Lecture Notes in Mathematics

Edited by A. Dold and B. Eckmann

598

J. Hoffmann-Jørgensen
T. M. Liggett
J. Neveu

Ecole d'Eté de Probabilités de Saint-Flour VI-1976

Edité par P.-L. Hennequin

Springer-Verlag
Berlin · Heidelberg · New York 1977

Authors
J. Hoffmann-Jørgensen
Matematisk Institut
Universitetsparken Ny Munkegade
8000 Aarhus/Denmark

T. M. Liggett
University of California
Department of Mathematics
Los Angeles, California 90024/USA

J. Neveu
Université de Paris VI
U.E.R. 48.00
Laboratoire de Probabilités
4. Place Jussieu, Tour 56
75230 Paris Cedex 05/France

Editor
P.-L. Hennequin
Université de Clermont
Complexe Scientifique des Cézeaux
Département de Mathématiques
Appliquées
B.P. 45
63170 Aubiere/France

Library of Congress Cataloging in Publication Data

Hoffman-Jørgensen, J
École d'été de probabilités de Saint-Flour VI-1976.

(Lecture notes in mathematics ; 598)
Includes bibliographies and index.
1. Probabilities—Congresses. 2. Measure theory—Congresses. 3. Point processes—Congressess. 4. Particles—Statistics—Congresses. I. Liggett, Thomas Milton, 1944- joint author. II. Neveu, Jacques, joint author. III. Ecole d'été de probabilitiés de Saint-Flour, 6th, 1976. IV. Series. Lecture notes in mathematics (Berlin) ; 598.
QA3.L28 no. 598 [QA273.A1] 510'.8s [519] 77-9882

AMS Subject Classifications (1970): 28A40, 46B99, 46E40, 60-02, 60B05, 60F05, 60G45, 60G50, 60K25, 60K35, 70-02

ISBN 3-540-08340-5 Springer-Verlag Berlin · Heidelberg · New York
ISBN 0-387-08340-5 Springer-Verlag New York · Heidelberg · Berlin

Printing and binding: Beltz Offsetdruck, Hemsbach/Bergstr.
2141/3140-543210

J. Hoffmann-Jørgensen

Probability in Banach Spaces

T. M. Liggett

The Stochastic Evolution of Infinite Systems of Interacting Particles

J. Neveu

Processus Ponctuels

INTRODUCTION

La sixième Ecole d'Eté de Calcul des Probabilités de Saint-Flour s'est tenue du 22 Août au 8 Septembre 1976 et a rassemblé, outre les conférenciers, soixante et un participants.

Les trois conférenciers, Messieurs Hoffmann-Jørgensen, Liggett et Neveu, ont tenu à élaborer une rédaction définitive de leurs cours qui complète sur certains points les exposés faits à l'école. Nous les en remercions vivement.

En outre les exposés suivants ont été faits par les participants durant leur séjour à Saint Flour :

A. BADRIKIAN	Lois stables et espaces du type p d'après De Acosta
J. BADRIKIAN	Films pour l'enseignement des probabilités
O. ADELMAN	Simple Markov chains
Ph. ARTZNER	Un modèle probabiliste de tâtonnement non-walrasien
R. CARMONA	Loi du logarithme itéré pour les suites de vecteurs gaussiens
E. DETTWEILER	Infinitely divisible measures on the cone of an orderes locally convex vector spaces
J. DUCHON	Fonctions-spline et espérances conditionnelles de champs gaussiens
X. FERNIQUE	Minorations de processus gaussiens non stationnaires
M. GHIDOUCHE	Récurrence au sens de Harris des GI/G/q
B. HEINKEL	Un théorème de la limite centrale dans C (S)
Y. ITOH	Random collision models for competition species
F. LAMBERT	Conditions d'existence et d'unicité de la solution pour une équation différentielle fonctionnelle stochastique
G. LETAC	Récurrence des librairies sur un arbre
B. MAISONNEUVE	Décomposition atomique de martingales de la classe $\mathcal{H}^1$
J. MEMIN	Conditions d'optimalité pour un problème de contrôle portant sur une famille de probabilités équivalentes
D. MOUCHTARI	La topologie du type Sazonov pour les Banach et les supports hilbertiens
D. NUALART-RODON	Intégrales stochastiques par rapport au processus de Wiener à deux paramètres

J. PELLAUMAIL	Une construction élémentaire de l'intégrale stochastique
M. ROBIN	Contrôle impulsionnel avec retard pour des processus de Markov
W. SCHACHERMAYER	Eberlein-compacts et espaces de Radon
R.L. TWEEDIE	Geometric ergodicity for a class of Markov chains

La plupart de ces exposés font l'objet du numéro 61 des Annales Scientifiques de l'Université de Clermont.

La frappe du manuscrit a été assurée par les Départements de Aarhus, Los Angeles et Clermont et nous félicitons, pour leur soin, les secrétaires qui se sont chargées de ce travail.

Nous exprimons enfin notre gratitude à la Société Springer Verlag qui permet d'accroître l'audience internationale de notre Ecole en accueillant ces textes dans la collection Lecture Notes in Mathematics.

P.L. HENNEQUIN

Professeur à l'Université de Clermont

B.P. 45

F-63170 AUBIERE

TABLE DES MATIERES

J. HOFFMANN-JØRGENSEN : "PROBABILITY IN BANACH SPACE"

T.M. LIGGETT : "THE STOCHASTIC EVOLUTION OF INFINITE SYSTEMS OF INTERACTING PARTICLES

LISTE DES AUDITEURS

Mr. ACQUAVIVA A.	Université de Brest
Mr. ADELMAN O.	Université de Paris VI
Mr. AL-BARHAWE A.	Université de Mossoul (Irak)
Mr. ALJ A.	Université de Rabat (Maroc)
Mr. AMIRA T.	Université de Paris VI
Mr. ARTZNER P.	Université de Strasbourg I
Mr. AZEMA J.	Université de Paris V
Mr. BADRIKIAN A.	Université de Clermont II
Mme BADRIKIAN J.	Université de Clermont I
Mr. BERNARD P.	Université de Clermont II
Mle BERTEIN F.	Université de Paris VI
Mr. BETHOUX P.	Université de Lyon I
Mr. BONNEMOY C.	Université de Clermont II
Mle BOYER DE BOUILLANE C.	E.N.S. de Fontenay-aux-Roses
Mr. BRANDOUY P.	Université de Pau
Mr. CARMONA R.	Université d'Aix-Marseille III
Mr. CHEMARIN P.	Université de Lyon I
Mle CHEVET S.	Université de Clermont II
Mr. CLAVILIER A.	Université de Clermont II
Mr. DAZORD J.	Université de Lyon I
Mr. DEMONGEOT J.	Université de Grenoble I
Mr. DETTWEILER E.	Université de Tübingen (R.F.A.)
Mr. DUCHON J.	Université de Grenoble I
Mle FARJOT P.	Université de Clermont II
Mme FAUCONNET A.	Université d'Aix-Marseille I
Mr. FERNIQUE X.	Université de Strasbourg I
Mr. FLIPO D.	Université de Lille III
Mr. FOURT G.	Université de Clermont II
Mr. GALVES A.	Ecole Polytechnique de Paris
Mle GAROT C.	C.E.N. de Saclay
Mr. GHIDOUCHE M.	Université de Rouen
Mr. GOLDBERG J.	I.N.S.A. de Lyon
Mr. HEINKEL B.	Université de Strasbourg I
Mr. HELM W.	Université Technique de Berlin (R.F.A.)

Mr. HENNEQUIN P.L.	Université de Clermont II
Mr. ITOH Y.	IRIA-LABORIA de Rocquencourt
Mr. KHELLADI A.	Université d'Alger (Algérie)
Mr. KIPNIS C.	Université de Paris VII
Mr. LAMBERT F.	Ecole de l'Air de Salon de Provence
Mr. LETAC G.	Université de Toulouse III
Mle LIRON M.	E.N.S. de Fontenay-aux-Roses
Mr. MAISONNEUVE B.	Université de Grenoble II
Mr. MEMIN J.	Université de Rennes I
Mr. MENALDI J.L.	IRIA-LABORIA de Rocquencourt
Mr. MEZIANI L.	Université d'Alger (Algérie)
Mr. MOGHA G.	Université de Paris VI
Mr. MOUCHTARI D.	Université de Kazan (U.R.S.S.)
Mr. NANOPOULOS C.	Université de Strasbourg I
Mr. NOBELIS P.	Université de Strasbourg I
Mr. NUALART-RODON D.	Université de Barcelone (Espagne)
Mr. NURNBERG R.	Université de Göttingen (R.F.A.)
Mle OLIVARES M.	Université de Paris VI
Mr. PELLAUMAIL	I.N.S.A. de Rennes
Mme PICQUE M.	E.N.S.E.T. de Cachan
Mr. ROBIN M.	IRIA-LABORIA de Rocquencourt
Mr. ROUSSIGNOL M.	Université de Paris VI
Mr. SCHACHERMAYER W.	Université de Clermont II
Mr. SCHEFFER C.	Ecole Polytechnique de Delft (Pays-Bas)
Mr. TORFS P.	Université de Nimègue (Pays-Bas)
Mr. TWEEDIE R.	CSIRO de Canberra (Australie)
Mr. VAN DEN BOOGERD Y.	Université Technique de Delft (Pays-Bas)

PROBABILITY IN BANACH SPACE

PAR J. HOFFMANN-JØRGENSEN

CHAPTER I

Prerequisites

1. Introduction

Probability in Banach spaces was first studied by M.R. Fortet and E. Mourier, [29], [30] and [91], their studies was later continued by Chatterji, [15] and [16], K. Ito and M. Nisio, [45], A. Beck, [4] and others. However the subject first became popular after the inspiring nomograph by J.-P. Kahane, [57], which has been the inspiration of many probabilist in the field.

The recent years has given us a wealth of information and most of the questions, which was asked in the late sixties has now been answered. The subject has now reached maturity, and I have here tried to the best of my ability to describe some important parts of the theory. It is with some sadness, that I have to admit, that to my opion the subject is almost closed, in the sense that there are almost no important problems left open. The only opening I can see, is to study similar results in spaces like the right continuous functions or the cad-lag functions or the measurable functions, which are closer related to probability theory than the Banach spaces. However my feeling is that this study will need completely different methods than those treated here.

There is only few references in the main text, so let me comment the different sections below. I have tried to make the references as complete as possible, but I may have forgotten some, if so this is certainly not intended.

Chapter §.2. This section contains the basic notion of probability in linear spaces.

In connection with (I.2.4) we have the following open problem:

(1.1) Does there exist nonseparable a Banach space $(E,\|\cdot\|)$, such that $(E,\mathcal{B}(E))$ is a measurable linear space?

Or more specific

(1.2) Is ℓ^∞ a measurable linear space under its $\|\cdot\|$-Borel structure?

My guess is that these questions are connected with certain additional axioms of set theory like the continuum hypothesis or Martin's axiom.

I am sure that the spaces $(L_E^{(\varphi)}, \lambda_\varphi)$ has been treated in the litterature, but have not been able to find any references. Note that $L^{(\varphi)}$ is closely related to the Laurent spaces.

In connection with (I.2.22) I refer to reader to [110].

Chapter I.3 This section contains those parts of geometry in Banach spaces, which is needed in the sequel, but this is only the top of the iceberg. For more information I refer the reader to Lindenstraus and Tzafriri [80], Maurey [88], Maurey and Pisier [90].

Chapter I.4 This section contains the standard results of weak convergence of measures on topological spaces. The results may be found in any standard textbook in the subject, e.g. [9], [94] or [111].

Chapter I.5 The results of this section are inspired by V. Strassen [108]. In this section we study, when it is possible to find a probability measure on a product space in a given class with given marginals, and the results are extensions of the results in [108]. They will later be applied in connection with the contraction principle.

Chapter II.1 Here er study integrability of $\sup_n |X_n|$ where (X_n) are independent random variables. Lemma 1.2 has been proved in [53] (Lemma 3.1), however the result here is more informative and our proof is much simpler.

Chapter II.2 The first convex version of the contraction principle was proved in [57], and was later generalized in [41], [42] and [53]. Note the convex versions of the contraction principle (Theorems II.2.9 - II.2.12) are merely just consequences of the definition of $\xi \models \eta$. The non-convex version (see Theorem II.2.15) is a good deal more involved, and the idea is taken from some papers of Musial, Woyczynski and Ryll-Nardzewski, [92] and [105].

Chapter II.3 The theory of convex measures has been developed by C. Borell [10] - [13]. And I refer the reader to these for more information of this beautiful subject. The zero-one law is known for other classes of measures (see e.g. [59], [49] and [42])

Chapter II.4 Lemma 4.1 and Theorem 4.2. are known as the Lévy symmetrization inequalities. Note that Theorem 4.2 (b) does not hold for φ-substituted by φ, take for example $E = \mathbb{R}$ and

$$X_n = (1-2^{-n})\varepsilon_n$$

$$q(x) = \sup_n |x_n|$$

$$\varphi = 1_{[1,\infty[}$$

Then $\varphi(M) = 1$ and $\varphi(S_n) = 0$.

In connection with Theorem 4.7 (see also Corollary 4.8) one may ask the following question:

(13) Does $P(N > t) \leq Ke^{-at}$ imply $P(M > t) \leq ce^{-bt}$?

The space (CLT,c(•)) was introduced by Pisier in [95] and Theorem 4.10 is due to him and to N.C. Jain [46].

Chapter II.5 The first version of Theorem II.5.3 (case (6)) was proved by Ito and Nisio [45], and later extended in [57] and [41]. The version here seems to me to be the most satisfactory version possible. Theorems II.5.4 and II.5.5 can be found in [42] and [53].

The theorems II.5.6 - II.5.9 goes under the name "the comparison principle" and was first proved by Kahane in [57] and later extended in [53] and [41].

Chapter II.6 The type and cotype was introduced by Maurey, Pisier and myself simultaneously in [40], [87], [88], and [99]. For further information I refer the reader to these papers.

Chapter II.7 Theorem II.7.1 is due to S. Kwapien [75] and myself [41]. The rest of this section is taken from Maurey and Pisier, [89], to which I refer the reader for more information.

Chapter III.1 The law of large number was the start of probability and in probability we have ever since tried to understand this theorem, and we have proved it over and over, again and again.

(III.1.1) is the classical version of the law large numbers and has its roots back to Chebyshev, Markov and Lyapounov. The extension to Banach spaces was first made by Fortet and Mourier [29] and [91] (see also A. Beck [4])

(III.1.2) is Kolmogorov's law of large numbers. It was first extended to Banach spaces by A. Beck [4] (the Equivalence of (a) and (g) in Theorem III.1.2), and later Pisier [101] proved the other equivalences of Theorem III.1.2.

(III.1.3) is Chung's law of large numbers (see [17]). It was extended by Pisier and myself in [44].

Chapter III.2 The central limit theorem is the second pearl of probability. It was first proved by de Moivre (1733) for the Bernoulli sequence, Laplace proved in 1812 for skew Bernoulli sequences and it was later generalized to real i.i.d.'s big Lyapounov.

This beautiful theorem is even more astonishing than the law of large numbers, and with full right this theorem is still studied intensively in modern probability.

Fortet and Mourier was the first to consider the theorem in Banach spaces, [29], [30] and [91]. And a large number of probabilist has studied the theorem in C(S), see e.g. [2], [24], [32], [33], [35], [36], [38], [39], [54], [83] and [117].

Corollary III.2.4 and Example III.2.5 are due to Pisier [96] (see also Jain [46] - [48]).

Chapter III.3 Stochastic integration of vectorvalued function in the form presented here was first introduced by myself (see [44]).

It has commonly been beleived that the "right" way of proving central limit theorem is to prove tightness of $\{(X_1+...+X_n)/\sqrt{n}\}$. This is to my opinion a wrong track, which actually mislead the investigation of the central limit theorem. And the method of stochastic integration (see Chapter III.4 or [44]) or direct use of the Prohorov metric (see [96]) has proved to much more efficient.

Chapter III.4 Theorem III.4.1 was proved by Pisier and myself in [44]. Theorem III.4.2 was partly proved by Pisier in [96].

<u>Chapter III.5</u> The result of this section is in essence due to J. Zinn [117].

Let me at last comment on those sections which are not present in this exposition.

The third pearl of probability is the law of the iterated logarithm. This law has only been mentioned én passant in Chapter II.4. The law of the iterated logarithm belongs to modern probability and have not yet had an impact on probability, that can be compared with that of the law of large number or the central limit theorem. One reason may be, that even though the law of the iterated logarithm has a great theoretical importance, it can not be observed even in a large number of data, and that it contradiets, at least in intuitive terms, the Arcsine-law, which can be observed even in a small number of data.

Readers who are interested in the law of the iterated logarithm in Banach spaces is referred to [47], [61] - [71], [78], [95] - [97] and [107].

Another section which is not present is the theory of vector-valued martingales. For this important and exciting subject I refer the reader to [14] - [16], [44], [95], [98], [100] and [116].

2. Random variables

Let (Ω,F,P) be a probability space, which will be fixed once for all in these notes, except for the fact, that we will allow ourselves to think of (Ω,F,P) to be so rich that we may define any random variable or sequence of random variables we would like. Kolmogorov's consistency theorem justifies this attitude.

A random variable is an measurable function $X:\Omega \to \mathbb{R}$. More general if $(E,\mathcal{B})$ is a linear space with a σ-algebra $\mathcal{B}$, satisfying

(2.1) $(x,y,\lambda) \rightarrow \lambda x + y: E\times E \times \mathbb{R} \to E$ is a measurable map, if we equip $E \times E \times \mathbb{R}$ with its product σ-algebra,

then $(E,\mathcal{B})$ is called a measurable linear space. And a measurable function $X:\Omega \to E$ is called an E-valued random vector.

If T is a topological space, then $\mathcal{B}(T)$ denotes the Borel σ-algebra on T, that is the σ-algebra generated by all closed subsets of T. If T is a set, and (S,Σ) a measurable space, and H a class of functions: $T \to S$, then $\sigma(H)$ denotes the least σ-algebra, which makes all functions in H measurable. With This notation we have

(2.2) $(E,\mathcal{B}(E))$ is a measurable linear space, if $(E,\|\cdot\|)$ is a separable normed space

(2.3) If F is a subset of E^*(=the algebraic dual of E), then $(E,\sigma(F))$ is a measurable linear space.

However if $(E,\|\cdot\|)$ is a non-separable Banach space then

$(E,\mathcal{B}(E))$ is not in general a measurable linear space, which is seen from the following proposition:

(2.4) If $(E,\mathcal{B})$ is a measurable linear space, so that $\{0\} \in \mathcal{B}$, then there exists an injective measurable map $f: E \to \mathbb{R}$. In particular we have, that the cardinal of E is at most that of the continuum.

For this reason we shall call a Banach space valued function, $X:\Omega \to E$, an E-valued random vector, if X is Borel measurable, and has separable range.

Let $(E,\mathcal{B})$ be a measurable linear space, and X an E-valued random vector, the the law of X, denoted $L(X)$, is the probability measure on $(E,\mathcal{B})$ given by

$$L(X) = P \circ X^{-1}$$

A random vector X is called even if $L(X) = L(-X)$. If $X = (X_n)$ is a finite or infinite sequence of E_n-valued random vectors ($(E_n,\mathcal{B}_n)$ is a measurable linear space for all $n \geq 1$), then (X_n) is said to be symmetric if $L(X) = L(Y)$, whenever $Y = (\pm X_n)$ for some choice of signs $\pm$.

If X is an E-valued random vector, then X^s is said to be a symmetrization of X, if $X^s = X' - X''$, where

$$X',\ X'' \text{ are independent, } L(X) = L(X') = L(X'')$$

And if $X = (X_n)$ is a random sequence then $X^s = (X_n^s)$ is called a symmetrization of X, if $X_n^s = X_n' - X_n''$ where

(X_n') and (X_n'') are independent and both has the same distribution as (X_n).

Then we have

(2.5) If (X_n) is a sequence of independent random vectors, and $X^s = (X_n^s)$ is a symmetrization of (X_n), Then X^s is symmetric.

Note that the symmetrization of a random sequence is always even, but not necessarily symmetric. Note also:

(2.6) If $X_1, X_2, \dots$ are independent and even, then (X_n) is symmetric.

We shall now introduce some notions for functions $q: E \to \bar{\mathbb{R}} = \mathbb{R} \cup \{\pm\infty\}$, where E is a linear space. We shall say that q is

<u>even</u> : $q(x) = q(-x)$

<u>convex</u> : $q(tx+(1-t)y) \leq tq(x) + (1-t)q(y) \quad \forall t \in [0,1]$

<u>quasiconvex</u>: $q(tx+(1-t)y) \leq \max\{q(x), q(y)\} \quad \forall t \in [0,1]$

<u>subadditive</u>: $q(x+y) \leq q(x) + q(y)$

<u>homogenuous</u>: $q(tx) = |t| q(x) \quad \forall t \in \mathbb{R}$

And if $q: E_1 \times \dots \times E_n \to \bar{\mathbb{R}}$, then q is said to be <u>symmetric</u> if

$$q(x_1, \dots, x_n) = q(\pm x_1, \dots, \pm x_n)$$

for all choices of signs $\pm$ and all $x_1 \in E_1, \dots, x_n \in E_n$.

We shall use $\mathbb{E}(X)$ for the expectation of a real random variable X, i.e.

$$\mathbb{E} X = \int_\Omega X dP$$

and we allow $\mathbb{E}X$ to assume the values $\pm\infty$. If X is a Banach space valued random vector, then $\mathbb{E}X$ denotes the Bochner integral of X (see e.g. chap. II §2 in [19]).

Let $(E, \mathcal{B})$ be a measurable linear space, and let q be a measurable map: $E \to \bar{\mathbb{R}}_+ = [0, \infty]$, satisfying:

(2.7) $\exists a \in \mathbb{R}\ \exists A > 0: \quad q(x+y) \le A(q(ax) + q(ay)) \quad \forall x, y \in E$

(e.g. convex function: $a = A = \frac{1}{2}$, or subadditive function: $a = A = 1$). Then we have

(2.8) $\mathbb{E}\, q(2Y) \le 2A\ \mathbb{E}\, q(a(X+Y))$ whenever X and Y are independent and X is symmetric.

Which follow easily from

$$2Y = (Y+X) + (Y-X), \quad L(Y+X) = L(Y-X)$$

Now let $(E, \|\cdot\|)$ be a Banach space, then $L^0(\Omega, F, P, E)$ denotes the set of all E-valued random vectors, and if $0 < p \le \infty$, then $L^p(\Omega, F, P, E)$ denote the set of $X \in L^0(\Omega, F, P, E)$, such that $\|X(\cdot)\| \in L^p(\Omega, F, P)$. We shall use the shorthand:

$$L^p_E = L^p(\Omega, F, P, E), \quad L^p = L^p_{\mathbb{R}}$$

for $0 \le p \le \infty$.

If $0 \le p < 1$, we can introduce a Fréchet metric $\|\cdot\|_p$ in L^p_E by

$$\|X\|_0 = \mathbb{E}\left\{\frac{\|X\|}{1+\|X\|}\right\} \qquad \text{if } p = 0$$

$$\|X\|_p = \mathbb{E}\|X\|^p \qquad \text{if } 0 < p < 1$$

If $1 \le p \le \infty$, we can introduce a Banach norm in L^p_E by

$$\|X\|_p = \{\mathbb{E}\|X\|^p\}^{1/p} \qquad \text{if } 1 \le p < \infty$$

$$\|X\|_\infty = \operatorname{ess\,sup}_\omega \|X(\omega)\| \qquad \text{if } p = \infty$$

A function $q\colon E \to \bar{\mathbb{R}}_+$ is called <u>measure</u> <u>convex</u>, if q is Borel measurable and

$$q(\mathbb{E}\,X) \leq \mathbb{E}\,q(X) \qquad \forall X \in L^1_E$$

it is well known (Jensen's inequality) that we have

(2.9) If $\dim E < \infty$ then any measurable convex function $q\colon E \to \bar{\mathbb{R}}_+$ is measure convex.

(2.10) If $q = \varphi \circ p$, where $p\colon E \to \bar{\mathbb{R}}_+$ is lower semicontinuous and convex and $\varphi\colon \bar{\mathbb{R}}_+ \to \bar{\mathbb{R}}_+$ is increasing and convex, then q is measure convex

---

Let $\varphi\colon \mathbb{R}_+ \to \bar{\mathbb{R}}_+$ be decreasing and assume that $\varphi \not\equiv \infty$, then we define

$$\lambda_\varphi(X) = \inf\{a \geq 0 \mid P(\|X\| > at) \leq a\varphi(t) \qquad \forall t \geq 0\}$$

with the usual convention: $\inf \emptyset = \infty$. Then we have

(2.11) $a \geq \lambda_\varphi(X) \iff P(\|X\| > t) \leq a\varphi(t/a) \qquad \forall t \geq 0$

(2.12) $\lambda_\varphi(X) < \infty \iff P(\|X\| > t) = 0(\varphi(\varepsilon t))$ as $t \to \infty$ for some $\varepsilon > 0$

(2.13) λ_φ is subadditive on L^0_E

(2.14) $\lambda_\varphi(sX) \leq \max\{1, |s|\}\lambda_\varphi(X) \qquad \forall s \in \mathbb{R} \;\; \forall X \in L^0_E$

(2.15) $\lambda_\varphi(X) = 0 \iff X = 0$ a.s.

So the space

$$L_E^{(\varphi)} = L^{(\varphi)}(\Omega,F,P,E) = \{X \in L_E^0 \mid \lambda_\varphi(X) < \infty\}$$

is a linear subspace of L_E^0, and $\lambda_\varphi(X-Y)$ defines a metric on $L_E^{(\varphi)}$, which is stronger than $\|\cdot\|_0$, and it is easily checked that we have

(2.16) $(L_E^{(\varphi)}, \lambda_\varphi(\cdot))$ is a Fréchet space.

If $\varphi(t) = t^{-p}$ $(0 < p < \infty)$, we use the notation

$$L_E^{(\varphi)} = L_E^{(p)}, \quad \lambda_\varphi = \lambda_p$$

If $\varphi(t) = \infty$ on $[0,1[$ and $\varphi(t) = 1$ on $[1,\infty[$, then $L_E^{(\varphi)}$ and λ_φ is denoted by $L_E^{(0)}$ and λ_0, and λ_0 is given by

$$\lambda_0(X) = \inf\{a > 0 \mid P(\|X\| > a) \le a\}$$

If $\varphi(t) = \infty$ on $[0,1[$ and $\varphi(t) = 0$ on $[1,\infty[$, then $L_E^{(\varphi)}$ and λ_φ is denoted by $L_E^{(\infty)}$ and λ_∞, and λ_∞ is given by

$$\lambda_\infty(X) = \inf\{a > 0 \mid P(\|X\| > a) = 0\}$$

It is easy to check that we have

(2.16) If $0 < p < \infty$ then $\lambda_p(X)^{p+1} \le \mathbb{E}\|X\|^p$ and $L_E^p \subseteq L_E^{(p)}$

(2.17) $L_E^{(0)} = L_E^0$ and we have

$$\frac{\|X\|_0}{2-\|X\|_0} \le \lambda_0(X) \le \|X\|_0 + \sqrt{\|X\|_0}$$

(2.18) $L_E^{(\infty)} = L_E^\infty$ and $\lambda_\infty(X) = \|X\|_\infty$

(2.19) If $0 < p < q < \infty$ then $L_E^{(q)} \subseteq L_E^p$ and $\mathbb{E}\|X\|^p \le \frac{q}{q-p}\lambda_q(X)^{p(1+1/q)}$

(2.20) $\lambda_p(aX) = |a|^{p/(p+1)}\lambda_p(X) \quad \forall a \in \mathbb{R} \quad \forall X \in L_E^{(p)}$

If $(E, \|\cdot\|)$ is a normed space, then on E^∞ (the countable product of E) we define

$$\|x\|_0 = \sum_{n=1}^{\infty} 2^{-n} \frac{\|x_n\|}{1+\|x_n\|}$$

$$\|x\|_p = \sum_{n=1}^{\infty} \|x_n\|^p \quad \text{if} \quad 0 < p \leq 1$$

$$\|x\|_p = \left(\sum_{n=1}^{\infty} \|x_n\|^p\right)^{1/p} \quad \text{if} \quad 1 \leq p < \infty$$

$$\|x\|_\infty = \sup_n \|x_n\|$$

$$\ell_E^p = \{x \in E^\infty \mid \|x\|_p < \infty\}$$

$$c_0(E) = \{x \in E^\infty \mid \lim_{n\to\infty} x_n = 0\}$$

where $x = (x_n)$ is a vector in E^∞. If $E = \mathbb{R}$ we use the notation ℓ^p and c_0.

Let $\varepsilon_1, \varepsilon_2, \ldots$ be independent random variables with $P(\varepsilon_j = \pm 1) = \frac{1}{2}$, then (ε_n) is called a <u>Bernouilli sequence</u>. The Bernouilli sequence will play an essential role in many contexts later on. A very simple, but extremely useful, observation is:

(2.21) If (X_n) is a symmetric sequence, and (ε_n) is a Bernouilli sequence independent of (X_n). Then (X_n) and $(\varepsilon_n X_n)$ has the same law.

Using the multinormal formula and the monotonicity of $\{\mathbb{E}|X|^p\}^{1/p}$, one easily derives the important <u>Khinchine inequalities</u>:

(2.22) $$k(p)\left(\sum_1^n t_j^2\right)^{p/2} \le \mathbb{E}\left|\sum_{j=1}^n t_j\varepsilon_j\right|^p \le K(p)\left(\sum_1^n t_j^2\right)^{p/2}$$

for all $p>0$, $n\ge 1$ and $t_1,\dots,t_n \in \mathbb{R}$. Here $k(p)$ and $K(p)$ are universal constants.

If $(\xi_1,\dots,\xi_n)$ is a symmetric n-dimensional random vector with finite p-th moment $(p\ge 2)$, then

$$\mathbb{E}\left|\sum_{j=1}^n \xi_j\right|^p = \int_{\mathbb{R}^n} \mathbb{E}\left|\sum_1^n t_j\varepsilon_j\right|^p \mu(dt_1,\dots,dt_n)$$

where $\mu = L(\xi_1,\dots,\xi_n)$. Hence

$$\left\{\mathbb{E}\left|\sum_{j=1}^n \xi_j\right|^p\right\}^{2/p} \le K(p)\left\{\mathbb{E}\left|\sum_{j=1}^n \xi_j^2\right|^{p/2}\right\}^{2/p}$$

$$\le K(p)\sum_{j=1}^n \left(\mathbb{E}|\xi_j|^p\right)^{2/p}$$

by Minkowski's inequality. That is

(2.23) $$\mathbb{E}\left|\sum_{j=1}^n \xi_j\right|^p \le K(p)\left\{\sum_{j=1}^n \left(\mathbb{E}|\xi_j|^p\right)^{2/p}\right\}^{p/2} \qquad \forall p\ge 2$$

whenever $(\xi_1,\dots,\xi_n)$ is symmetric and has finite p-th moment.

If $\xi_1\dots\xi_n$ are independent with $\mathbb{E}\xi_j = 0$ and $\mathbb{E}|\xi_j|^p < \infty$ where $p\ge 2$. Then by symmetrizing $(\xi_1,\dots,\xi_n)$ and applying Jensen's inequality we find

(2.24) $$\mathbb{E}\left|\sum_{j=1}^n \xi_j\right|^p \le 2^p K(p)\left\{\sum_{j=1}^n \left(\mathbb{E}|\xi_j|^p\right)^{2/p}\right\}^{p/2} \qquad \forall p\ge 2$$

Let $(E,\mathcal{B})$ be a measurable linear space where $\mathcal{B}$ is given by

$$\mathcal{B} = \sigma(E')$$

for some linear subset $E' \subseteq E^*$ (= the algebraie dual of E), where E' separates points in E. If μ is a probability measure on E we define its <u>Fourier transform</u> by

$$\hat{\mu}(x') = \int_E e^{i<x',x>} \mu(dx) \qquad \forall x' \in E'$$

and its <u>covariance function</u> by

$$R(x',y') = \int_E <x',x><y',x> \mu(dx) \qquad \forall x',y' \in E'$$

whenever $E' \subseteq L^2(\mu)$. Finally if $E' \subseteq L^1(\mu)$ we define the <u>Gelfand mean</u> of μ by:

$$E(x') = \int_E <x',x> \mu(dx)$$

which is a linear functional on E'.

A probability measure μ on $(E,\mathcal{B})$ is called <u>gaussian</u> if we have

(2.25) $\qquad \hat{\mu}(x') = \exp(i\,E(x') - R(x',x'))$

where E is the Gelfand mean of μ and R is the covariance function of μ. (2.25) is evident equivalent to:

(2.26) $\mu_{x'_1,\dots,x'_n}$ is a Gauss measure on $\mathbb{R}^n$ for all $x'_1 \dots x'_n \in E'$ $\forall n \geq 1$ where $\mu_{x'_1,\dots,x'_n}$ is the image measure of μ under the map

$$x \longrightarrow (<x'_1,x>,\dots,<x'_n,x>) : E \longrightarrow \mathbb{R}^n$$

If $(E,\|\cdot\|)$ is a Banach space and $G \subseteq L^0_E$, then G is said to <u>gaussian</u>, if $L(X)$ is gaussian and $\mathbb{E}\,X = 0$ for all $X \in G$. We shall need the following two elementary facts about Gaussian measures and spaces:

Proposition 2.1. *Let* $G \subseteq L^0$ *be a gaussian linear space, and* $G_0 \subseteq G$. *If* $F_0 = \sigma(G_0)$, *then we have*

$$\mathbb{E}(X | F_0) = \pi_0 X \qquad \forall X \in G$$

where $\pi_0 : L^2 \to L^2$ *is the orthogonal projection of* L^2 *onto* $\overline{\text{span}}\,(G_0)$.

Proposition 2.2. *If* (X,Y) *is an* $E \times E$*-valued gaussian random vector with* $\mathbb{E} X = \mathbb{E} Y = 0$, *then* X *and* Y *are independent if and only if*

$$\mathbb{E}\langle x',X\rangle\langle y',Y\rangle = 0 \quad \forall x',y' \in E'$$

3. Geometry in Banach spaces.

I shall assume that the reader is accustomed to the introductory part of the theory of Banach spaces and Fréchet spaces e.g. the closed graph theorem, principle of uniform boundedness, and interior mapping principle (see e.g. [19])

However in connection with types and cotypes of a Banach space we shall need some of the more exotic parts of the theory of Banach space, which I will describe below, but without proofs.

Let E and F be normed spaces, then we define their distance, $d(E,F)$ by:

$$d(E,F) = \inf\{\|T\|\,\|T^{-1}\| \mid T \text{ isomorphism of } E \text{ onto } F\}$$

Hence $d(E,F) < \infty$ if and only if E and F are isomorphic and we have

(3.1) $$d(E,F) \geq 1 \qquad \forall E,F$$

And we say that E is finitely λ-representable in F where λ is a number ≥ 1, if they satisfies

(3.2) $\forall E_0$ finite dimensional subspace of E, $\exists F_0$, a finite dimensional subspace of F with $d(E_o,F_o) \leq \lambda$.

If E is finitely λ-representable in F for some $\lambda < \infty$, we just say that F parodies E. And if E is finitely λ-representable in F for all $\lambda > 1$, then we say that F mimics E.

There is a similar notion for a bounded linear operator $T: E \to G$, where E,F and G are normed linear spaces. We say that T is finitely λ-representable through F, if for every subspace E_0 of E with $\dim E_0 < \infty$, there exists a subspace F_0 of F, and bounded linear maps $U:E_0 \to F_0$, and $V:F_0 \to G$, so that

(3.3) $$VU = T_0 \quad \text{and} \quad \|V\|\,\|U\| \leq \lambda \|T_0\|$$

where T_0 is the restriction of T to E_o. Hence E is finitely λ-representable in F, if and only if the identity: $E \to E$ is finitely λ-representable through F. As before we say that T is parodied through F if T is finitely λ-representable through F for some $\lambda < \infty$. And we say that T is mimicked through F is T is finitely λ-representable through F for all $\lambda > 1$.

Clearly a bounded operator $T:E \to G$ is finitely λ-representable through F, if and only if:

(3.4) $\forall x_1 \ldots x_n \in E$, $\exists y_1, \ldots, y_n \in F$, so that $\forall t_1 \ldots t_n \in \mathbb{R}$:

$$\|\sum_{j=1}^{n} t_j T x_j\| \leq \|\sum_{1}^{n} t_j y_j\| \leq \lambda \|T_0\| \, \|\sum_{1}^{n} t_j x_j\|$$

where T_0 is the restriction of T to $\overline{\text{span}}\{x_1,\ldots,x_n\}$.

And as above we may define $d(T,F)$, by

$$d(T,F) = \inf\{\|U\|\|V\| \mid T = VU,\ U:E \to F,\ V:F \to G\}$$

whenever $T:E \to G$ is a bounded linear operator.

From (3,4) it is fairly simple to prove the following 3 propositions:

Proposition 3.1. *Let $T:E \to G$ be a bounded linear operator, then T is mimicked through any dense linear subspace $E_0 \subseteq E$.* □

Proposition 3.2. *Let $T:E \to G$ be a bounded linear operator, and let $\{E_\alpha \mid \alpha \in A\}$ be a family of subspaces of E, satisfying:*

(a) $\forall \alpha, \beta \in A \quad \gamma \in A$ *so that* $E_\alpha \cup E_\beta \subseteq E_\gamma$

(b) $\bigcup_\alpha E_\alpha$ *is dense in* E

(c) $d(T_\alpha, F) \leq \lambda \quad \forall \alpha \in A$

where T_α is the restriction of T to E_α. Then T is finitely λ-representable through F. □

Proposition 3.3. *If F parodies E, then there exists a renorming of E (i.e. a norm on E which is equivalent to the original norm on E), such that F mimics E, if E is equipped with this new norm.* □

Since every finite dimensional Banach space is isometric to a subspace of c_0, we have

Proposition 3.4. The space c_o mimics any normed space E, and any bounded linear operator $T:E \to G$ is mimicked through c_o. □

By the socalled principle of local reflexivity (see e.g. [80]) we have

Theorem 3.5. If E is a normed space, then E mimics E''.

And the famous and very deep Dvoretsky-Rogers' lemma states:

Theorem 3.5. If E is an infinite dimensional normed space, then E mimics any Hilbert space. □

See e.g. [80]

For the purpose of these notes, we shall only need to consider representability of the injections:

$$\ell^1 \to \ell^p \quad \text{and} \quad \ell^p \to \ell^\infty$$

for $1 \leq p \leq \infty$. And we note that by Proposition 3.2 and (3.5) we have

Proposition 3.6. The injection $\ell^1 \to \ell^p$ $(1 \leq p < \infty)$ is finitely λ-representable through E, if and only if there exists $\{x_j^n \mid j \leq n\} \subseteq E$, so that

(a) $\|x_j^n\| \leq \lambda \qquad \forall j \leq n$

(b) $\left(\sum_1^n |t_j|^p\right)^{1/p} \leq \left\|\sum_{j=1}^n t_j x_j^n\right\| \qquad \forall t_1 \ldots t_n \in \mathbb{R}^n$

The injection $\ell^p \to \ell^\infty$ $(1 \le p \le \infty)$ is finitely λ-representable through E, if and only if there exists $\{x_j^n | j \le n\} \subseteq E$, so that

(c) $$\max_{1\le j\le n} |t_j| \le \|\sum_{j=1}^n t_j x_j^n\| \le \lambda(\sum_{j=1}^n |t_j|^p)^{1/p} \quad \forall t_1 \ldots t_n \in \mathbb{R}^n \quad \square$$

And from Theorem 3.5 it follows that we have

Theorem 3.7. If $1 \le p \le 2 \le q \le \infty$, then the injection $\ell^p \to \ell^q$ is mimicked through any infinite dimensional normed space E. $\square$

Moreover for the cases we are considering here there is no difference between "mimicked" and "parodied". To see this we return for a while to the general case.

Now let E, F and G be Banach spaces and T a bounded linear operator: $F \to G$. Let $(f_n) \subseteq F$ be a sequence of linearly independent vectors with

$$\overline{\operatorname{span}}\{f_n | n \ge 1\} = F$$

And put $g_n = Tf_n$. Then we may define

$$p_n(x_1,\ldots,x_n) = \max\{\|\sum_1^n t_j x_j\| : \|\sum_1^n t_j f\| \le 1\}$$

$$q_n(x_1,\ldots,x_n) = \min\{\|\sum_1^n t_j x_j\| : \|\sum_1^n t_j g_j\| \ge 1\}$$

And obviously we have

(3.6) $$q_n(x_1,\ldots,x_n)\|\sum_1^n t_j g_j\| \le \|\sum_1^n t_j x_j\| \le p_n(x_1\ldots x_n)\|\sum_1^n t_j f_j\|$$

Now we put

$$\alpha(n) = \inf\{p_n(x_1,\ldots,x_n) \mid q_n(x_1,\ldots,x_n) \geq \|T_n\|^{-1}\}$$

$$\beta(n) = \sup\{q_n(x_1,\ldots,x_n) \mid p_n(x_1,\ldots,x_n) \leq \|T_n\|\}$$

where T_n is T restricted to $F_n = \text{span}\{f_1,\ldots,f_n\}$. Then it is easily checked, that we have

(3.7) $\beta(n) \leq 1 \leq \alpha(n)$ $\forall n \geq 1$

(3.8) $\beta(n) \cdot \alpha(n) = 1$ $\forall n \geq 1$

(3.9) T is finitely $(\lambda+\varepsilon)$-representable through E for all $\varepsilon > 0$, if and only if $\alpha(n) \leq \lambda$ for all $n \geq 1$.

(3.10) $\alpha(1) = \beta(1) = 1$, and if $\|T_n\| = \|T\|$ for all $n \geq 1$, then $\alpha(\cdot)$ is increasing and $\beta(\cdot)$ is decreasing.

For the injections $\ell^1 \to \ell^p$ or $\ell^p \to \ell^\infty$ we of course take (f_n) and (g_n) to be the cannonical unit vector bases, and it is then fairly easy to see that $\alpha(\cdot)$ become supermultiplicative (i.e. $\alpha(nk) \geq \alpha(n)\alpha(k)$), and since α is increasing we have either $\alpha(n) \equiv 1$ or $\alpha(n) \to \infty$ at least as fast as some power of n. So we have (see e.g. [89]).

<u>Theorem 3.8</u>. <u>If</u> E <u>is a Banach space and</u> $1 \leq p \leq \infty$, <u>then we have</u>

(a) $\ell^1 \to \ell^p$ <u>is parodied through</u> E, <u>if and only if it is mimicked through</u> E.

(b) $\ell^p \to \ell^\infty$ <u>is parodied through</u> E, <u>if and only if it is mimicked through</u> E. □

It is clear that if $\ell^1 \to \ell^p$ is mimicked through E then so is $\ell^1 \to \ell^r$ for all $r \geq p$. And if $\ell^p \to \ell^\infty$ is mimicked through E then so is $\ell^r \to \ell^\infty$ for all $1 \leq r \leq p$. Hence the sets

$$\{p \in [1,\infty]: \ell^1 \to \ell^p \text{ is mimicked through } E\}$$

$$\{p \in [1,\infty]: \ell^p \to \ell^\infty \text{ is mimicked through } E\}$$

are intervals containing $[2,\infty]$ respectively $[1,2]$. What is not so clear is that these intervals are actually closed (see [89]). That is:

<u>Theorem 3.9</u>. <u>Let</u> E <u>be a Banach space, then the set</u>:

$$\{p \in [1,\infty] \mid \ell^1 \to \ell^p \text{ is mimicked through } E\}$$

<u>is a closed interval containing</u> $[2,\infty]$. <u>And the set</u>:

$$\{p \in [1,\infty] \mid \ell^p \to \ell^\infty \text{ is mimicked through } E\}$$

<u>is a closed interval containing</u> $[1,2]$. □

The next theorem may be found in [89] (p.68), and is an application of the Pietsch' factorization theorem

Theorem 3.10. Let $2<p<\infty$, and suppose that $\ell^p \to \ell^\infty$ is not mimicked through E. Then there exists constant $K>0$, satisfying

(a) $\forall x_1,\dots,x_n \in E,\ \exists \alpha_1,\dots,\alpha_n \geq 0,$ with $\sum_1^n \alpha_j = 1$ and

$$\left\| \sum_{j=1}^n t_j x_j \right\| \leq K \max_\pm \left\| \sum_{j=1}^n \pm x_j \right\| \left(\sum_{j=1}^n |t_j|^p \alpha_j \right)^{1/p}$$

for all $t_1,\dots,t_n \in \mathbb{R}$. □

Even though the proof of Theorem 3.10 applies the theory of cotype developed in Chapter II §6, I shall not give the proof here but refer the reader to [89]. However, let us deduce some consequences of Theorem 3.10.

Suppose that $\ell^p \to \ell^\infty$ is not mimicked through E, and that (x_j) is a sequence with unconditionally bounded partial sums, that is

$$\sup_n \max_\pm \left\| \sum_{j=1}^n \pm x_j \right\| < \infty \tag{3.6}$$

Then by Theorem 3.10 there exist $\alpha_{jn} \geq 0$ for $1 \leq j \leq n < \infty$, so that

$$\sum_{j=1}^n \alpha_{jn} = 1 \qquad \forall\, n \geq 1$$

$$\left\| \sum_{j=1}^n t_j x_j \right\| \leq C \left(\sum_{j=1}^n |t_j|^p \alpha_{jn} \right)^{1/p} \qquad \forall\, t_1,\dots,t_n \in \mathbb{R}$$

where C is a constant independent of n. Using this for $t = (0,\dots,0,1,0,\dots)$ we find

$$\|x_j\|^p \leq C^p \alpha_{jn} \qquad \forall\, j \leq n$$

So we find

$$\sum_{j=1}^{n} \|x_j\|^p \leq C^p \sum_{j=1}^{n} \alpha_{jn} = C^p \qquad \forall\ n \geq 1$$

Hence we have

Corollary 3.11. If $\ell^p \to \ell^\infty$ is not mimicked through E, and (x_j) is a sequence in E with unconditionally bounded partial sums, then we have $(x_j) \in \ell^p_E$. □

If E and F are Banach spaces, we shall say that F contains E if F contains a subspace isomorphic to E.

The final results we shall need from the geometric theory of Banach spaces are the following:

Theorem 3.12. Let (S,Σ,μ) be a positive measure space, and E a Banach space. Then we have

(a) If $\ell^p \to \ell^\infty$ is not mimicked through E, then neither is it mimicked through $L^q(\mu,E)$ for any $1 \leq q < p$.

(b) If $\ell^1 \to \ell^p$ is not mimicked through E, then neither is it mimicked through $L^q(\mu,E)$ for any $p < q < \infty$.

(c) If c_o is not contained in E, then neither is c_o contained in $L^q(\mu,E)$ for any $1 \leq q < \infty$. □

This theorem is a deep result and the proof of all 3 statements requires stochastic methods (see [89],[40] and [75]). But we shall sustain from the proof here.

The property of not containing c_o, may be expressed in similar facon as in Corollary 3.11 (see [8]):

Theorem 3.13. Let E be a Banach space not containing c_o, and (x_j) a sequence with unconditionally bounded partial sums. Then we have $x_j \to 0$ (i.e. $(x_j) \in c_o(E)$), and $\sum_{j=1}^{\infty} x_j$ is unconditionally convergent. □

4. Weak convergence

Let T be a completely regular Hausdorff space in all of this section, and let

$$C(T) = \{f \mid f \text{ continuous bounded: } T \to \mathbb{R}\}$$

$$Pr(T) = \{\mu \mid \mu \text{ is a Radon probability on } T\}$$

On $Pr(T)$ we define the weak topology, $w = \sigma(Pr,C)$, by

$$\mu_\alpha \xrightarrow{w} \mu \Leftrightarrow \int_T f\, d\mu_\alpha \longrightarrow \int_T f\, d\mu \qquad \forall f \in C(T)$$

If T is metrized by a metric d, we define the Prohorov metric d_o by

$$d_o(\mu,\nu) = \inf \left\{ \varepsilon > 0 \;\middle|\; \begin{array}{ll} \mu(F) \le \nu(F^\varepsilon) + \varepsilon & \forall F \in \mathcal{F} \\ \nu(F) \le \mu(F^\varepsilon) + \varepsilon & \forall F \in \mathcal{F} \end{array} \right\}$$

where $\mathcal{F}$ is the class of all closed sets and

$$A^\varepsilon = \{x \in T \mid \exists y \in A: \; d(x,y) < \varepsilon\} = \bigcup_{y \in A} b^o(y,\varepsilon)$$

The following propositions are well known

(4.1) If d is a metric for T, then d_o is a metric for $(Pr(T),w)$

(4.2) If (T,d) is complete, then so is $(Pr(T),d_o)$

(4.3) $\mu_\alpha \xrightarrow{w} \mu$ if and only $\mu(U) \le \liminf_\alpha \mu_\alpha(U)$ for all open sets $U \subseteq T$.

Let X,Y be random variables with valued in a separable subset of (T,d), and let $\varphi: \mathbb{R}_+ \to \overline{\mathbb{R}}_+$ be decreasing and $\not\equiv \infty$, then we define

$$\Lambda_\varphi(X,Y) = \lambda_\varphi(d(X,Y))$$

(cf. §2). Now using the inequality:

$$P(X \in F) \le P(d(X,Y) \ge \varepsilon) + P(Y \in F^\varepsilon)$$

we find

$$d_o(L(X), L(Y)) \le \Lambda_\varphi(X,Y)\max\{\varphi(t), t\} \qquad \forall\, t \ge 0 \tag{4.4}$$

Lemma 4.1. Let T be a completely regular Hausdorff space and F a subset of $C(T)$, so that

(a) $f \cdot g \in F \quad \forall\, f, g \in F$

(b) F separates points in T

If (μ_n) and μ are Radon probabilities on T, so that

(c) $\int_T f\, d\mu_n \to \int_T f\, d\mu \quad \forall\, f \in F$

then $\mu_n \overset{w}{\to} \mu$ if T is equipped with the $\sigma(T,F)$-topology.

Remark. The $\sigma(T,F)$-topology is the weakest topology on T which makes all functions $f \in F$ continuous, i.e.

$$t_\alpha \to t \text{ in } \sigma(T,F) \Leftrightarrow f(x_\alpha) \to f(x) \qquad \forall\, f \in F$$

Proof. Let H be the class of functions $f \in C(T)$, for which

$$\lim_{n\to\infty} \int_T f\, d\mu_n = \int_T f\, d\mu$$

Then H is a $\|\cdot\|_\infty$-closed linear subspace of $C(T)$ containing F and all constant functions. So if $F_o = \overline{\text{span}(F \cup \{1_T\})}$, then $F_o \subseteq H$, and F_o is a uniformly closed algebra in $C(T)$.

Let U be a $\sigma(T,F)$-open set, then for every $x \in U$ there exist a finite set $F(x) \subseteq F$ and $\varepsilon(x) > 0$, so that

$$U(x) = \{y \in T \mid \max_{f \in F(x)} |f(y) - f(x)| < \varepsilon(x)\} \subseteq U$$

Let $g_x = \varepsilon(x)^{-1} \max_{f \in F(x)} |f(\cdot) - f(x)|$, then $g_x \in F_o$ and

$$x \in U(x) = \{g_x < 1\} \subseteq U$$

Since μ is a Radon measure and $U = \bigcup_{x \in U} U(x)$, there exists a finite set $\pi \subseteq U$, so that

$$\mu(\bigcup_{x \in \pi} U(x)) \geq \mu(U) - \varepsilon$$

where ε is any given positive number. Let

$$h = \min_{x \in \pi} g_x$$

then $h \in F_o$ and $\{h < 1\} = \bigcup_{x \in \pi} U(x) \subseteq U$.

Now let $\nu_n = \mu_n o h^{-1}$ and $\nu = \mu o h^{-1}$, then ν_n and ν are probability measures on $\mathbb{R}$. And since $\varphi o h \in F_o$ for all $\varphi \in C(\mathbb{R})$ (Stone-Weierstrass' theorem) we have $\nu_n \overset{w}{\to} \nu$. So by (4.3) we find

$$\begin{aligned} \mu(U) - \varepsilon &\leq \mu(h < 1) = \nu(-\infty, 1) \\ &\leq \liminf_{n\to\infty} \nu_n(-\infty, 1) \\ &= \liminf_{n\to\infty} \mu_n(h < 1) \\ &\leq \liminf_{n\to\infty} \mu_n(U) \end{aligned}$$

Hence the lemma follows from (4.3). □

<u>Proposition 4.2</u>. <u>Let</u> f <u>be a continuous map</u>: $T \to \mathbb{R}$, <u>and</u> $\mu_n \in Pr(T)$, <u>so that</u> $\mu_n \overset{w}{\to} \mu$. <u>If</u> f <u>is equiintegrable with</u>

<u>respect to</u> $\{\mu_n\}$, <u>that is if</u>

(a) $\int_{\{|f|>a\}} |f| d\mu_n \underset{a\to\infty}{\Longrightarrow} 0$ uniformly in $n\geq 1$

<u>Then we have</u>

(b) $$\lim_{n\to\infty} \int_T f\, d\mu_n = \int_T f\, d\mu$$

<u>Proof</u>. The proof is simple and we shall leave the verification to the reader. □

Let T and S be completely regular spaces, and F a continuous map: $T \to S$, then F induces a map: $\tilde{F}: Pr(T) \to Pr(S)$, by

$$\tilde{F}(\mu) = \mu \circ F^{-1}$$

i.e. $\tilde{F}(\mu)$ is the distribution law of F under μ or the image measure of μ under F. It is clear that we have:

(4.5) If $F: T \to S$ is continuous, then $\tilde{F}: Pr(T) \to Pr(S)$ is weakly continuous.

Another very important continuous map is the product measure:

(4.6) If T and S are completely regular Hausdorff spaces then $(\mu,\nu) \to \mu \times \nu$ is a continuous map: $Pr(T) \times Pr(S) \to Pr(S\times T)$.

The proof of this proposition is straight forward but requires some work. We shall leave the verification to the reader.

Let E be a Banach space and (X_n) and X be E-valued random vectors. Then $L(X)$ is a Radon measure, since X has separable range. And we say that $\{X_n\}$ <u>converges in law to</u> X if $L(X_n) \xrightarrow{w} L(X)$.

A family $M \subseteq Pr(T)$ is called uniformly tight, if M satisfies

(4.7) $\forall\, \varepsilon > 0 \;\; \exists\, K$ compact $\subseteq T$: $\mu(K) \geq 1-\varepsilon$, $\forall\, \mu \in M$

It is easy to see that

(4.8) If $M \subseteq Pr(T)$ is uniformly tight, then M is conditionally w-compact.

The converse is not true in general, but we have Pro orov's theorem:

(4.9) If T is a complete metric space, and $M \subseteq Pr(T)$ is w-compact, then M is uniformly tight.

5. Measures with given marginals

Let S and T denote completely regular spaces in all of this section. If $\sigma: S \to \mathbb{R}_+$ is continuous and bounded below away from 0 we define

$$M(S) = \{\mu \mid \mu \text{ is a finite Radon measure on } S\}$$
$$M_\sigma(S) = \{\mu \in M(S) \mid \sigma \text{ is } \mu\text{-integrable}\}$$
$$Pr_\sigma(S) = M_\sigma(S) \cap Pr(S)$$
$$C_\sigma(S) = \{f \mid f/\sigma \in C(S)\}$$

Then $f \in L_1(\mu)$ for all $f \in C_\sigma(S)$ and $\mu \in M_\sigma(S)$, so we can define the w_σ-topology on $Pr_\sigma(S)$ by

$$\mu_\alpha \to \mu \text{ in } w_\sigma \iff \int_S f d\mu_\alpha \to \int_S f d\mu \quad \forall f \in C_\sigma(S)$$

Clearly we have $w = w_1$ and

(5.1) $\quad \mu \to \int \sigma d\mu$ is a homeomorphism of $(M_\sigma(S), w_\sigma)$ onto $(M(S), w)$

If $f: S \to \mathbb{R}$ and $g: T \to \mathbb{R}$ are maps, then $f \oplus g$ denotes the function:

$$f \oplus g(s,t) = f(s) + f(t) \quad \forall (s,t) \in S \times T$$

<u>Lemma 5.1</u>. <u>Let</u> (γ_α) <u>be a net in</u> $M_+(S\times T)$ <u>and let</u> μ_α <u>and</u> ν_α <u>be the marginals of</u> γ_α. <u>If</u>

(a) $$\mu_\alpha \xrightarrow{w} \mu, \quad \nu_\alpha \xrightarrow{w} \nu$$

<u>for some</u> $\mu \in M_+(S)$ <u>and some</u> $\nu \in M_+(T)$. <u>Then</u> (γ_α) <u>has a limit point</u> $\gamma \in M_+(S\times T)$ <u>with marginals</u> μ <u>and</u> ν.

Proof. Clearly any limit point γ for (γ_α) has marginals μ and ν, so it suffices to prove that (γ_α) has a limit point. To this we shall use a compactness criteria by F. Topsøe (Theorem 9.1 of [111]), which state that is suffices to prove:

(i) $\limsup_\alpha \gamma_\alpha(S\times T) < \infty$

(ii) $\forall \varepsilon > 0 \ \exists G \in \mathcal{G}: \limsup_\alpha \gamma_\alpha(G^c) \leq \varepsilon$

whenever $\mathcal{G}$ is a family of open subsets of $S \times T$, such that

(iii) $\forall K$ compact $\exists G \in \mathcal{G}$ so that $K \subseteq G$

Since $\gamma_\alpha(S\times T) = \mu_\alpha(S) = \nu_\alpha(T)$, (i) is fullfilled by (a). So let $\mathcal{G}$ satisfy (iii) and let $\varepsilon > 0$ be given, then we can find compact sets $K \subseteq S$, $L \subseteq T$, so that

$$\mu(K^c) \leq \tfrac{1}{2}\varepsilon, \quad \nu(L^c) \leq \tfrac{1}{2}\varepsilon$$

Let $G \in \mathcal{G}$ be chosen so that $G \supseteq K \times L$, then by compactness of K and L we can find open sets $U \supseteq K$ and $V \supseteq L$ so that $G \supseteq U \times V$. Hence

$$\limsup_\alpha \mu_\alpha(U^c) \leq \mu(U^c) \leq \mu(K^c) \leq \tfrac{1}{2}\varepsilon$$

$$\limsup_\alpha \nu_\alpha(V^c) \leq \nu(V^c) \leq \nu(L^c) \leq \tfrac{1}{2}\varepsilon$$

and since

$$G^c \subseteq (U\times V)^c \subseteq (U^c\times T) \cup (S\times V^c)$$

we have

$$\lim_{\alpha}\sup \gamma_{\alpha}(G^{c}) \leq \varepsilon$$

and so (ii) holds and the lemma is proved. □

Theorem 5.2. Let $\sigma : S \to \mathbb{R}_{+}$ and $\tau : T \to \mathbb{R}_{+}$ be continuous and bounded below away from 0, and let $\rho = \sigma \oplus \tau$. Suppose that $\mu \in Pr_{\sigma}(S)$, $\nu \in Pr_{\tau}(T)$, $\Lambda \subseteq Pr_{\rho}(S\times T)$ and that

(a) $$\int_S f d\mu + \int_T g d\nu \leq p(f \oplus g) \quad \forall f \in C_{\sigma}(S)\, \forall g \in C_{\tau}(T)$$

where

(b) $$p(\varphi) = \sup_{\lambda\in\Lambda}\int_{S\times T} \varphi d\lambda \quad \text{for } \varphi \in C_{\rho}(S\times T)$$

Then there exist $\gamma \in Pr_{\rho}(S\times T)$ with marginals μ and ν, which satisfies

(c) $$\int_{S\times T} \varphi d\gamma \leq p(\varphi) \qquad \forall \varphi \in C_{\rho}(S\times T)$$

Remark. If Λ is convex and w_{ρ}-closed, then any $\gamma \in M_{\rho}(S\times T)$ satisfying (c) belongs to Λ.

Proof. Let Γ be the set of all measures in $Pr_{\rho}(S\times T)$, which satisfies (c). Then Γ is convex and w_{ρ}-closed, and $\Lambda \subseteq \Gamma$, moreover the definition of p implies:

(i) $$p(\varphi) = \sup_{\gamma \in \Gamma} \int \varphi d\gamma \qquad \forall \varphi \in C\rho(S\times T)$$

Now let Δ be the set of all $(\alpha,\beta) \in Pr_\sigma(S) \times Pr_\tau(T)$, for which there exists a $\gamma \in \Gamma$ with marginals α and β. Then Δ is a convex subset of $M_\sigma(S) \times M_\tau(T)$, which is the dual of $C_\sigma(S) \times C_\tau(T)$ under the pairing:

$$\langle(\alpha,\beta), (f,g)\rangle = \int_S fd\alpha + \int_T gd\beta$$

(i) and (a) implies that

$$\langle(\mu,\nu), (f,g)\rangle \le \sup_{(\alpha,\beta)\in\Delta} \langle(\alpha,\beta), (f,g)\rangle$$

So $(\mu,\nu) \in cl(\Delta)$ (closure in $w_\sigma \times w_\tau$).

But then there exists $\gamma_\alpha \subseteq \Gamma$ with marginals μ_α and ν_α, so that $\mu_\alpha \to \mu$ in w_σ and $\nu_\alpha \to \nu$ in w_τ. So by Lemma 5.1 we have that (γ_α) has a w-limit point γ with marginals μ and ν.

Now we note that

$$\int_{S\times T} \rho\, d\gamma_\alpha = \int_S \sigma\, d\mu_\alpha + \int_T \tau\, d\nu_\alpha$$

$$\underset{\alpha}{\to} \int_S \sigma\, d\mu + \int_T \tau\, d\nu$$

$$= \int_{S\times T} \rho\, d\gamma$$

But then the theorem follows from the following easy proposition

(ii) If $\mu_\alpha \overset{w}{\to} \mu$ and $\lim_\alpha \int_S \sigma\, d\mu_\alpha = \int_S \sigma\, d\mu$, then $\mu_\alpha \to \mu$ in w_σ. □

<u>Corollary 5.3</u>. <u>Let</u> σ,τ <u>and</u> $\rho = \sigma \oplus \tau$ <u>be as in Theorem 5.2, let</u> Λ <u>be a convex</u> w_ρ<u>-closed subset of</u> $Pr_\rho(S\times T)$ <u>and</u> $M(s)$ <u>a subset of</u> $Pr_\tau(T)$ <u>for all</u> $s \in S$, <u>so that</u>

(a) $\delta_s \times \alpha \in \Lambda \quad \forall s \in S \quad \forall \alpha \in M(s)$

<u>If</u> $g \in C_\tau(T)$ <u>we define</u>

$$g^*(s) = \sup_{\alpha \in M(s)} \int_T g\,d\alpha$$

$$g_*(s) = \inf_{\alpha \in M(s)} \int_T g\,d\alpha$$

<u>Let</u> $\mu \in Pr_\sigma(S)$ <u>and</u> $\nu \in Pr_\tau(T)$, <u>then the following two statements are equivalent:</u>

(b) $$\int_S f d\mu + \int_T g d\nu \le \sup_{s\in S}(f(s) + g^*(s)) \quad \forall f \in C_\sigma(S) \forall g \in C_\tau(T)$$

(c) $$\int_S f d\mu + \int_T g d\nu \ge \inf_{s\in S}(f(s) + g_*(s)) \quad \forall f \in C_\sigma(S) \forall g \in C_\tau(T)$$

<u>And if one of them holds then there exists a</u> $\gamma \in \Lambda$ <u>with marginals</u> μ <u>and</u> ν.

<u>Proof</u>. The equivalence of (b) and (c) follows from the equation: $(-g)_* = -(g^*)$.

Suppose that (b) holds and define

$$p(\varphi) = \sup_{\gamma\in\Lambda} \int_{S\times T} \varphi d\gamma$$

$$\varphi^*(s) = \sup_{\alpha \in M(s)} \int \varphi(s,t)\alpha(dt)$$

for $\varphi \in C_\rho(S\times T)$. Then $\varphi^*(s) \le p(\varphi)$ for all $s \in S$, since

$\delta_s \times \alpha \in \Lambda$ for $\alpha \in M(s)$, hence

$$\sup_{s \in S} \varphi^*(s) \leq p(\varphi)$$

and since $(f \oplus g)^* = f + g^*$ the corollary is a consequence of Theorem 5.2 and the remark to Theorem 5.2. □

Remark. Note that

$$(5,2) \qquad \int_T g d\nu \leq \int^* g^* d\mu \qquad \forall g \in C_\tau(T)$$

implies (b), where $\int^*$ denotes the outer integral. And note that

$$(5,3) \qquad \int_T g d\nu \geq \int_* g_* d\mu \qquad \forall g \in C_\tau(T)$$

implies (c), where $\int_*$ denote the inner integral.

Corollary 5.4. Let D be a closed subset of $S \times T$ and $\mu \in Pr(S)$, $\nu \in Pr(T)$ such that

$$\int_S f d\mu + \int_T g d\nu \leq \sup_{(s,t) \in D} (f(s) + g(t))$$

for all $f \in C(S)$ and all $g \in C(T)$. Then there exists $\gamma \in Pr(D)$ with marginals μ and ν

Proof. Let $D(s) = \{t \mid (s,t) \in D\}$, and put $\sigma = \tau \equiv 1$, $\Lambda = Pr(D)$ and

$$M(s) = \{\delta_t \mid t \in D(s)\}$$

in Corollary 5.3. □

In order to apply Corollary 5.3 and its remark it is important to know measurability or continuity properties of g^* and g_*. To do this we need the following definitions:

A map $R: S \to \{\text{the subsets of } T\}$, is called lower continuous if

$$\{s \mid R(s) \cap V \neq \emptyset\} \quad \text{is open} \quad \forall V \text{ open } \subseteq T$$

R is called upper continuous if

$$\{s \in S \mid R(s) \subseteq V\} \quad \text{is open} \quad \forall V \text{ open } \subseteq T$$

And R is called continuous if R is upper and lower continuous at the same time.

Theorem 5.5. Let $M(s) \subseteq Pr_\tau(T)$, where $\tau: T \to \mathbb{R}_+$ is continuous and bounded below away from 0. Let g^* and g_* be defined by

$$g^*(s) = \sup_{\alpha \in M(s)} \int_T g\,d\alpha$$

$$g_*(s) = \inf_{\alpha \in M(s)} \int_T g\,d\alpha$$

for $g \in C_\tau(T)$.

If M is lower continuous, then g^* is lower semicontinuous and g_* is upper semicontinuous for all $g \in C_\tau(T)$.

If M is upper continuous, then g^* is upper semicontinuous and g_* is lower semicontinuous for all $g \in C_\tau(T)$

If M is continuous, then g^* and g_* are continuous for all $g \in C_\tau(T)$.

Proof. Suppose that M is lower continuous and $g \in C_\tau(T)$. If $g^*(s_0) > a$, then

$$V = \{\alpha \in Pr_\tau(T) \mid \int_T g d\alpha > a\}$$

is an open set with $V \cap M(s_o) \neq \emptyset$. Now let

$$U = \{s \mid M(s) \cap V \neq \emptyset\}$$

then U is an open neighbourhood of s_0, and clearly $g^*(s) > a$ for $s \in U$. Hence g^* is lower semicontinuous and so $g_* = -(-g)^*$ is upper semicontinuous.

Suppose that M is upper continuous and $g \in C_\tau(T)$. If $g^*(s_0) < a$, then we choose $b \in \mathbb{R}$ so that $g^*(s_0) < b < a$. Let

$$V = \{\alpha \in Pr_\tau(T) \int_T g d\alpha < b\}$$

then V is open and $M(s_0) \subseteq V$, so we have

$$U = \{s \in S \mid M(s) \subseteq V\}$$

is an open neighbourhood of s_0 with $g^*(s) \leq b < a$ for all $s \in U$. Hence g^* is upper semicontinuous and $g_* = -(-g)^*$ is lower semicontinuous.

The last statement follows immediately from the previous two statements. □

Theorem 5.6. Let $q\colon C(S \times T) \to \mathbb{R} \cup \{-\infty\}$ be superadditive (i.e. $q(f+g) \geq q(f) + q(g)$) and positive homogenuous (i.e.

$q(af) = aq(f)$ *if* $a \geq 0$). *If* $\mu \in Pr(S)$ *and* $\nu \in Pr(T)$ *satisfies*

(a) $$\int_S f d\mu + \int_T g d\nu \geq q(f \oplus g) \qquad \forall f \in C(S)\, \forall g \in C(T)$$

(b) $$q(\varphi) \geq 0 \qquad \forall \varphi \in C_+(S\times T)$$

Then there exists $\gamma \in Pr(S\times T)$ *with marginals* μ *and* ν *satisfying*

(c) $$\int_{S\times T} \varphi d\gamma \geq q(\varphi) \qquad \forall \varphi \in C(S\times T)$$

Proof. Let $p(\varphi) = -p(-\varphi)$, then p is subadditive and positively homogenuous. Let

$$L = \{f \oplus g \mid f \in C(S), g \in C(T)\}$$

$$\bar{\gamma}(f \oplus g) = \int_S f d\mu + \int_T g d\nu = \int_{S\times T} (f \oplus g) d(\mu\times\nu)$$

Then L is a linear subspace of $C(S\times T)$ and $\bar{\gamma}$ is a linear functional on L dominated by p on L. Hence by Hahn-Banach's theorem (which is also valid, in our case, due to (b) and the structure of L) $\bar{\gamma}$ may be extended to a linear functional $\bar{\bar{\gamma}}$ on $C(S\times T)$, satisfying

(i) $$q(\varphi) \leq \bar{\bar{\gamma}}(\varphi) \leq p(\varphi) \qquad \forall \varphi \in C(S\times T)$$

Hence by (b) we have that $\bar{\bar{\gamma}}$ is a positive linear functional and $\bar{\bar{\gamma}}(1) = 1$. We shall now show that $\bar{\bar{\gamma}}$ is a Radon measure, i.e.

(ii) $\forall \varepsilon > 0 \;\; \exists\, C$ compact, so that $\bar{\bar{\gamma}}(\varphi) \leq \varepsilon$ whenever $0 \leq \varphi \leq 1$ and $\varphi = 0$ on C.

So let $\varepsilon > 0$ be given, and choose compact sets $K \subseteq S$ and $L \subseteq T$ with

$$\mu(K^c) \le \varepsilon/4 \quad \text{and} \quad \nu(L^c) \le \varepsilon/4$$

Now put $C = K \times L$ and let $\varphi \in C(S\times T)$ with $0 \le \varphi \le 1$ and $\varphi = 0$ on C. Then $K \times L \subseteq \{\varphi < \frac{1}{2}\varepsilon\}$, and so there exists open sets $W_1 \supseteq K$ and $W_2 \supseteq L$, so that

$$K \times L \subseteq W_1 \times W_2 \subseteq \{\varphi < \tfrac{1}{2}\varepsilon\}$$

And by complete regularity there exists $f \in C(S)$ and $g \in C(T)$, so that $0 \le f \le 1$, $0 \le g \le 1$ and

$$f(s) = \begin{cases} 1 & \forall s \notin W_1 \\ 0 & \forall s \in K \end{cases}$$

$$g(t) = \begin{cases} 1 & \forall s \notin W_2 \\ 0 & \forall s \in L \end{cases}$$

It is now easily checked that

$$\varphi \le \tfrac{1}{2}\varepsilon + f \oplus g$$

So by positivity of $\overline{\overline{\gamma}}$ we find

$$\begin{aligned} \overline{\overline{\gamma}}(\varphi) &\le \tfrac{1}{2}\varepsilon + \int_S f d\mu + \int_T g d\mu \\ &\le \tfrac{1}{2}\varepsilon + \mu(K^c) + \nu(L^c) \\ &\le \varepsilon \end{aligned}$$

so (ii) is proved. □

Example 5.7. Let me describe a typical situation to which the results applies:

Let S be a completely regular Hausdorff space and $\sigma: S \to \mathbb{R}_+$ be continuous and bounded below away from 0 (in the following we take $T = S$, $\sigma = \tau$ and $\rho = \sigma \oplus \sigma$). Let $\Lambda \subseteq \Pr_\rho(S\times S)$ and $M(s) \subseteq \Pr(S)$ $(s \in S)$ satisfy:

(5.4) $\qquad \delta_s \in M(s) \quad \forall s \in S$

(5.5) $\qquad \Lambda$ is convex and w_ρ-closed

(5.6) $\qquad \delta_s \times \alpha \in \Lambda \quad \forall \alpha \in M(s)\ \forall s \in S$

As above we define g^* and g_* by

$$g^*(s) = \sup_{\alpha\in M(s)} \int g d\alpha$$

$$g_*(s) = \inf_{\alpha\in M(s)} \int g d\alpha$$

for $g \in C_\sigma(S)$. Now let H be a class of functions, $h: S \to \mathbb{R}_+$, satisfying

(5.7) $\qquad g_*(s) \in H \quad \forall g \in C_\sigma(S)$ with $0 \le g \le \sigma$

A function $f: S \to \bar{\mathbb{R}}$ is called M-convex if

(5.8) $\qquad f(s) \le a\, f(t) + (1-a) f(u)$, whenever $0 \le a \le 1$ and $aM(t) + (1-a)M(u) \subseteq M(s)$

The set of M-convex functions is denoted M.

With these assumptions and notation we clearly have:

(5.9) $\qquad g_* \le g \le g^* \qquad \forall g \in C_\sigma(S)$

(5.10) $\qquad g_* \in M \qquad \forall g \in C_\sigma(S)$

(5.11) If $f, g \in M$ and $a \geq 0$, then $f+g$, af and $\max\{f,g\}$ all belongs to M

(5.12) Every constant function is M-convex

And we have

<u>Theorem 5.8</u>. <u>Under the assumptions</u> (5.4) - (5,7), <u>we have that if</u> $\mu, \nu \in Pr_\sigma(S)$, <u>satisfies</u>

(a) $$\int_* g d\mu \leq \int_* g d\nu \qquad \forall g \in H \cap M$$

<u>Then there exist</u> $\gamma \in \Lambda$ <u>with marginals</u> μ <u>and</u> ν.

<u>Proof</u>. Let $g \in C_\sigma(S)$ be bounded below, then there exists $a,b > 0$ so that $0 \leq ag+b \leq \sigma$, hence

$$(ag+b)_* = ag_* + b \in H \cap M$$

and so

$$\begin{aligned} a\int_* g_* d\mu + b &= \int_* (ag+b)_* d\mu \\ &\leq \int_* (ag+b)_* d\nu \\ &= a\int_* g_* d\nu + b \end{aligned}$$

from which we deduce that

(i) $$\int_* g_* d\mu \leq \int_* g_* d\nu$$

whenever $g \in C_\sigma(S)$ and g is bounded below.

Let $g \in C_\sigma(S)$, and put $g_n = \max\{g,-n\}$ then $g_n \in C_\sigma(S)$ and g_n is bounded below. Hence by (i) we have

(ii) $$\int_* (g_n)_* d\mu \le \int_* (g_n)_* d\nu$$

Now $g_n \downarrow g$ so $(g_n)_*$ decreases to some function $h: S \to \mathbb{R} \cup \{-\infty\}$, with $h \ge g_*$. Let $a > g_*(s)$ then there exists $\alpha \in M(s)$, so that

$$\int_S g d\alpha < a$$

But since $g_n \downarrow g$ there exists $n_0 \ge t$, so that

$$\int_S g_n d\alpha < a \qquad \forall n \ge n_0$$

Hence $(g_n)_*(s) < a$ $\forall n \ge n_o$, and so $h(s) \le a$ for any $a > g_*(s)$. That is $(g_n)_* \downarrow g_*$ and $(g_n)_* \le g_0$, $g_0 \in L^1(\mu) \cap L^1(\nu)$. Since the <u>decreasing</u> convergence theorem holds for the inner integral we have by (ii):

$$\int_* g_* d\mu \le \int_* g_* d\nu$$

So the theorem follows from Corollary 5.3 and the remark (5.3). □

<u>Example 5.9</u>. Let S be a Banach space with norm $\|\cdot\|$, and put $\sigma = \|\cdot\| + 1$, $\rho = \sigma \oplus \sigma$. We can then take

$$M(s) = \{\alpha \in Pr_\sigma(S) \mid \int_S x d(dx) = s\}$$

and Λ to be the set of all $\gamma \in Pr_\rho(S\times S)$ satisfying

$$\int_{S\times S} y f(x) \gamma(dx,dy) = \int_{S\times S} x f(x) \gamma(dx,dy)$$

for all $f \in C(S)$. Then clearly (5.4), (5.5) and (5.6) are satisfied. Now

$$aM(u) + (1-a)M(v) \subseteq M(s)$$

if and only if $au + (1-a)v = s$.

So M-convexity is ordinary convexity. Let g be continuous with $0 \le g \le \|\cdot\| + 1$. Then $0 \le g_* \le \sigma$ and g_* is convex. Suppose that $g_*(s_0) < a$, then there exist $\alpha \in M(s)$ so that

$$\int_S g d\alpha < a$$

Let $\alpha_x(A) = \alpha(A-x)$, then

$$p(x) = \int_S g d\alpha_x = \int_S g(y+x)\alpha(dy)$$

is continuous and $p(0) = \int_S g d\alpha < a$. Hence there exists $\delta > 0$ so that $p(x) < a$ for $\|x\| < \delta$. But $\alpha_x \in M(s_0+x)$, and so $g_*(s_0+x) < a$ for $\|x\| < \delta$. Hence g_* is lower semicontinuous.

However any lower semicontinuous convex finite function on a Banach space is continuous. So if we take H to be the set of all continuous functions, f, satisfying: $0 \le f \le \sigma$, then by Theorem 5.8, and the definition of Λ we get:

<u>Theorem 5.10</u>. <u>If</u> μ <u>and</u> ν <u>are Radon measures on a Banach space</u> $(S, \|\cdot\|)$, <u>so that</u>

(a) $\int_S \|x\|\mu(dx) < \infty$, $\int_S \|x\|\nu(dx) < \infty$

(b) $\int_S f d\mu \le \int_S f d\nu$ $\forall f: S \to \mathbb{R}$ <u>convex continuous and with</u> $0 \le f \le 1 + \|\cdot\|$

Then there exist a probability space (Ω, F, P) and two E-valued random vectors, X and Y, with

(c) $L(X) = \mu$ and $L(Y) = \nu$

(d) $X = \mathbb{E}(Y|X)$ □

Example 5.11. Let $\leq$ be a relation on S satisfying

(5.13) $s \leq s \quad \forall s \in S$

(5.14) $s \leq t,\ t \leq u \Rightarrow s \leq u$

(5.15) $D = \{(s,t) \mid t \leq s\}$ is closed in $S \times S$

If $D(s) = \{t \mid t \leq s\}$ is lower continuous, upper continuous or continuous, we say that $\leq$ is lower continuous, upper continuous or continuous.

Put $\sigma \equiv 1$, and

$$M(s) = \{\delta_t \mid t \leq s\}$$

$$\Lambda = \{\gamma \in Pr(S \times S) \mid \gamma(D) = 1\}$$

Then clearly (5.4) and (5.6) holds.

Now $aM(s) + (1-a)M(t) \subseteq M(u)$, if and only if $a = 0$ and $t \leq u$, or $a = 1$ and $s \leq u$, so M-convexity is equivalent to:

$$f(u) \leq f(s) \qquad \forall u \geq s$$

Any function f satisfying this inequality is called decreasing, and any function f satisfying the converse inequality:

$$f(u) \geq f(s) \qquad \forall u \geq s$$

is called increasing.

A subset $A \subseteq S$ is called a <u>right interval</u> (<u>left interval</u>) if 1_A is increasing (decreasing)

<u>Theorem 5.12. Let</u> $\mu, \nu \in Pr(S)$ <u>and let</u> $\leq$ <u>be a relation satisfying</u> (5.13), (5.14) <u>and</u> (5.15). <u>Then the following 3 statements are equivalent</u>

(a) $\int^* f d\nu \leq \int^* g d\mu \quad \forall f,g: S \to \mathbb{R}$, <u>satisfying</u> $f(t) \leq g(s)$ $\forall t \geq s$, $f \geq f_0$ <u>and</u> $g \geq g_0$ <u>for some</u> $f_0 \in L^1(\nu)$ <u>and some</u> $g_0 \in L^1(\mu)$

(b) $\int^* f d\nu \leq \int^* f d\mu \quad \forall f: S \to [0,1]$, <u>increasing</u>

(c) $\exists \gamma \in Pr(D)$ <u>with marginals</u> μ <u>and</u> ν

<u>If</u> $\leq$ <u>is lower continuous, then</u> (a)-(c) <u>is equivalent to</u>

(d) $\nu(A) \leq \mu(A) \quad \forall A$ <u>a closed right interval</u>.

<u>If</u> $\leq$ <u>is upper continuous, then</u> (a)-(c) <u>is equivalent to</u>

(e) $\nu(A) \leq \mu(A) \quad \forall A$ <u>an open right interval</u>.

<u>If</u> $\leq$ <u>is continuous, then</u> (a)-(c) <u>is equivalent to</u>

(f) $\int_S f d\nu \leq \int_S f d\mu \quad \forall f: S \to [0,1]$ <u>continuous and increasing</u>.

<u>Proof</u>. The implications (a) $\Rightarrow$ (b), (a) $\Rightarrow$ (d), (a) $\Rightarrow$ (e) and (a) $\Rightarrow$ (f) are all evident.

(b) $\Rightarrow$ (c): Let $H = \{f \mid f: S \to [0,1]\}$, then (5,7) is satisfied. If $g \in H$ is M-convex then g is decreasing, and $f = 1-g$ is increasing and $0 \leq f \leq 1$. Hence

$$\int_* g d\mu = \int_* (1-f) d\mu = 1 - \int^* f d\mu$$
$$\leq 1 - \int^* f d\nu = \int_* (1-f) d\nu = \int_* g d\nu$$

So (c) follows from Theorem 5.8.

(c) $\Longrightarrow$ (a): Follows from the following fact:

(*) If $\alpha \in Pr(S)$ and $p: S \to T$ is continuous and $\beta = \alpha \circ p^{-1}$, then

$$\int^* h d\beta = \int^* (h \circ p) d\alpha$$

$\forall h: T \to \mathbb{R}$, so that $h \geq h_0$ for some $h_0 \in L^1(\beta)$

which implies

$$\int^* f d\nu = \int^* (f \circ q) d\gamma \leq \int^* (g \circ p) d\gamma = \int^* g d\mu$$

where $q(s,t) = t$ and $p(s,t) = s$, since $f \circ q \leq g \circ p$.

(*) is not as evident as it looks, and it is essential that p is <u>continuous</u> and α is a <u>Radon</u> measure. I shall however leave the verification to the reader.

Now suppose that $\leq$ is lower continuous, and that (d) holds. If $f: S \to [0,1]$ is decreasing and upper semicontinuous, then $\{g \geq t\}$ is a closed right interval for all $t \geq 0$, where $g = 1 - f$. So we have

$$\int_S f d\mu = 1 - \int_S g d\mu = 1 - \int_0^\infty \mu(g \geq t) dt$$
$$\leq 1 - \int_0^\infty \nu(g \geq t) dt = \int_S f d\nu$$

since $g \geq 0$. Hence (c) follows from Theorem 5.8 by putting $\mathcal{H} = \{f | f \text{ l.s.c. } : S \to [0,1]\}$ (cf. Theorem 5.5.)

The two last statements follows similarly. □

CHAPTER II

The tail distribution of convex functions of infinitely many variables

1. Supremum of independent random variables

The study of convergence or boundedness of $\sum_1^\infty X_n$ in L_E^p, where (X_n) is an independent E-valued sequence, may often be reduced to study $N = \sup_n \|X_n\|$ (see §4 and §5). This section is devoted to the study of such supremas.

On $\mathbb{R}_+^n$ we introduce coordinatewise ordering, i.e.

$$(1.1) \qquad (x_1,\dots,x_n) \le (y_1,\dots,y_n) \Leftrightarrow x_j \le y_j \quad \forall\, j = 1,\dots,n$$

which is a continuous ordering on $\mathbb{R}_+^n$ (cf. Example I.5.11).

On $M_+(\mathbb{R}_+^n)$ we can introduce the ordering:

$$(1.2) \qquad \mu \vdash \nu \Leftrightarrow \int f d\mu \le \int f d\nu \qquad \forall\, f \in H(\mathbb{R}_+^n)$$

where $H(\mathbb{R}_+^n)$ is the set of increasing measurable functions $f: \mathbb{R}_+^n \to \mathbb{R}$ which are bounded below. From Theorem I.5.12 we have that $\mu \vdash \nu$ if and only if one of the following 3 statements holds

(1.3) $\quad \mu(U) \le \nu(U) \quad \forall\, U$ an open right interval

(1.4) $\quad \mu(F) \le \nu(F) \quad \forall\, F$ a closed right interval

(1.5) $\quad \int f d\mu \le \int f d\nu \quad \forall\, f$ bounded, non negative, continuous and increasing

If $\xi = (\xi_1,\dots,\xi_n)$ and $\eta = (\eta_1,\dots,\eta_n)$ are n-dimensional random vectors, we define

$$\xi \vdash \eta \Leftrightarrow \mathbb{E} f(|\xi|) \leq \mathbb{E} f(|\eta|) \qquad f \in H(\mathbb{R}_+^n)$$

where $|\xi| = (|\xi_1|,\ldots,|\xi_n|)$ and similarly for $|\eta|$. That is

$$\xi \vdash \eta \Leftrightarrow L(|\xi|) \vdash L(|\eta|)$$

Notice that if $n = 1$, then this is equivalent to (by (1.3))

$$P(|\xi| > t) \leq P(|\eta| > t) \qquad \forall t \geq 0.$$

Lemma 1.1. Let ξ and η be n-dimensional random vectors, such that

(a) $\xi_1,\ldots,\xi_n$ are independent and $\eta_1 \cdots \eta_n$ are independent

(b) $\xi_j \vdash \eta_j \qquad \forall j = 1,\ldots,n$

where ξ_j respectively η_j is the j-th coordinate of ξ respectively η. Then $\xi \vdash \eta$, so if $\varphi: \mathbb{R}_+^n \to \mathbb{R}_+$ is increasing on $\mathbb{R}_+^n$ we have $\mathbb{E}\tilde{\varphi}(\xi) \leq \mathbb{E}\tilde{\varphi}(\eta)$ where $\tilde{\varphi}$ is given by:

(c) $$\tilde{\varphi}(t_1,\ldots,t_n) = \varphi(|t_1|,\ldots,|t_n|) \quad \forall (t_1,\ldots,t_n) \in \mathbb{R}^n$$

Proof. We may assume without loss of generality that $\xi_j \geq 0$ and $\eta_j \geq 0$ for all $j = 1,\ldots,n$. If $n = 1$ then the statement is just the definition of $\xi_1 \vdash \eta_1$. Now suppose that the lemma holds for $n-1$. Let

$$\psi_o(t) = \mathbb{E}\varphi(\xi_1,\ldots,\xi_{n-1},t) \qquad t \in \mathbb{R}_+$$

$$\psi_1(t) = \mathbb{E}\varphi(\eta_1,\ldots,\eta_{n-1},t) \qquad t \in \mathbb{R}_+$$

then by induction hypothesis we have $\psi_o \leq \psi_1$, and since ψ_o and ψ_1 are increasing on $\mathbb{R}_+$, we have

$$\mathbb{E}\,\psi_1(\xi_n) \leq \mathbb{E}\psi_1(\eta_n)$$

and from (a) it then follows that

$$\mathbb{E}\varphi(\xi) = \mathbb{E}\psi_o(\xi_n) \leq \mathbb{E}\psi_1(\xi_n) \leq \mathbb{E}\psi_1(\eta_n) = \mathbb{E}\varphi(\eta)$$

Hence $\xi \vdash \eta$. □

<u>Lemma 1.2</u>. <u>Let</u> (ξ_n) <u>be a sequence of independent random variables, so that</u>

(a) $$N = \sup|\xi_n| < \infty \qquad \text{a.s.}$$

<u>If</u> $\varphi: \mathbb{R}_+ \to \mathbb{R}_+$ <u>is increasing and</u> $\psi: \mathbb{R}_+ \to \mathbb{R}_+$ <u>is left continuous, then we have</u>

(b) $$P(N \leq a) \sum_1^\infty \int_{|\xi_n|>a} \varphi(|\xi_n|)\,dP \leq \mathbb{E}\varphi(N) \qquad \forall\ a \geq 0$$

(c) $$\mathbb{E}\psi(N) \leq \sum_1^\infty \mathbb{E}\psi(|\xi_n|)$$

<u>Proof</u>. (b): Let $T = \inf\{n \mid |\xi_n| > a\}$, then $\varphi(|\xi_T|) \leq \varphi(N)$ for $T < \infty$, so we have

$$\mathbb{E}\varphi(N) \geq \mathbb{E}\varphi(|\xi_T|)\, 1_{\{T<\infty\}} = \sum_{n=1}^\infty \int_{T=n} \varphi(|\xi_n|)\,dP$$

$$= \sum_{n=1}^\infty P(|\xi_1| \leq a, \ldots, |\xi_{n-1}| \leq a) \int_{|\xi_n|>a} \varphi(|\xi_n|)\,dP$$

$$\geq P(N \leq a) \sum_{1}^\infty \int_{|\xi_n|>a} \varphi(|\xi_n|)\,dP$$

since $N \leq a$ implies $|\xi_j| \leq a\ \forall j$.

(c): By left continuity of ψ we have

$$\psi(N) \leq \sup_n \psi(|\xi_n|) \leq \sum_1^\infty \psi(|\xi_n|).$$ □

If (ξ_n) are independent and $N = \sup_n |\xi_n|$, then it follows from Lemma 1.2 and the Borel-Cantelli lemma that

$$(1.6)\qquad \begin{aligned} N<\infty \text{ a.s.} &\Leftrightarrow \exists t: \sum_1^\infty P(|\xi_n|>t) < \infty \\ &\Leftrightarrow \lim_{t\to\infty} \sum_1^\infty P(|\xi_n|>t) = 0 \end{aligned}$$

Moreover we have that if $N<\infty$ a.s. then

$$(1.7)\qquad P(N>t) \sim \sum_1^\infty P(|\xi_n|>t) \qquad \text{as} \quad t\to\infty$$

where $f\sim g$ as $t\to\infty$ means $f(t)/g(t) \xrightarrow[t\to\infty]{} 1$.

Lemma 1.3. Let (ξ_n) be independent random variables, satisfying

(a) $N = \sup_n |\xi_n| < \infty$ a.s.

(b) $P(|\xi_n| > ts) \le K(s)\varphi(t)\, P(|\xi_n|>t) \quad \forall t,s \ge a \ \forall n\ge 1$

where $a>0$, and K and φ are functions: $\mathbb{R}_+ \to \mathbb{R}_+$. Then there exists constants $K_o>0$ and $\varepsilon>0$, so that

(c) $P(N>t) \le K_o\, \varphi(\varepsilon t) \quad \forall t\ge b$

Proof. Since $N<\infty$ a.s. we can find $s_o>a$, so that $\sum_1^\infty P(|\xi_n|>s_o)<\infty$. Now let $t\ge b = as_o$, then we have

$$\sum_1^\infty P(|\xi_n|>t) \le K(s_o)\varphi(t/s_o) \sum_1^\infty P(|\xi_n|>s_o) .$$

So (c) follows from Lemma 1.2. □

Lemma 1.4. Let (ξ_n) be independent identical distributed random variables, $\{a_n\} \subseteq \mathbb{R}_+$ and φ an even function: $\mathbb{R} \to \mathbb{R}_+$, which increases on $\mathbb{R}_+$. Now suppose that there exists $K < \infty$, so that

(a) $a_n \uparrow \infty$ and $\varphi(a_n) \leq K n \quad \forall\, n \geq 1$

(b) $N = \sup\{|\xi_n|/a_n\} < \infty$ a.s.

Then $\mathbb{E}\,\varphi(\varepsilon\xi_1) < \infty$ for some $\varepsilon > 0$.

Moreover if $\{a_n\}$ and φ in addition to (a) and (b) satisfies

(c) $a_{nm} \leq C\, a_n a_m \quad \forall\, n,m \geq 1$ for some $C > 0$

(d) $k\, n \leq \varphi(a_n) \quad \forall\, n \geq 1$ for some $k > 0$

Then for some $K_o < \infty$ and $\delta > 0$ we have

(e) $P(N>t) \leq K_o \varphi(\delta t)^{-1} \quad \forall\, t \geq 0$

Proof. Since $N < \infty$ a.s. we can find $T > 0$, so that $\sum_1^\infty P(|\xi| > a_j T) < \infty$. (Since ξ_j all have the same law we may drop the j). For $a > 0$ we define

$$f_a(x) = \sum_{j=1}^{\infty} 1_{]ja,\infty[}(x)$$

The one easily checks that f_a satisfies

(i) $a\, f_a(x) \leq x \leq a\, f_a(x) + a \quad \forall\, x \geq 0 \quad \forall\, a > 0$

Let $\varepsilon = T^{-1}$, then by use of (i) for $a = K$ (the constant appearing in (a)) we find

$$\begin{aligned}\mathbb{E}\varphi(\varepsilon\xi) &\leq K + K \sum_1^\infty P(\varphi(\varepsilon\xi) > jK) \\ &\leq K + K \sum_1^\infty P(\varphi(\varepsilon\xi) > \varphi(a_j)) \\ &\leq K + K \sum_1^\infty P(|\xi| > Ta_j) < \infty\end{aligned}$$

which proves the first part of the lemma.

Now suppose that (c) and (d) hold. Then by Lemma 1.2 and (i) with $a = kp$, (where $k > 0$ is the constant appearing in (d), and $p \geq 1$ is a given integer) we have:

$$\begin{aligned} P(N > a_p t) &\leq \sum_{j=1}^{\infty} P(|\xi| > a_j a_p t) \\ &\leq \sum_{j=1}^{\infty} P(|\xi| > C^{-1} t\, a_{jp}) \\ &\leq \sum_{j=1}^{\infty} P(\varphi(s\xi) \geq \varphi(a_{jp})) \\ &\leq \sum_{j=1}^{\infty} P(\varphi(s\xi) \geq k\,pj) \\ &\leq k^{-1} p^{-1}\, \mathbb{E}\varphi(s\xi) \end{aligned}$$

where $s = C\,t^{-1}$. So if $\delta = \varepsilon C^{-1}$, then we have

$$P(\delta N > a_p) \leq k^{-1} p^{-1}\, \mathbb{E}\varphi(\varepsilon\xi) = A/(p+1)$$

where A is a finite constant. Put $a_o = 0$ if $t \geq 0$ then we can find $p \geq 0$ so that $a_p \leq \delta t \leq a_{p+1}$ and we have

$$P(N>t) \leq P(\delta N > a_p) \leq A(p+1)^{-1}$$

$$\varphi(\delta t) \leq \varphi(a_{p+1}) \leq K(p+1)$$

So

$$P(N>t) \leq A K \varphi(\delta t)^{-1} \qquad \forall\ t \geq 0 \qquad \square .$$

Suppose that $\{X_n\}$ are independent identical distributed random vectors with values in a Banach space: $(E, \|\cdot\|)$. If X_1 satisfies the <u>strong law of large numbers</u>, that is if

$$(1.8) \qquad \lim_{n\to\infty} \frac{1}{n} \sum_{1}^{n} X_j \quad \text{exists a.s.}$$

then $\sup_n \frac{1}{n} \|X_n\| < \infty$ a.s. So applying Lemma 1.4, to

$$\xi_n = \|X_n\|, \quad a_n = n, \quad \varphi(t) = |t|$$

we find that

$$(1.9) \qquad \mathbb{E}\|X_1\| < \infty, \qquad P(\sup_n \frac{1}{n} \|X_n\| > t) \leq \frac{K}{t}$$

Similarly if X_1 satisfies the law of the iterated logarithm, that is if

$$(1.10) \qquad \limsup_{n\to\infty} \frac{\|\sum_1^n X_j\|}{\sqrt{n \log\log n}} < \infty \text{ a.s.}$$

Then $\sup_n (\|X_n\| / a_n) < \infty$ a.s. where

$$a_n = \sqrt{n \log\log n} \qquad \text{for } n \geq 3$$

$$a_m = 1 \qquad \text{for } n = 1,2$$

and it is easily checked that $\{a_n\}$ satisfies (c), and if is defined by

$$\varphi(t) = \frac{t^2}{\log\log t} \qquad |t| \geq e^e$$

$$\varphi(t) = 1 \qquad |t| < e^e$$

Then it is easily checked that (a) and (d) are satisfied so we have

$$(1.11) \qquad \mathbb{E}\{\frac{\|X_1\|^2}{LL(X_1)}\} < \infty$$

where $LL(t)$ is defined by:

$$(1.12) \qquad LL(t) = \begin{cases} \log\log|t| & \text{for } |t| \geq e^e \\ 1 & \text{for } |t| \leq e^e \end{cases}$$

Collecting these remarks we have proved:

Theorem 1.5. Let (X_n) be independent identical distributed random vectors with values in a Banach space $(E, \|\cdot\|)$.

If X_1 satisfies the strong law of large numbers, then we have

(a) $$\mathbb{E}\|X\| < \infty \quad \text{and} \quad P(\sup_n (\|X_n\|/n) > t) \leq K/t$$

If X_1 satisfies the law of the iterated logarithm, then we have

(b) $$\mathbb{E}\{\frac{\|X\|^2}{L\,L(X)}\} < \infty$$

(c) $$P(\sup_n \frac{\|x_n\|}{\sqrt{n\,L\,L(n)}} > t) \leq \frac{K\,t^2}{L\,L(t)} \qquad \forall t$$ □

2. The contraction principle

We shall in this chapter study the socalled contraction principle, which states that

$$\mathbb{E}q(X_1,\dots,X_n) \le \mathbb{E}q(Y_1,\dots,Y_n)$$

whenever X_j is "smaller" than Y_j and q belongs to a suitable class of functions.

If $x=(x_1,\dots,x_n)\in\mathbb{R}^n$ and $y=(y_1,\dots,y_n)\in\mathbb{R}^n$, then we define

$$|x| = (|x_1|,\dots,|x_n|)$$
$$xy = (x_1y_1,\dots,x_ny_n)$$
$$x\le y \iff x_1\le y_1,\dots,x_n\le y_n$$

Let $\xi=(\xi_1,\dots,\xi_n)$ and $\eta=(\eta_1,\dots,\eta_n)$ be n-dimensional random vectors, then we defined $\xi\vdash\eta$ in §1, and it is clear that $\xi\vdash\eta$ if and only if

(2.1) $\mathbb{E}\,\varphi(\xi)\le\mathbb{E}\varphi(\eta)$ $\forall\varphi\colon \mathbb{R}^n\to\bar{\mathbb{R}}_+$ symmetric Borel function, which increases on $\mathbb{R}^n_+$

We shall introduce another ordering among the n-dimensional random vectors: We say that $\xi\models\eta$ if we have

(2.2) $\mathbb{E}\,\varphi(\xi)\le\mathbb{E}\varphi(\eta)$ $\forall\varphi\colon \mathbb{R}^n\to\bar{\mathbb{R}}_+$ symmetric, convex Borel function.

Since a symmetric convex function necessarily increases on $\mathbb{R}^n_+$, we have that (2.2) is equivalent to

(2.3) $\mathbb{E}\varphi(|\xi|)\le\mathbb{E}\varphi(|\eta|)$ $\forall\varphi\colon \mathbb{R}^n_+\to\bar{\mathbb{R}}_+$ increasing, convex Borel function.

And so we have

(2.4) $\xi \vdash \eta \Rightarrow \xi \models \eta$

If φ is a map: $\mathbb{R}^n \to \bar{\mathbb{R}}_+$, then we define its symmetrization, $\tilde{\varphi}$, by

$$\tilde{\varphi}(t) = 2^{-n} \sum_{\varepsilon} \varphi(\varepsilon t)$$

where we sum over all $\varepsilon = (\pm 1, \pm 1, \dots, \pm 1)$. Then clearly we have

(2.5) $\tilde{\varphi}$ is symmetric

(2.6) If φ is convex then so is $\tilde{\varphi}$

(2.7) $\tilde{\varphi} = \varphi$ if φ is symmetric

(2.8) $\mathbb{E}\,\tilde{\varphi}(\xi) = \mathbb{E}\varphi(\xi)$ if ξ is symmetric.

It is also evident that (2.2) is equivalent to

(2.9) $\mathbb{E}\tilde{\varphi}(\xi) \le \mathbb{E}\tilde{\varphi}(\eta) \quad \forall \varphi\colon \mathbb{R}^n \to \bar{\mathbb{R}}_+$ convex Borel function.

By definition (1.2) and Jensen's inequality (I.2.9) we have

(2.10) If $|\xi| \le |\eta|$ then $\xi \vdash \eta$

(2.11) If $|\xi_j| \le \mathbb{E}(|\eta_j| \,|\, G) \ \forall_j$, then $\xi \models \eta$

Lemma 2.1 If μ and ν are probability measures on $\mathbb{R}^n$ so that $\mu \vdash \nu$, then there exists random variables ξ and η with distributions μ and ν, so that $|\xi| \le |\eta|$

And if μ and ν are probability measures on $\mathbb{R}^n$ so that

(a) $\int \|x\| d\mu < \infty \,, \quad \int \|x\| d\nu < \infty$

(b) $\int f d\mu \leq \int f d\nu \qquad \forall f: \mathbb{R}^n \to \mathbb{R}_+$, convex continuous, symmetric with $f(x) \leq \|x\| + 1$

Then there exist random vectors ξ and η with distributions μ and ν and satisfying $|\xi_j| \leq \mathbb{E}(|\eta_j| \| \xi)$ for all $j = 1, \ldots, n$.

Proof. The first part follows from Theorem I. 5.12. For the second part we use Example I. 5.7 with

$$\sigma = \|\cdot\| + 1$$

$$M(s) = \{\alpha \mid \int |t| d\alpha \geq |s|\}$$

$$\Lambda = \{\gamma \mid \int |t| f(s) d\gamma \geq \int |s| f(s) d\gamma \quad \forall f \in C_+(\mathbb{R}^n)\}$$

Note that $\int |t| d\alpha$ and $\int |t| f(s) d\gamma$ are vector integrals. Then

$$aM(u) + (1-a)M(v) \subseteq M(s) \qquad (0 \leq a \leq 1)$$

if and only if $a|u| + (1-a)|v| \geq |s|$. So any M-convex function is convex and symmetric. Moreever if $f: \mathbb{R}^n \to \mathbb{R}_+$ is convex then f is continuous. So we can take H to be the set of all continuous functions: $\mathbb{R}^n \to \mathbb{R}_+$ and the lemma follows from Theorem I. 5.8. □

Corollary 2.2. If $\xi \vdash \eta$ and φ and ψ are measurable maps: $\mathbb{R}^n \to \bar{\mathbb{R}}_+$, satisfying

(a) $\varphi(t) \leq \psi(s)$ if $|t| \leq |s|$

Then we have

(b) $$\mathbb{E}\varphi(\xi) \le \mathbb{E}\psi(\eta) \qquad \square$$

Corollary 2.3. If $\mathbb{E}\|\xi\| < \infty$ *and* $\mathbb{E}\|\eta\| < \infty$ *and*

(a) $$\mathbb{E}\varphi(\xi) \le \mathbb{E}\varphi(\eta)$$

for all continuous convex symmetric function with $0 \le \varphi \le \|\cdot\| + 1$, *then* $\xi \models \eta$. $\square$

Lemma 2.4. Let ξ *and* η *be real random variables, so that*

(a) $$aP(b|\xi| > t) \le P(|\eta| > t) \quad \textit{if} \quad t \ge t_0$$

where $a \in]0,1]$, $b \in \mathbb{R}_+$ *and* $t_0 \in \mathbb{R}_+$. *Let* α *and* η_0 *be random variables so that*

(b) α *and* ξ *are independent*

(c) $P(|\alpha| = b) = a, \quad P(\alpha = 0) = 1-a$

(d) $|\eta_0| \ge \max(|\eta|, t_0)$

Then if $\varphi: \mathbb{R}_+^n \to \mathbb{R}_+$ *is an increasing Borel map we have*

(e) $\alpha\xi \vdash \eta_0$

(f) $a\,\mathbb{E}\varphi(b|\xi|) \le \mathbb{E}\varphi(|\eta_0|)$

(g) $ab\xi \models \eta_0$

Proof (a). We have

$$\begin{aligned} P(|\alpha\xi| > t) &= P(|\alpha| = b,\ b|\xi| > t) \\ &= a\,P(b|\xi| > t) \\ &\le \begin{cases} 1 & \text{if } t < t_0 \\ P(|\eta| > t) & \text{if } t \ge t_0 \end{cases} \\ &\le P(|\eta_0| > t) \end{aligned}$$

since $|\eta_0| > t$ a.s. if $t < t_0$ and $|\eta_0| \ge |\eta|$ if $t \ge t_0$. Hence $\alpha\xi \vdash \eta_0$ by (1.3)

(b): By independence of α and ξ we have

$$\mathbb{E}\,\varphi(|\alpha\xi|) = a\,\mathbb{E}\varphi(b|\xi|)$$

so (b) follows from (a).

(c): Let φ be convex increasing: $\mathbb{R}_+ \to \bar{\mathbb{R}}_+$, and note that $\mathbb{E}|\alpha| = ab$. Then we have

$$\varphi(abt) = \varphi(t\mathbb{E}|\alpha|) \le \mathbb{E}\varphi(t|\alpha|)$$

by Jensen's inequality (cf(I. 2.9)), and so by independence of α and ξ we have

$$\mathbb{E}\varphi(ab|\xi|) \le \mathbb{E}\varphi(|\alpha\xi|)$$

Hence $ab\xi \models \alpha\xi$, and $\alpha\xi \vdash \eta_0$, so $ab\xi \models \eta_0$ by (2.4). □

In analogy with Lemma 1.1 (and with an analogous proof) we have

Lemma 2.5. *Let* ξ *and* η *be* n-*dimensional random vectors, satisfying*

(a) $\xi_1,\ldots,\xi_n$ *are independent*

(b) $\eta_1,\ldots,\eta_n$ *are independent*

(c) $\xi_j \models \eta_j \quad \forall j=1,\ldots,n$

Then $\xi \models \eta$. □

Lemma 2.6. *Let* ξ', ξ'', η' *and* η'' *be* n-*dimensional vectors, so that*

(a) ξ' *and* ξ'' *are independent*

(b) η' *and* η'' *are independent*

Then we have

(c) $\xi' \vdash \eta', \quad \xi'' \vdash \eta' \Longrightarrow \xi'\xi'' \vdash \eta'\eta''$

(d) $\xi' \models \eta', \quad \xi'' \models \eta'' \Longrightarrow \xi'\xi'' \models \eta'\eta''$

Proof (c): Let $\varphi: \mathbb{R}_+^n \to \bar{\mathbb{R}}_+$ be increasing then $\psi(t) = \mathbb{E}\varphi(t|\xi'|)$ is increasing on $\mathbb{R}_+^n$, and so

$$\mathbb{E}\varphi(|\xi'\xi''|) = \mathbb{E}\psi(|\xi''|) \leq \mathbb{E}\psi(|\eta''|)$$

and since $\psi(t) \leq \mathbb{E}\varphi(t|\eta'|)$ we have

$$\mathbb{E}\varphi(|\xi'\xi''|) \leq \mathbb{E}\varphi(|\eta'\eta''|)$$

(d) Similarly! □

Lemma 2.7. Let $\eta = (\eta_1, \ldots, \eta_n)$, so that $a_j = \mathbb{E}|\eta_j| < \infty$, then $a \models \eta$, where $a = (a_1, \ldots, a_n)$.

Proof. Let $\varphi: \mathbb{R}^n \to \bar{\mathbb{R}}_+$ be a symmetric convex Borel function, then by Jensen's inequality we have

$$\varphi(a) = \varphi(\mathbb{E}|\eta|) \leq \mathbb{E}\varphi(|\eta|) = \mathbb{E}\varphi(\eta) \quad \square$$

Corollary 2.8. Let α, ξ and η be n-dimensional random vectors, so that

(a) α and ξ are independent

(b) $\alpha\xi \models \eta$

(c) $a_j = \mathbb{E}|\alpha_j| < \infty \quad \forall_j = 1, \ldots n$

Then $a\xi \models \eta$ where $a = (a_1, \ldots, a_n)$. $\square$

Let us now apply these results to random vectors with values in a linear space. So let $(E_j, \mathcal{B}_j)$ be measurable linear spaces, and put

$$E^n = \prod_1^n E_j \qquad \mathcal{B}^n = \bigoplus_1^n \mathcal{B}_j$$

If $q: E^n \to \bar{\mathbb{R}}_+$ is a measurable function, we define its symmetrization, $\tilde{q}$, as before:

$$\tilde{q}(x_1, \ldots, x_n) = 2^{-n} \sum_{\pm} q(\pm x_1, \ldots, \pm x_n)$$

If $Y = (Y_1, \ldots, Y_n)$ is a E^n-valued random vector and Q is a class of measurable maps: $E^n \to \bar{\mathbb{R}}_+$, then we say that Y

admits a Q-central symmetrization, Y^s, if

$$(2.12)\quad \begin{cases} Y^s = Y' - Y'' \quad \text{where} \quad L(Y') = L(Y'') = L(Y) \\ Y^s = (Y_1^s,\dots,Y_n^s) \quad \text{is symmetric} \\ \mathbb{E}q(Y) \le \mathbb{E}q(Y^s) \qquad \forall q \in Q \end{cases}$$

Note that we do not require Y' and Y'' are independent. Actually they may very well be dependent.

For example let $E_1,\dots,E_n$ be Banach spaces and $Y_j = \gamma_j Z_j$ where γ_j is a real γ and Z are independent. If Q is the class of all measure convex functions: $E \to \bar{\mathbb{R}}_+$, then Y admits a Q-central symmetrization on each of the following cases:

(2.13) $\gamma_1,\dots,\gamma_n$ are independent and have mean 0

(2.14) $Z_1,\dots,Z_n$ are independent and have mean 0

(In the first case $(\gamma_1^s Z_1,\dots,\gamma_n^s Z_n)$ is a Q-central symmetrization, where γ^s is a symmetrization of γ which is independent of Z, and similar in the second case).

If $t = (t_1,\dots,t_n) \in \mathbb{R}^n$ and $x = (x_1,\dots,x_n) \in E^n$, then we define

$$tx = (t_1x_1,\dots,t_nx_n) \in E^n$$

with the notation we have.

Theorem 2.9. Let ξ and η be n-dimensional random vectors, and X an E^n-valued random vector such that

(a) ξ and X are independent

(b) η and X are independent

(c) $\xi \models \eta$

If $q: E^n \to \bar{\mathbb{R}}_+$ is convex and measurable, then we have

(d) $\mathbb{E}\,\tilde{q}(\xi X) \leq \mathbb{E}\,\tilde{q}(\eta X)$

Proof. Apply (2.9) to ξ, η and $\varphi(t) = \mathbb{E}\,q(tX)$ for $t \in \mathbb{R}^n$. □

Theorem 2.10. Let ξ and η be n-dimensional random vectors, X an E^n-valued random vector and q a convex measurable function: $E^n \to \bar{\mathbb{R}}_+$, such that $\xi \models \eta$ and

(a) ξ and X are independent

(b) η and X are independent

(c) ξX admits a q-central symmetrization

(d) ηX admits a r-central symmetrization

where $r(t) = \tilde{q}(2t)$. Then we have

(e) $\mathbb{E}\,q(\xi X) \leq \tfrac{1}{2}\mathbb{E}\,q(4\eta X) + \tfrac{1}{2}\mathbb{E}\,q(-4\eta X)$

Proof. Let $Y = \xi X$ and $Z = \eta X$, and let $Y^s = Y' - Y''$ and $Z^s = Z' - Z''$ be a q-central respectively r-central symmetrization of Y and Z. Then by (2.12) and (2.8) we have

$$\mathbb{E}\,q(Y) \leq \mathbb{E}\,q(Y^s) = \mathbb{E}\,\tilde{q}(Y^s)$$

And by convexity and symmetry of $\tilde{q}$ we have

$$\mathbb{E}\,\tilde{q}(Y^s) \leq \tfrac{1}{2}\mathbb{E}\,\tilde{q}(2Y') + \tfrac{1}{2}\mathbb{E}\,\tilde{q}(-2y'') = \mathbb{E}\,\tilde{q}(2Y)$$

And similarly

$$\mathbb{E}\,\tilde{q}(2Z) \le \mathbb{E}\,\tilde{q}(2Z^s) = \mathbb{E}\,q(2Z^s)$$
$$\le \tfrac{1}{2}\mathbb{E}\,q(4Z) + \tfrac{1}{2}\mathbb{E}\,q(-4Z)$$

And by Theorem 2.9 we have $\mathbb{E}\,\tilde{q}(2Y) \le \mathbb{E}\,\tilde{q}(2Z)$, so (e) is proved. □

Now let E be a banach space, and let Q be the class of functions, $q:E^n \to \mathbb{R}_+$, of the form:

$$q(x_1,\dots,x_n) = \varphi(\|\sum_{j=1}^{n} x_j\|)$$

where φ is an increasing convex function: $\mathbb{R}_+ \to \mathbb{R}_+$. If $Y = (Y_1,\dots,Y_n)$ admits a Q-central symmetrization, we say that Y admits a <u>central symmetrization</u>. Since q is measure convex for all $q \in Q$ (cf. (I.2.10)) we have

(2.15) $Y = \gamma Z$, where γ and Z are independent, admits a central symmetrization if either (2.13) or (2.14) holds.

And as an immediate consequence of Theorem 2.10 we have

<u>Theorem 2.11</u>. <u>Let</u> ξ <u>and</u> η <u>be</u> n-<u>dimensional random vectors and</u> $(X_1,\dots,X_n) = X$ E-<u>valued random vectors, such that</u>

(a) ξ <u>and</u> X <u>are independent</u>

(b) η <u>and</u> X <u>are independent</u>

(c) ξX <u>and</u> ηX <u>admit central symmetrizations</u>

(d) $\xi \models \eta$

Then for every increasing convex function: $\mathbb{R}_+ \to \mathbb{R}_+$:

(e) $$\mathbb{E}\varphi(\| \sum_{j=1}^{n} \xi_j X_j \|) \le \mathbb{E}\varphi(4 \| \sum_{j=1}^{n} \eta_j X_j \|). \quad \square$$

And as an immediate consequence of Theorem 2.9 we have

Theorem 2.12. Let ξ and η be n-dimensional random vectors, and $X = (X_1, \dots, X_n)$ a symmetric random sequence, such that

(a) ξ and X are independent

(b) η and X are independent

(c) $\xi \models \eta$

Then for every increasing convex function $\varphi: \mathbb{R}_+ \to \mathbb{R}_+$:

(d) $$\mathbb{E}\varphi(\| \sum_{1}^{n} \xi_j X_j \|) \le \mathbb{E}\varphi(\| \sum_{1}^{n} \eta_j X_j \|). \quad \square$$

A sequence (ξ_n) of real random variables is said to be stochastically bounded away from 0, if

(2.19) The sequence $\{\xi_n\} \in L^0$, do not have 0 as a limit point in L^0

Let $f: \mathbb{R}_+ \to \mathbb{R}_+$ be bounded, increasing, continuous, and such that $f(x) > 0$ for $x > 0$ and $f(0) = 0$, then clearly (2.19) is equivalent to each of the following 4 statements:

(2.20) $\inf_n \mathbb{E} f(|\xi_n|) > 0$

(2.21) δ_0 is not a limit point of $\{ L(\xi_n) \}$

(2.22) $\exists a,b > 0$ so that $P(|\xi_n| > a) \geq b \quad \forall n \geq 1$

(2.23) $\exists a,b > 0$ and random variables (α_n) independent of (ξ_n), so that $P(|\alpha_n| = a) = b$, $P(\alpha_n = 0) = 1 - b$

$\alpha_n \vdash \xi_n \quad \forall n$

Let (ξ_n) be a sequence of random variables and let (ξ_n^s) be a symmetrization of (ξ_n), if (ξ_n^s) is stochastically bounded away from 0 we say that (ξ_n) is <u>totally non degenerated</u>. It is fairly easy to see that (ξ_n) is totally non degenerated if and only if any one of the following 2 conditions holds

(2.24) $\exists a,b > 0$: $P(|\xi_n - x| > a) \geq b \quad \forall n \geq 1 \ \forall x \in \mathbb{R}$

(2.25) $\{\xi_n - x_n\}$ is stockastically bounded away from 0 for all sequences $(x_n) \in \mathbb{R}^\infty$

The theorems 2.9-2.12 are all versions of the contraction principle. Note that in each of these theorems we require that q or φ is convex, and that they are all easy consequences of the definition of "$\xi \models \eta$". We shall now show a version of Theorem 2.12, where φ is only increasing. The non-convex case is however a good deal more complicated, and we shall need a couple of lemmas.

<u>Lemma 2.13</u>. <u>Let</u> ξ <u>be a non negative random variable with</u> $0 < \mathbb{E}\xi^2 < \infty$, <u>then we have</u>

$$P(\xi > \lambda \mathbb{E}\xi) \geq (1-\lambda)\frac{2(\mathbb{E}\xi)^2}{\mathbb{E}\xi^2} \qquad \forall 0 \leq \lambda \leq 1$$

Proof. Let $A=\{\xi>\lambda\mathbb{E}\,\xi\}$, $\eta=\xi 1_A$ and $\rho=\xi-\eta=\xi 1_{A^c}$, then $\rho\leq\lambda\mathbb{E}\,\xi$, so we have $\mathbb{E}\,\rho\leq\lambda\mathbb{E}\xi$ and

$$\mathbb{E}\eta = \mathbb{E}\xi - \mathbb{E}\rho \geq (1-\lambda)\mathbb{E}\xi$$

And by Cauchy-Schwarz' inequality we have

$$(\mathbb{E}\eta)^2 \leq (\mathbb{E}\xi^2)(\mathbb{E}1_A^2) = P(A)\mathbb{E}\xi^2$$

From which the lemma follows. □

Lemma 2.14. Let E be a Banach space and let t and s belong to $\mathbb{R}^n$ so that $|t|\leq|s|$. Then there exists random variables $\alpha_1,\dots,\alpha_n$, so that

(a) $\alpha_j=\pm 1$ a.s. $\forall j=1,\dots n$

(b) $P(5\|\sum_1^n \alpha_j s_j x_j\| \geq \|\sum_1^n t_j x_j\|) \geq 1/6 \quad \forall x_1\dots x_n\in E$

Proof. Choose $\lambda_j\in[-1,1]$, such that $t_j=\lambda_j s_j$, and let $\varepsilon_1,\dots,\varepsilon_n,\beta_1\dots,\beta_n$ be independent random variables only assuming the values $+1$ and -1, and such that

$$P(\varepsilon_j=1) = P(\varepsilon_j=-1) = \tfrac{1}{2}$$

$$P(\beta_j=1) = \tfrac{1}{2}(1+\lambda_j),\ P(\beta_j=-1) = \tfrac{1}{2}(1-\lambda_j)$$

Let $x_1,\dots,x_n\in E$ and choose $x'\in E'$ so that

$$\|x'\| = 1 \qquad \|\sum_1^n t_j x_j\| = \langle x', \sum_1^n t_j x_j\rangle = a$$

Let $r_j = s_j\langle x',x_j\rangle$ and suppose that $\sum_{j=1}^{n} r_j^2 > \frac{8a^2}{25}$. We shall then prove:

(i) $$P(\|\sum_1^n \varepsilon_j s_j x_j\| > \frac{a}{5}) \geq \frac{1}{3}(1-\frac{1}{8})^2 = 0.2552 > \frac{1}{4}$$

Let $S = \sum_1^n \varepsilon_j r_j$, then $S^2 \leq \|\sum_1^n \varepsilon_j s_j x_j\|^2$, and so

$$P(\|\sum_1^n \varepsilon_j s_j x_j\| > \frac{a}{5}) \geq P(S^2 > \frac{a^2}{25}) \geq P(S^2 > \frac{1}{8}\mathbb{E}S^2)$$

since $\mathbb{E}\, S^2 = \sum_1^n r_j{}^2 > \frac{8a^2}{25}$. An easy computation show that $K(4) \leq 3$, where $K(4)$ is the constant in Khinchine's inequality (I.2.22). Hence $\mathbb{E}\, S^4 \leq 3(\mathbb{E}\, S^2)^2$, and so (i) follows from Lemma 2.13.

Let us now assume that $\sum_1^n r_j{}^2 \leq \frac{8a^2}{25}$. We shall then show

(ii) $$P(\|\sum_1^n \beta_j s_j x_j\| > \frac{a}{5}) \geq \tfrac{1}{2}$$

Let $T = \sum_1^n \beta_j r_j$, then

$$\mathbb{E}\, T = \sum_1^n \lambda_j r_j = \sum_1^n \lambda_j s_j \langle x',x_j\rangle = \sum_1^n t_j\langle x',x_j\rangle = a$$

$$\operatorname{Var} T = \sum_1^n (1-\lambda_j^2) r_j^2 \leq \frac{8a^2}{25}$$

since $\mathbb{E}\,\beta_j = \lambda_j$ and $\operatorname{Var}\beta_j = (1-\lambda_j^2)$. Now we note that $|T| \leq \|\sum_1^n \alpha_j s_j x_j\|$ and $a-|T| \leq |T-a|$, so by Chebyshev's inequality we find

$$\begin{aligned} P(\|\sum_1^n \alpha_j s_j x_j\| > \frac{a}{5}) &\geq P(|T| > \frac{a}{5}) \\ &\geq P(|T-a| < \frac{4a}{5}) \\ &\geq 1 - (\frac{5}{4a})^2 \operatorname{Var} T \\ &\geq 1 - \frac{25}{16}\,\frac{8}{25} = \tfrac{1}{2} \end{aligned}$$

So (ii) is proved.

Now let β_0 be independent of $\varepsilon_1 \dots \varepsilon_n, \beta_1 \dots \beta_n$ with $P(\beta_0 = 1) = \frac{2}{3}$, $P(\beta_0 = -1) = \frac{1}{3}$, and put

$$\alpha_j = \begin{cases} \varepsilon_j & \text{if } \beta_0 = 1 \\ \beta_j & \text{if } \beta_0 = -1 \end{cases}$$

Then

$$P(\|\sum_1^n \alpha_j s_j x_j\| > \frac{a}{5}) = \frac{2}{3}P(\|\sum_1^n \varepsilon_j s_j x_j\| > \frac{a}{5}) + \frac{1}{3}P(\|\sum_1^n \beta_j s_j x_j\| > \frac{a}{5})$$

and since either (i) of (ii) holds we see that (b) holds for any choice of $x_1 \dots x_n$. □

Theorem 2.15. *Let $(X_1, \dots, X_n)$ be a symmetric sequence of E-valued random vectors, where E is a Banach space, and let ξ and η be n-dimensional random variables, so that*

(a) ξ *and* X *are independent*

(b) η *and* X *are independent*

(c) $\xi \vdash \eta$

Then for every increasing function $\varphi : \mathbb{R}_+ \to \bar{\mathbb{R}}_+$, *we have*

(d) $$\mathbb{E}\,\varphi(\|\sum_1^n \xi_j X_j\|) \le 6\,\mathbb{E}\,\varphi(5\|\sum_1^n \eta_j X_j\|)$$

Proof. First we show:

(i) $$P(\|\sum_1^n t_j X_j\| > c) \le 6P(5\|\sum_1^n s_j X_j\| > c) \qquad \text{if } |t| \le |s|$$

So let $|t| \le |s|$ and $c \ge 0$, then we choose a random

vector $\alpha = (\alpha_1, \ldots x_n)$ independent of X satisfying (a) and (b) in Lemma 2.13. Let $T = \|\sum_1^n t_j X_j\|$, then by symmetry of X and independence of α and X, we have $L(X) = L(\alpha X)$, and so we find

$$P(5\|\sum_1^n s_j X_j\| > c) = P(5\|\sum_1^n \alpha_j s_j X_j\| > c)$$

$$\geq P(5\|\sum_1^n \alpha_j s_j X_j\| > T, \quad T > c)$$

$$= \int_A P(5\|\sum_1^n \alpha_j s_j x_j\| > \|\sum_1^n t_j x_j\|)\ \mu(dx_1, \ldots, dx_n)$$

where $\mu = L(X)$ and

$$A = \{(x_1 \ldots, x_n) \mid \|\sum_1^n t_j x_j\| > c\}$$

Then by Lemma 2.13 we have

$$6P(5\|\sum_1^n s_j X_j\| > c) \geq \mu(A) = P(\|\sum_1^n t_j X_j\| > c)$$

and (i) is proved.

Let $\psi(t_1, \ldots, t_n) = \mathbb{E}\,\varphi(\|\sum_1^n t_j X_j\|)$, then by (i) and Lemma 2.4 we have

$$\psi(t) \leq 6\psi(5s) \qquad \forall |t| \leq |s|$$

And so by Corollary 2.2 we have

$$\mathbb{E}\,\psi(\xi) \leq 6\mathbb{E}\,\psi(5\eta)$$

But by (a) and (b) we have

$$\mathbb{E}\,\psi(\xi) = \mathbb{E}\,\varphi(\|\sum_1^n \xi_j X_j\|)$$

$$\mathbb{E}\psi(5\eta) = \mathbb{E}\,\varphi(5\|\sum_1^n \eta_j X_j\|)$$

and the theorem is proved. □

3. Convex measures

We shall now very briefly study some measures on linear spaces which were introduced by C. Borell. Let $(E,\mathcal{B})$ be a measurable linear space, and let μ be a probability measure on $(E,\mathcal{B})$. If μ_* denotes the inner measure, and if $\alpha \geq -\infty$ then we say that μ is <u>α-convex</u> if it satisfies

(3.1) $$\mu_*(\lambda A + (1-\lambda)B) \geq \{\lambda\mu(A)^\alpha + (1-\lambda)\mu(B)^\alpha\}^{1/\alpha}$$

for all $A,B \in \mathcal{B}$, where the right hand side should be interpreted by continuity for $\alpha = -\infty$ and $\alpha = 0$, that is:

$$\{\lambda x^\alpha + (1-\lambda)y^\alpha\}^{1/\alpha} = \begin{cases} x^\lambda y^{1-\lambda} & \text{for } \alpha = 0 \\ \min\{x,y\} & \text{for } \alpha = -\infty \end{cases}$$

If $E = \mathbb{R}^n$, μ = Lebesgue measure, then (3.1) holds for $\alpha = \frac{1}{n}$ (the Brun-Minkowski inequality), and actually any measure on $\mathbb{R}$ satisfying (3.1) for $\alpha = \frac{1}{n}$ is proportional to the Lebesgue measure. Moreover in $\mathbb{R}^n$ there exist no non-zero α-convex measure for $\alpha > \frac{1}{n}$. Hence in an infinite dimensional space α-convexity is only of interest if $\alpha \leq 0$.

Let $M_\alpha(E)$ denote the space of all α-convex probability measures on $(E,\mathcal{B})$, it is then clear that we have

(3.2) $$M_0(E) \subseteq M_\alpha(E) \subseteq M_\beta(E) \subseteq M_{-\infty}(E) \quad \forall\ 0 \geq \alpha \geq \beta \geq -\infty$$

So $M_{-\infty}(E)$ is the largest of these spaces, and a measure in $M_{-\infty}(E)$ will simply be called a <u>convex measure</u>.

If $-\infty \leq \alpha \leq 0$, then $\mu \in M_\alpha(\mathbb{R}^n)$ if and only if there exists an affine subspace $H \subseteq \mathbb{R}^n$ of dimension $k \leq n$, with the properties:

$$\mu << \lambda_H \qquad (\lambda_H = \text{Lebesgue measure on } H)$$

$$\left(\frac{d\mu}{d\lambda_H}\right)^{\alpha/(1-\alpha k)} \quad \text{is convex if} \quad -\infty < \alpha < 0$$

$$\left(\frac{d\mu}{d\lambda_H}\right)^{-1/k} \quad \text{is convex if} \quad \alpha = -\infty$$

$$\log\left(\frac{d\mu}{d\lambda_H}\right) \quad \text{is concave if} \quad \alpha = 0$$

In general we have that if $\mathcal{B} = \sigma(F)$ for some linear subspace F of the (algebraic) dual E^* then $\mu \in M_\alpha(E)$ if and only if $\mu_{x_1',\ldots,x_n'} \in M_\alpha(\mathbb{R}^n)$ for all $x_1' \ldots x_n' \in F$, where $\mu_{x_1' \ldots x_n'}$ is the image of μ under the map:

$$x \to (x_1'(x),\ldots,x_n'(x)) : E \to \mathbb{R}^n$$

So in principle this solves the problem of determining whether or not a given measure belongs to $M_\alpha(E)$. I shall not prove any of these facts here, but refer to [10] and [11].

Theorem 3.1. Let $q: E \to \mathbb{R} \cup \{+\infty\}$ be measurable and convex, and let μ be an α-convex probability measure on $(E,\mathcal{B})$. Let $F(t)$ denote the distribution function of q under μ:

$$F(t) = \mu(q \leq t) \qquad t \in \mathbb{R}$$

If $-\infty < \alpha < 0$, then $F(t)^\alpha$ is convex (here we define $0^\alpha = \infty$). If $\alpha = 0$ then $\log F(t)$ is concave (we define $\log 0 = -\infty$).

Let $a = \inf\{t \mid F(t) > 0\}$, then F may be written in the form

$$F(t) = F(a) + \int_a^t F'(s)ds \qquad \forall\, t \geq a$$

So F is absolutely continuous apart from a possible jump at a.

Proof. By convexity of q we have

$$\lambda\{q\leq t\} + (1-\lambda)\{q\leq s\} \subseteq \{q\leq \lambda t + (1-\lambda)s\}$$

So $F(t)^{\alpha}$ is convex if $-\infty < \alpha < 0$, and $\log F(t)$ is concave if $\alpha = 0$ (recall that $\alpha < 0$, so raising both sides of (3.1) to the power α, reverse the inequality). Now a convex function is absolutely continuous on the interior of the set where it is finite, and since $F(t)^{\alpha} < \infty$ for $t > a$ and similar $\log F(t) > -\infty$ for $t > a$, the last part of the theorem follows. □

Theorem 3.2. Let $q: E \to \overline{\mathbb{R}}_+$ be a measurable function satisfying

(a) $q(tx+sy) \leq (|t|+|s|)\max\{q(x),q(y)\} \quad \forall t,s \in \mathbb{R}, \forall x,y \in E$

Let μ be an α-convex probability measure on $(E,\mathcal{B})$ where $-\infty < \alpha \leq 0$, so that

(b) $\mu(q = \infty) < \frac{1}{2}$

Then we have $q < \infty$ μ- a.s., and for $t > 0$

(c) $$\mu(q > t) \leq \begin{cases} K t^{1/\alpha} & \text{if } -\infty < \alpha < 0 \\ K e^{-\varepsilon t} & \text{if } \alpha = 0 \end{cases}$$

for some constants $K < \infty$ and $\varepsilon > 0$.

Note that (a) is equivalent to q being quasiconvex and satisfying: $q(tx) \leq |t|\, q(x)$.

Proof. Let $a > 0$ be chosen so that $\theta = \mu(q\leq a) > \frac{1}{2}$. Let us then show that

(i) $$\{q>a\} \supseteq \frac{2}{t+1}\{q > ta\} + \frac{t-1}{t+1}\{q \leq a\} \quad \forall t \geq 1$$

So let $q(x) > ta$ and $q(y) \leq a$, and consider

$$z = \frac{2}{t+1} x + \frac{t-1}{t+1} y$$

Then we find

$$x = \tfrac{1}{2}(t+1)z + \tfrac{1}{2}(1-t)y$$

So by (a) we have:

$$ta < q(x) \leq t \max\{q(y), q(z)\}$$

and since $q(y) \leq a$, we find that $q(z) > a$, and (i) is proved.

If $-\infty < \alpha < 0$, then (i) and (3.1) gives

$$1-\theta \geq \{\frac{2}{t+1} \mu(q>ta)^{\alpha} + \frac{t-1}{t+1} \theta^{\alpha}\}^{1/\alpha}$$

Isolating $\mu(q>ta)$ we find (recall that α is negative):

$$\mu(q>ta) \leq (\beta t+\gamma)^{1/\alpha} \quad \forall\ t \geq 1$$

where $\beta = \tfrac{1}{2}(1-\theta)^{\alpha} - \tfrac{1}{2}\theta^{\alpha}$ and $\gamma = \tfrac{1}{2}(1-\theta)^{\alpha} + \tfrac{1}{2}\theta^{\alpha}$. Now since $\theta > \tfrac{1}{2}$ and $\alpha < 0$ we have that $\beta > 0$, so the first part of (b) and (c) are proved.

If $\alpha = 0$ then (i) and (3.1) gives

$$\log(1-\theta) \geq \frac{2}{t+1} \log \mu(q>ta) + \frac{t-1}{t+1} \log \theta$$

and isolating $\log \mu(q>ta)$ gives

$$\log \mu(q>ta) \leq -\varepsilon t + K$$

where $\varepsilon = \tfrac{1}{2} \log \frac{\theta}{1-\theta}$ and $K = \tfrac{1}{2} \log \theta(1-\theta)$. Now since $\theta > \tfrac{1}{2}$ we have $\varepsilon > 0$ and the second parts of (b) and (c) are proved. □

<u>Theorem 3.3</u>. <u>Let</u> E <u>be a linear topological Hausdorff space and</u> μ <u>a convex Radon measure. If</u> G <u>is any additive subgroup</u>

of E (that is $x,y \in G$ implies $x-y \in G$), then $\mu_*(G) = 0$ or $\mu_*(G) = 1$, where μ_* is the inner measure corresponding to μ.

Note. Even when μ is a gaussian Radon measure, $E = \mathbb{R}^\infty$ and G is a linear subspace, it is unknown whether the zero-one law above holds for the outer measure, μ^*.

Proof. Suppose that $\mu_*(G) > 0$, then we can find a symmetric compact set $K_o \subseteq G$, so that $\mu(K_o) > 0$. Hence

$$G_o = \bigcup_{n=1}^{\infty} (K_o + \cdots + K_o)$$

is a σ-compact additive subgroup of E, with $K_o \subseteq G_o \subseteq G$. So it suffices to show that $\mu(G_o) = 1$.

Suppose in contrary, that $\mu(G_o) < 1$, then we choose $\varepsilon > 0$ so that

$$\mu(G_o) < 1-\varepsilon \qquad \text{and} \qquad \mu(K_o) > \varepsilon$$

Further we choose a compact subset $K_1 \subseteq E \setminus G_o$ with

$$\mu(G_o) + \mu(K_1) > 1-\varepsilon$$

Now let $K_n = (n-1)K_o + nK_1$ for $n \geq 0$, then we have

$$\text{(i)} \qquad E\setminus(G_o \cup K_1) \supseteq \frac{1}{n}\{E\setminus(G_o \cup K_n)\} + (1-\frac{1}{n})K_o$$

To see this let $x \in E\setminus(G_o \cup K_n)$ and $y \in K_o$, and put $z = \frac{1}{n}x + (1-\frac{1}{n})y$, then

$$x = nz + (n-1)(-y)$$

and $(-y) \in K_o \subseteq G_o$ so $(n-1)(-y) \in (n-1)K_o \subseteq G_o$. Now $x \notin G_o$ hence $z \notin G_o$, moreover since $x \notin K_n = (n-1)K_o + nK_1$ we have $z \notin K_1$. That is $z \in E\setminus(G_o \cup K_1)$, and (i) is proved.

From (i) and (3.1) we find

$$1 - \mu(G_o) - \mu(K_1) \geq \min\{\mu(E\setminus(G_o \cup K_n)), \mu(K_o)\}$$

By the choice of K_1 and ε we have

$$1 - \mu(G_o) - \mu(K_1) < \varepsilon < \mu(K_o)$$

and so

$$1 - \mu(G_o) - \mu(K_n) \leq \mu(E\setminus(G_o \cup K_n)) < \varepsilon$$

Hence we conclude

(ii) $\qquad \exists\, a > 0: \quad \mu(K_n) \geq a \quad \forall\, n \geq 1$

Let C be a compact subset of E, then is it easy to check that for some $n \geq 1$ we have $K_n \cap C = \emptyset$ (notice that $0 \notin K_o + K_1$ and use compactness of K_o, K_1 and C). So by (ii) we have

$$\mu(E\setminus C) \geq a \quad \forall\, C \text{ compact} \subseteq E$$

which contradicts our assumption that μ is a Radon measure. □

Let E be a linear space and E' a linear subspace of the algebraic dual of E, and suppose that E' separates points of E.

If we put $\mathcal{B} = \sigma(E')$, then $(E,\mathcal{B})$ is a measurable linear space and we can define a <u>gaussian measure</u> to be a probability measure on $(E,\mathcal{B})$, with the property:

(3.2) $\qquad \mu_{x_1' \cdots x_n'}$ is gaussian on $\mathbb{R}^n$ $\forall\, x_1' \cdots x_n' \in E'$

In terms of the <u>Fourier transform</u>:

$$\hat{\mu}(x') = \int_E e^{i<x',x>} \mu(dx) \qquad x' \in E'$$

and the <u>convariance function</u>

$$R(x',y') = \int_E \langle x',x\rangle \langle y',x\rangle \ \mu(dx) \quad x',y' \in E'$$

and the Gelfand-mean

$$\alpha(x') = \int_E \langle x',x\rangle \ \mu(dx)$$

(3.2) is equivalent to

$$(3.3) \qquad \hat{\mu}(x') = \exp(i\ \alpha(x') - \tfrac{1}{2}\mathbb{R}(x',x')) \qquad \forall x' \in E'$$

Now any gaussian measure on $\mathbb{R}^n$ has a density of the form $C \exp(-Q(x-a))$, with respect to the Lebesgue measure on some affine subspace of $\mathbb{R}^n$, where Q is a quadratic form. So it follows from the remarks in the beginning of this §, that we have

(3.4) any gaussian measure belongs to $M_o(E)$.

So if X is an E-valued gaussian random vector and q is a measurable seminorm, with $q(X) < \infty$ a.s. then for some $\varepsilon > 0$ we have

$$\mathbb{E}\{e^{\varepsilon q(X)}\} < \infty$$

(see Theorem 3.2). Actually a theorem of X. Fernique (see [26]), states that we have:

Theorem 3.4. Let E be a locally convex space and $\mathcal{B} = \sigma(E')$. If X is an E-valued gaussian random variable, and q is a measurable seminorm: $E \to \overline{\mathbb{R}}_+$. Then $q(X) = \infty$ a.s. or $q(X) < \infty$ a.s., and in the latter case we have

$$\mathbb{E}\{e^{\varepsilon q(X)^2}\} < \infty$$

for some $\varepsilon > 0$.

Proof. Since $\{q < \infty\}$ is a linear space it has probability 0 or 1 by Theorem 3.3.

Now suppose that $q(X) < \infty$, and choose $\varepsilon > 0$ so that

$$K = \mathbb{E}\{e^{\varepsilon q(X)}\} < \infty$$

Let $X_1, X_2, \ldots$ be independent copies of X, and assume that $\mathbb{E}X = 0$. Put

$$S_n = \sum_1^n X_j \quad \text{and} \quad Z_n = n^{-\frac{1}{2}} S_n$$

Then $Z_n \sim X$, and by Chebyshev's inequality we find:

$$\begin{aligned} P(q(X) > a\sqrt{n}) &= P(q(Z_n) > a\sqrt{n}) = P(q(S_n) > an) \\ &\leq \mathbb{E}\{\exp(\varepsilon q(S_n))\} \; e^{-\varepsilon an} \\ &\leq \mathbb{E}\{\exp(\varepsilon \sum_1^n q(X_j))\} \; e^{-\varepsilon an} \\ &= r^n \end{aligned}$$

where $r = K\,e^{-\varepsilon a}$. Hence we may choose a so large that $r \leq e^{-1}$, and then we have

$$P(q(X) > a\sqrt{n}) \leq e^{-n} \quad \forall\; n \geq 0$$

If $t \geq 0$ we can find $n \geq 1$ with $\sqrt{n} \leq t \leq \sqrt{n+1}$, and then we have

$$P(q(X) > at) \leq P(q(X) > a\sqrt{n}) \leq e^{-n} \leq e^{-t^2+1}$$

So we find

$$P(q(X) > t) \leq \exp(-(t/a)^2 + 1) \quad \forall\; t \geq 0$$

which proves the theorem. □

4. Integrability of seminorms

In §3 we saw that if $\mathcal{L}(X)$ is α-convex $(-\infty < \alpha \le 0)$ then $q(X) \in L^p$ for all $p < -1/\alpha$ for any measurable seminorm $q: E \to \overline{\mathbb{R}}_+$ with $q(X) < \infty$ a.s. We shall now show some similar results for

$$q^*(X) = \sup q(X_1,\ldots,X_n,0,0,\ldots)$$

where (X_n) are independent or symmetric and q an appropriate function: $E^\infty \to \overline{\mathbb{R}}_+$.

Let $(E_n,\mathcal{B}_n)$ be a measurable linear spaces for all $n\ge 1$, and put

$$E^\infty = \prod_1^\infty E_j \qquad \mathcal{B}^\infty = \bigotimes_1^\infty \mathcal{B}_j$$

Let Π_n and p_n be the projections respectively injections defined by

$$\pi_n(x) = (x_1,\ldots,x_n,0\cdots) \quad \text{for } x = (x_j) \in E^\infty$$

$$p_n(y) = (0,\ldots,0,y,0\ldots) \quad \text{for } y \in E_n$$

If q is a map: $E^\infty \to \overline{\mathbb{R}}_+$, we define

$$q^*(x) = \sup_n q(\Pi_n x) = \sup_n q(x_1,\ldots,x_n,0,\ldots)$$

<u>Lemma 4.1</u>. <u>Let</u> $(E,\mathcal{B})$ <u>be a measurable linear space and</u> $q: E \to \overline{\mathbb{R}}_+$ <u>a measurable subadditive function</u>. <u>If</u> X <u>is an</u> E-<u>valued random vector and</u> X^s <u>a symmetrization of</u> X, <u>then we have</u>

(a) $$P(q(X) \le a)\, \mathbb{E}\varphi(q(X)) \le \mathbb{E}\varphi(q(X^s) + a)$$

(b) $$\mathbb{E}\varphi(q(X^s)) \le \mathbb{E}\varphi(2q(X)) + \mathbb{E}\varphi(2q(-X))$$

<u>for any</u> $a\ge 0$ <u>and any increasing function</u> $\varphi: \mathbb{R}_+ \to \mathbb{R}_+$.

Proof (a). Let $A = \{x \in E \mid q(x) \leq a\}$ and let $\mu = L(X)$ then we have

$$\mathbb{E}\,\varphi(q(X^s) + a) \geq \int_A \mathbb{E}\,\varphi(q(X-x) + a)\mu(ds)$$

so there exist $x_o \in A$ (that is $q(x_o) \leq a$), such that

$$\begin{aligned}\mathbb{E}\,\varphi(q(X^s) + a) &\geq \mu(A)\ \mathbb{E}\,\varphi(q(X-x_o) + a)\\ &\geq \mu(A)\ \mathbb{E}\,\varphi(q(X-x_o) + q(x_o))\\ &\geq \mu(A)\ \mathbb{E}\,\varphi(q(X))\end{aligned}$$

since q is subadditive.

(b) By subadditivity of q and monotonicity of φ we have

$$\begin{aligned}\varphi(q(X^s)) &= \varphi(q(X) + q(-X))\\ &\leq \varphi(2q(X)) + \varphi(2q(-X)) \qquad \square\end{aligned}$$

Theorem 4.2. Let $X = (X_n)$ be a symmetric E^∞-valued random sequence, and $q: E^\infty \to \overline{\mathbb{R}}_+$ a measurable quasiconvex function. Let

$$S_n = q(\pi_n X), \qquad M_n = \max_{1 \leq j \leq n} S_j, \qquad M = \sup_j S_j$$

If $\varphi: \mathbb{R}_+ \to \mathbb{R}_+$ is increasing and $\varphi^-(x) = \sup_{y<x} \varphi(y)$ then we have

(a) $\mathbb{E}\,\varphi(M_n) \leq 2\ \mathbb{E}\,\varphi(S_n) \quad \forall\ n \geq 1$

(b) $\mathbb{E}\,\varphi^-(M) \leq 2 \liminf_{n\to\infty} \mathbb{E}\,\varphi(S_n)$

(c) $\{S_n\}$ stochastic bounded $\Leftrightarrow$ $M < \infty$ a.s.

Moreover if $S_n \to S$ in law, then we have

(d) $\mathbb{E}\,\varphi(S) \leq \mathbb{E}\,\varphi(M) \leq 2\ \mathbb{E}\,\varphi(S)$.

<u>Proof</u>. Let $t \geq 0$ be given, and define the stopping time T (w.r.t. (X_n)), by

$$T = \inf\{j \mid S_j > t\} \qquad (\inf \emptyset = \infty)$$

Then $\{T \leq n\} = \{M_n > t\}$. For $1 \leq j \leq n$ we put

$$Z_{nj} = 2\pi_j X - \pi_n X = (X_1, \ldots, X_j, -X_{j+1}, \ldots, -X_n, 0, 0 \ldots)$$

and we have $Z_{jj} = \pi_j Z = \frac{1}{2} Z_{nj} + \frac{1}{2} Z_{nn}$, so by quasiconvexity of q we have

$$S_j = q(Z_{jj}) \leq \max \{q(Z_{nj}), S_n\}$$

If $T = j$, then $S_j > t$, so either $q(Z_{nj}) > t$ or $S_n > t$, and hence we find

(i) $$P(T = j) \leq P(T = j,\ S_n > t) + P(T = j, q(Z_{nj}) > t)$$

Now we notice that $Z_{nj} \sim Z_{nn}$ and that there exists a set $A_j \subseteq E^\infty$ so that

$$\{T = j,\ S_n > t\} = \{Z_{nn} \in A_j\}$$

$$\{T = j,\ q(Z_{nj}) > t\} = \{Z_{nj} \in A_j\}$$

Hence the two probabilities on the right side of (i) are equal and summing over $j = 1, \ldots, n$, gives

(ii) $$P(M_n > t) = \sum_1^n P(T=j) \leq 2\, P(S_n > t)$$

So (a) follows from (1.3).

Since φ^- is left continuous we have, $\varphi^-(M) = \lim_{n\to\infty} \varphi^-(M_n)$, so (b) follows from (a), since $\varphi^- \leq \varphi$.

(c): If $\{S_n\}$ is stochastic bounded, then (ii) shows that $M < \infty$ a.s., and the converse is trivial.

(d): By (a) with $\varphi = 1_{[t,\infty[}$ we have

$$P(S>t) \leq \liminf_{n\to\infty} P(S_n>t) \leq P(M>t) \leq \limsup_{n\to\infty} P(M_n \geq t)$$

$$\leq 2 \limsup_{n\to\infty} P(S_n \geq t) \leq 2P(S \geq t)$$

From which (d) follows partial integration. □

Theorem 4.3. Let $X = (X_n)$ be an E^∞-valued sequence of independent random vectors, and let $q: E^\infty \to \overline{\mathbb{R}}_+$ be an even, subadditive, quasiconvex, measurable function. Let

$$S_n = q(\pi_n X), \qquad M_n = \max_{1\leq j\leq n} S_j, \qquad M = \sup_j S_j$$

If $\varphi: \mathbb{R}_+ \to \mathbb{R}_+$ is increasing and $\varphi^-(x) = \sup_{y<x} \varphi(y)$, then we have

(a) $$P(M_n \leq a)\ \mathbb{E}\varphi(M_n) \leq 4\ \mathbb{E}\varphi(2S_n+a)$$

(b) $$P(M \leq a)\ \mathbb{E}\varphi^-(M) \leq 4 \liminf_{n\to\infty} \mathbb{E}\varphi(2S_n+a)$$

(c) $\{S_n\}$ stochastic bounded $\Leftrightarrow$ $M<\infty$ a.s.

Moreover if $S_n \to S$ in law, then we have

(d) $$P(M\leq a)\ \mathbb{E}\varphi(M) \leq 4\ \mathbb{E}\varphi(2S+a)$$

(e) $$\mathbb{E}\varphi(S) \leq \mathbb{E}\varphi(M)$$

for any $a \geq 0$.

Proof. Let $X^s = (X_n^s) = (X_n - X_n')$ be a symmetrization of X, and let S_n^s, M_n^s and M^s be defined as above with X substituted by X^s. Then by Lemma 4.1 and Theorem 4.2 we have

$$P(M_n \leq a)\ \mathbb{E}\varphi(M_n) \leq \mathbb{E}\varphi(M_n^s+a) \leq 2\ \mathbb{E}\varphi(S_n^s+a)$$

$$\leq 4\ \mathbb{E}\ \varphi(2S_n+a)$$

since q is even. This proves (a) and (b) follows similarly.

(c): Clearly $M < \infty$ implies that $\{S_n\}$ is stochastic bounded, so suppose that $\{S_n\}$ is stochastic bounded. Then by Lemma 4.1 we have that $\{S_n^s\}$ is stochastic bounded and so by Theorem 4.2 we have:

$$\infty > P(M^s < \infty) = \int_{E^\infty} P(q^*(X-x) < \infty)\mu(dx)$$

where $\mu = L(X)$. So for some $x_o \in E$, we have that $q^*(X-x_o) < \infty$ a.s. Hence we can find $a > 0$, so that

$$P(q^*(X-x_o) \leq a) > \tfrac{1}{2}, \qquad P(S_n \leq a) > \tfrac{1}{2} \quad \forall\ n \geq 1$$

But then the event:

$$\{S_n \leq a\} \cap \{q^*(X-x_o) \leq a\}$$

has positive probability and so it contains at least one point ω_n for each $n \geq 1$. And we have

$$\begin{aligned} M = q^*(X) &\leq q^*(X-x_o) + q^*(x_o) \\ &\leq q^*(X-x_o) + \sup_n\{q(\pi_n(X(\omega_n) - x_o)) + S_n(\omega_n)\} \\ &\leq q^*(X-x_o) + 2a < \infty \quad \text{a.s.} \end{aligned}$$

since $S_n(\omega_n) \leq a$ and $q(\pi_n(X(\omega_n) - x_o) \leq q^*(X(\omega_n) - x_o) \leq a$.

(d): From (a) with $\varphi = 1_{[t,\infty[}$ we have

$$\begin{aligned} P(M \leq a)\ P(M > t) &\leq \limsup_{n\to\infty} P(M_n \leq a)\ P(M_n \geq t) \\ &\leq 4 \limsup_{n\to\infty} P(2S_n + a \geq t) \leq 4\ P(2S + a \geq t) \end{aligned}$$

so (d) follows by partial integration.

(e): Since $S_n \leq M$ for all n (e) follows as above. □

Lemma 4.4. Let $X = (X_n)$ be an independent symmetric E^{∞}-valued random sequence, and $q: E^{\infty} \to \overline{\mathbb{R}}_+$ a measurable, quasiconvex, subadditive function. Let

$$S_n = q(\pi_n X), \quad M_n = \max_{1 \le j \le n} S_j, \quad M = \sup_j S_j$$

$$T_n = q(p_n X_n), \quad N_n = \max_{1 \le j \le n} T_j, \quad N = \sup_j T_j$$

Then we have

(a) $$P(S_n > s + t + u) \le P(N_n > s) + 2P(S_n > u)P(M_n > t)$$

(b) $$P(M > s + t + u) \le 2P(N > s) + 4P(M > u)P(M > t)$$

for all $s, t, u \in \mathbb{R}_+$.

Proof. Again we consider the stopping time:

$$T = \inf\{j \mid S_j > t\} \qquad (\inf \emptyset = \infty)$$

Then $S_n > t + s + u$ implies $T \le n$, and so we have

(i) $$P(S_n > t + s + u) = \sum_{j=1}^{n} P(T = j,\ S_n > t + s + u)$$

Let $1 \le j \le n$, then $\pi_n X = (\pi_n X - \pi_j X) + \pi_{j-1} X + p_j X$, so by subadditivity of q we find

$$S_n = q(\pi_n X) \le q(\pi_n X - \pi_j X) + S_{j-1} + N_n$$

Hence if $T = j$, $N_n \le s$ and $S_n > t + s + u$, then we have $S_{j-1} < t$, and

$$Y_{nj} = q(\pi_n X - \pi_j X) > u$$

moreover Y_{nj} and $\{T = j\}$ are independent, and so we have

(ii) $$P(T = j, S_n > t + s + u) \le P(T = j, N_n > s) + P(T = j)P(Y_{nj} > u)$$

If we apply Theorem 4.2 to the function $q_o(x) = q(x_n \dots x_1, 0 \dots)$ and to the vector $Y = (X_n, \dots, X_1, 0, 0 \dots)$, we find:

(iii) $$P(Y_{nj} > u) \le 2P(S_n > u)$$

Now if we combine (i), (ii) and (iii) we find

$$\begin{aligned} P(S_n > t+s+u) &\le P(N_n > s) + 2P(S_n > u) \sum_{j=1}^{n} P(T = j) \\ &= P(N_n > s) + 2P(S_n > u) P(M_n > t) \\ &\le P(N > s) + 2P(M > u) P(M > t) \end{aligned}$$

So (a) is proved and (b) follows from Theorem 4.2(b) with $\varphi = 1_{]t+s+u,\infty[}$. □

<u>Corollary 4.5</u>. <u>Let</u> $X = (X_n)$ <u>be an independent</u> E^∞<u>-valued random sequence, and</u> $q\colon E^\infty \to \overline{\mathbb{R}}_+$ <u>a measurable, even, quasi-convex, subadditive function. Let</u>

$$S_n = q(\pi_n X), \quad M = \sup_j S_j, \quad N = \sup_n q(p_n X_n)$$

<u>Then we have</u>

(a) $$P(M \le a)\, P(M > 2s+2t+2u+a) \le 8P(N>s) + 32P(M>t)\, P(M>u)$$

<u>for all</u> $a,s,t,u \in \mathbb{R}_+$. <u>Moreover there exists a decreasing function</u> $F\colon \mathbb{R}_+ \to [0,1]$, <u>so that</u>

(b) $P(M \le a)\, P(M > t+a) \le F(t)$ $\quad \forall\, t,a \ge 0$

(c) $F(2s+t+u) \le 4P(N > s) + 4F(t)F(u)$ $\quad \forall\, t,s,u \ge 0$

(d) $F(t) \le 2P(M > \tfrac{1}{2}t)$ $\quad \forall\, t \ge 0$

Proof. Let (X_n^s) be a symmetrization of (X_n) and define M^s and N^s as above for the sequence (X_n^s). Let

$$F(t) = P(M^s > t) \qquad G(t) = P(N^s > t)$$

Then (b) and (d) holds by Lemma 4.1, and

$$F(t+2s+u) \leq 2G(2s) + 4F(t)F(u)$$

by Lemma 4.4, and since

$$G(2s) \leq 2P(N > s)$$

by Lemma 4.1 we see that (c) holds, and (a) is an easy consequence of (b), (c), and (d). □

Theorem 4.6. Let $X = (X_n)$ be an independent E^∞-valued random sequence and $q: E^\infty \to \overline{\mathbb{R}}_+$ a measurable, even, quasiconvex, subadditive function. Let

$$S_n = q(\pi_n X), \quad M = \sup_n S_n, \quad N = \sup_n q(p_n X_n)$$

and suppose that $M < \infty$ a.s.

If $\varphi: \mathbb{R}_+ \to \mathbb{R}_+$ is increasing and satisfies: $\varphi(2t) \leq K\varphi(t)$ for $t \geq t_o$, then the following statements are equivalent:

(a) $\mathbb{E}\,\varphi(M) < \infty$

(b) $\sup_n \mathbb{E}\,\varphi(S_n) < \infty$

(c) $\mathbb{E}\,\varphi(N) < \infty$

And if $S_n \to S$ in law then (a)-(c) is equivalent to:

(d) $\mathbb{E}\,\varphi(S) < \infty$

Proof. Take $a > 0$ so that $P(M \leq a) \geq \frac{1}{2}$ then by Theorem 4.3 (b) we have

$$\mathbb{E}\varphi^-(M) \le 8 \sup_n \mathbb{E}\varphi(2S_n+a)$$

And since the condition: $\varphi(2t) \le K\varphi(t)$ for $t \ge t_o$, implies $\varphi(2t+a) \le C\varphi(t) + C$ and $\varphi(t) \le C\varphi^-(t) + C$ for some $C > 0$, we find that (a) and (b) are equivalent.

It is obvious that (a) implies (c), so suppose that $\mathbb{E}\varphi(N) < \infty$. Now we can find a constant $K_o > 0$ so that

(i) $$\varphi(6t + a) \le K_o\varphi(t) + K_o \quad \forall\ t \ge 0$$

where a is chosen so hat $P(M \le a) \ge \frac{1}{2}$. Let us then choose $b \ge a$ so that $P(M>b) \le (64K_o)^{-1}$. If $\xi \equiv b$, and $L = (M-a)^+/6$, then we have

$$\begin{aligned} P(L > t) &\le P(M > 6t + a) \\ &\le 8P(N > t) + 32P(M > t)^2 \\ &\le 8P(N > t) + 32P(\xi > t) + \alpha P(M > t) \end{aligned}$$

where $\alpha = (2K_o)^{-1}$ (cf. Corollary 4.5 (a)). Now put $\varphi_c(x) = \varphi(x)$ for $x \le c$ and $\varphi_c(x) = \varphi(c)$ for $x \ge c$. Then φ_c is increasing, bounded and it is easily seen that φ_c satisfies (i). So by partial integration we find

$$\mathbb{E}\varphi_c(L) \le 8\mathbb{E}\varphi(N) + 32\varphi(b) + (2K_o)^{-1}\mathbb{E}\varphi_c(M)$$

And since $M \le 6L+a$ we find that

$$\begin{aligned} \mathbb{E}\varphi_c(M) &\le K_o\mathbb{E}\varphi_c(L) + K_o \\ &\le K_1 + \tfrac{1}{2}\mathbb{E}\varphi_c(M) \end{aligned}$$

where $K_1 = 8K_o\mathbb{E}\varphi(N) + 32K_o\varphi(b) + K_o$. So we have $\mathbb{E}\varphi_c(M) \le 2K_1$ for all $c > 0$. Now letting $c \to \infty$ we find $\mathbb{E}\varphi(M) \le 2K_1 < \infty$. Hence (c) implies (a).

If $S_n \to S$ in law, then it follows from Theorem 4.3 (d) + (e) that (a) and (d) are equivalent. □

Theorem 4.7. *Let* $X = (X_n)$ *be an independent* E^∞*-valued random sequence and* $q: E^\infty \to \overline{\mathbb{R}}_+$ *a measurable, even, subadditive, quasiconvex function. As before let*

$$M = \sup_n q(\pi_n X), \qquad N = \sup_n q(p_n X_n)$$

and suppose that $M < \infty$ *a.s.*

Let f *and* g *be increasing continuous functions:* $\mathbb{R}_+ \to \mathbb{R}_+$ *with* $f(\infty) = g(\infty) = \infty$ *, and satisfying:*

(a) $$P(N > t) \le K_1 \exp(-g(t)) \qquad \forall\, t \ge t_o$$

(b) $$f(I(t)) \le K_2\, g(t) + K_2 \qquad \forall\, t \ge t_o$$

where t_o, K_1 *and* K_2 *are positive finite constants, and* I *is given by*

(c) $$I(t) = \int_{t_o}^{t} \frac{g(t)}{g(s)}\, ds$$

Then there exists constants $K > 0$, $\varepsilon > 0$, $\delta > 0$, *so that*

(d) $$P(M > t) \le K \exp(-\varepsilon f(\delta t)) \qquad \forall\, t \ge t_1.$$

Proof. We may assume that t_o is taken so large that $P(M > \tfrac{1}{2}t_o) \le \tfrac{1}{2}$. Let $A(t) = 8F(t)$ and $B(t) = 64\, P(N > t)$, where F is the function from Corollary 4.5. If $t \ge t_o$ then $t - \tfrac{1}{2}t_o \ge \tfrac{1}{2}t$, so from Corollary 4.5 (with $a = \tfrac{1}{2}t_o$) we find

$$P(M > t) \le 2F(t - \tfrac{1}{2}t_o) \le \tfrac{1}{4}A(\tfrac{1}{2}t)$$

$$\begin{aligned} A(2t+2s) &\le 32P(N>t) + 32P(M>t)^2 \\ &= \tfrac{1}{2}B(t) + \tfrac{1}{2}A(t)^2 \\ &\le \max\{B(t),\ A(t)^2\} \end{aligned}$$

So if we put

$$\alpha(t) = -\log A(t), \quad \beta(t) = g(t) - \log(64K_1)$$

Then by assumption (a) we have

(i) $$\alpha(2t+2s) \geq \min\{\beta(s), 2\alpha(t)\} \quad \forall\, t,s \geq 0$$

(ii) $$P(M > t) \leq 2\exp(-\alpha(\tfrac{1}{2}t)) \quad \forall\, t \geq t_o$$

Since $g(\infty) = \infty$ we may assume that $g(t) \leq 2\beta(t)$ for $t \geq t_o$, and so by (b) we have

(iii) $$f(I(t)) \leq 2K_2\beta(t) + K_2 \quad \forall\, t \geq t_o$$

Since β increases to $+\infty$ and is continuous, we can find $t_o = \sigma_o < \sigma_1 \cdots$, so that

(iv) $$\beta(\sigma_n) = 2^n\beta(\sigma_o) \quad \forall\, n \geq 0$$

And since α increases to $+\infty$, we can find a $\tau_o \geq t_o$ so that

(v) $$\alpha(\tau_o) \geq \tfrac{1}{2}\beta(\sigma_o)$$

Then we define

(vi) $$\tau_n = 2^n(\tau_o + \sum_{o}^{n-1} 2^{-j}\sigma_j) = 2\tau_{n-1} + 2\sigma_{n-1}$$

So by successive use of (i) we find

$$\begin{aligned}\alpha(\tau_n) &\geq \min\{\beta(\sigma_{n-1}), 2\alpha(\tau_{n-1})\} \\ &\geq \min\{\beta(\sigma_{n-1}), 2\beta(\sigma_{n-2}), \ldots, 2^{n-1}\beta(\sigma_o), 2^n\alpha(\tau_o)\} \\ &= \min_{0\leq j\leq n} \{2^{n-j-1}\beta(\sigma_j)\}\end{aligned}$$

since $2^n\alpha(\tau_o) \geq 2^{n-1}\beta(\sigma_o)$ by (v). Now from the choice of σ_j it follows that we have (see (iv)

(vii) $$\alpha(\tau_n) \geq 2^{n-1}\beta(\sigma_o) = \tfrac{1}{2}\beta(\sigma_n)$$

Now let us estimate τ_n. Using summation by parts we obtain:

$$\sum_o^{n-1} 2^{-j}\sigma_j = 2(\sum_o^{n-1} (2^{-j}-2^{j-1})\sigma_j)$$

$$= 2(\sigma_o-\sigma_{n-1}2^{-n} + \sum_1^{n-1} 2^{-j}(\sigma_j-\sigma_{j-1}))$$

$$\leq K_3 + K_3 \sum_1^{n-1} \int_{\sigma_{j-1}}^{\sigma_j} g(s)^{-1}\, ds$$

$$= K_3 + K_3 \int_{\sigma_o}^{\sigma_{n-1}} g(s)^{-1}\, ds$$

since $g(s)^{-1} \geq \tfrac{1}{2}\beta(s)^{-1} \geq 2^{-j-1}\beta(\sigma_o)$ for $s \leq \sigma_j$ (cf. (iv)). Since $\sigma_1 > \sigma_o$, we then find for $n \geq 2$:

$$\tau_n = 2^n(\tau_o + \sum_o^{n-1} 2^{-j}\sigma_j) \leq K_4\, 2^n \int_{t_o}^{\sigma_{n-1}} g(s)^{-1}\, ds$$

and since $2^{n-1}\beta(\sigma_o) = \beta(\sigma_{n-1}) \leq g(\sigma_{n-1})$ we have

(viii) $$\tau_n \leq K_5\, I(\sigma_{n-1}) \quad \forall\, n \geq 2$$

Let $t \geq K_5\, I(\sigma_1)$, then we can find $n \geq 2$, so that $K_5 I(\sigma_{n-1}) \leq t \leq K_5 I(\sigma_n)$, so by (iii), (vii) and (viii) we have

$$\alpha(t) \geq \alpha(\tau_n) \geq \tfrac{1}{2}\beta(\sigma_n)$$

$$f(t/K_5) \leq f(I(\sigma_n)) \leq 2K_2\beta(\sigma_n) + K_2$$

which gives $\alpha(t) \geq \varepsilon f(\delta t) - K_6$ for some $\varepsilon > 0$ $\delta > 0$ and $K_6 < \infty$. So (d) is then a consequence of (ii). □

Corollary 4.8. Let $X = (X_n)$, $q: E^\infty \to \overline{\mathbb{R}}_+$, M and N be as in Theorem 4.7. And suppose that $M < \infty$ a.s. and

(a) $$P(N>t) \le K_1 \exp(-g(t)) \quad \forall\ t \ge t_o$$

<u>where</u> $g: \mathbb{R}_+ \to \mathbb{R}_+$ <u>is continuous and increasing.</u>

<u>If</u> p <u>is an upper exponent for</u> g, <u>that is if</u>:

(b) $$g(ts) \le K_2 s^p g(t) \quad \forall\ s \ge 1 \quad \forall\ t \ge t_o$$

<u>then we have</u>

(c) $$P(M>t) \le \begin{cases} K \exp(-\varepsilon g(t)) & \text{if } 0<p<1 \\ K \exp(-g(\frac{t}{\log t})) & \text{if } p=1 \\ K \exp(-\varepsilon g(t^{1/p})) & \text{if } p>1 \end{cases}$$

<u>In particular if</u> $g(t) = \alpha t^p$, <u>then we have</u>

(d) $$P(M>t) \le \begin{cases} K e^{-\varepsilon t^p} & \text{if } 0<p<1 \\ K \exp(-\frac{t}{\log t}) & \text{if } p=1 \\ K e^{-\varepsilon t} & \text{if } p>1 \end{cases}$$

<u>If</u> $\int_{t_o}^{t} g(s)^{-p} ds < \infty$, <u>then we have</u>

(e) $$P(M>t) \le \begin{cases} K e^{-\varepsilon t^p} & \text{if } 0<p\le 1 \\ K e^{-\varepsilon t} & \text{if } p\ge 1 \end{cases}$$

<u>Proof</u>. Suppose that (b) holds. Then

$$I(t) = \int_{t_o}^{t} \frac{g(t)}{g(s)} ds \le K_2 \int_{t_o}^{t_o} (\frac{t}{s})^p ds =$$

$$= \begin{cases} K_2 (1-p)^{-1} (t - t^p t_o^{\,1-p}) & p \neq 1 \\ K_2 t \log (t/t_o) & p = 1 \end{cases}$$

If $0 < p < 1$, then $(1-p) > 0$ and $I(t) \leq K_3 t$. So $f(t) = g(t/K_3)$ satisfies (b) in Theorem 4.7. If $p = 1$ then $I(t) \leq K_3 t \log t$, so $f(t) = g(\frac{t}{K_3 \log t})$ satisfies (b) in Theorem 4.7. If $p > 1$, then $(1-p) < 0$ and $I(t) \leq K_3 t^p$, so $f(t) = g(t^{1/p} K_3^{-1/p})$ satisfies (b) in Theorem 4.7. Hence (c) follows from Theorem 4.7 by noting that (b) implies

$$g(\delta t) \geq K_2^{-1} \delta^{-p} g(t) \quad \forall\ t \geq t_o \quad \forall\ 0 < \delta < 1$$

Now suppose that $K_3 = \int_{t_o}^{\infty} g(s)^{-p} ds < \infty$, then we have

$$I(t) \leq K_3\ g(t) \qquad \text{if } p \leq 1$$

so $f(t) = t$ satisfies (b) in Theorem 4.7. And if $p > 1$ then

$$I(t) = \int_{t_o}^{t} \frac{g(t)}{g(s)}\, ds \leq \int_{t_o}^{t} \left(\frac{g(t)}{g(s)}\right)^p ds \leq K_3 g(t)^p$$

So $f(t) = t^{1/p}$ satisfies (b) in Theorem 4.7. Hence (e) follows from Theorem 4.7. □

Lemma 4.9. *Let* $(E, \|\cdot\|)$ *be a Banach space and* $\{X_n\}$ *independent identical distributed random vectors in* E, *and let* $S_n = X_1 + \cdots + X_n$ *for* $n \geq 1$.

If $a_n \uparrow \infty$ *and* $\varphi: \mathbb{R}_+ \to \mathbb{R}_+$ *is an increasing function, such that*

(a) $\qquad \varphi(a_n) \leq Kn$

(b) $\qquad \{S_n/a_n\}$ *is stochastic bounded.*

Then for some constants $K_o > 0$, $a > 0$, $\varepsilon_o > 0$ *we have*

(c) $\qquad P(\|X_1\| > t + a) \leq \dfrac{K_o}{\varphi(\varepsilon_o t)} \qquad \forall\ t \geq 0$

If we in addition to (a) and (b) assume that $\{a_n\}$ is submultiplicative, that is:

(d) $$a_{n\cdot m} \leq C\, a_n a_m \quad \forall\, n,m \geq 1$$

for some constant $C > 0$. Then there exists constants $K_1 > 0$, $a > 0$ and $\varepsilon_1 > 0$ so that

(e) $$P(S_n > a_n(t+a)) \leq \frac{K_1}{\varphi(\varepsilon_1 t)} \quad \forall\, t \geq 0 \quad \forall\, n \geq 1$$

Proof. By use of a standard symmetrization procedure, we may assume that X_n is symmetric. Now according to (b) we may choose $T > 0$ so that

(i) $$P(\|S_n\| > a_n T) \leq \frac{1}{4} \quad \text{for } n \geq 1$$

Then by Theorem 4.2 (a) (with $\varphi = 1_{]t,\infty[}$) we have

$$P(\|X_1\| \leq t)^n = P(\max_{1\leq j\leq n} \|X_j\| \leq t)$$

$$\geq 1 - P(\max_{1\leq j\leq n} \|S_j\| > \tfrac{1}{2} t)$$

$$\geq 1 - 2P(\|S_n\| > \tfrac{1}{2}t)$$

So choosing $t = 2a_nT$ gives

(ii) $$P(\|X_1\| > 2a_nT) \leq 1-2^{-1/n} \leq \frac{\log 4}{n+1} \quad \forall\, n\geq 0$$

(where we put $a_o = 0$). If $t\geq 0$ then we can find $n\geq 1$ so that $2a_{n-1}T \leq t \leq 2a_nT$, and so

$$P(\|X_1\| > t) \leq P(\|X_1\| > 2Ta_{n-1}) \leq \frac{\log 4}{n}$$

$$\varphi(t/2T) \leq \varphi(a_n) \leq K\, n$$

So we find:

(iii) $$P(\|X_1\| > t) \leq \frac{K_o}{\varphi(\varepsilon_o t)} \quad \forall\ t \geq 0$$

where $K_o = K \log 4$ and $\varepsilon_o = (2T)^{-1}$.

Now suppose that (d) holds. Let $k \geq 1$ be given and define

$$Y_n^k = (S_{n\cdot k} - S_{(n-1)\cdot k})a_k \quad \forall\ n = 1,2,\ldots$$

Then $Y_1^k, Y_2^k \cdots$ are independent identical distributed random vectors in E, and

$$a_n^{-1} \|\sum_1^n Y_j\| = a_n^{-1} a_k^{-1} \|S_{n\cdot k}\| \leq C\, a_{nk}^{-1} \|S_{n\cdot k}\|$$

So if T is chosen as before (that is so that (i) holds) then we have

(iv) $$P(\|\sum_1^n Y_j^k\| > C\,T\,a_n) \leq \frac{1}{4} \quad \forall\ n \geq 1$$

Hence by (iii) we have

(v) $$P(\|Y_1^k\| > t) \leq \frac{K_o}{\varphi(\varepsilon_1 t)} \quad \forall\ t \geq 0 \quad \forall\ k \geq 1$$

where $K_o = K \log 4$ and $\varepsilon_1 = (2CT)^{-1}$. But

$$Y_1^k = S_k / a_k$$

and so the theorem is proved. □

Let $(E, \|\cdot\|)$ be a Banach space and X a random E-valued vector. Then X is said to belong to the domain of normal attraction if

$$Z_n = (X_1 + \cdots + X_n)/\sqrt{n}$$

converges in law, where $X_1, X_2, \ldots$ are independent copies of X; and for any random vector X, we define:

$$c(X) = \sup_n E\|Z_n\|$$

The set of $X \in L_E^0$ for which $c(X) < \infty$ is denoted CLT, and it is easily seen that

(4.1) $(\mathrm{CLT}, c(\cdot))$ is a Banach space

(4.2) If $X \in \mathrm{CLT}$, then $X \in L_E^1$ and $\mathbb{E}X = 0$.

Moreover from Lemma 4.9 we find

Theorem 4.10. Let $(E, \|\cdot\|)$ be a Banach space. Then there exists $K > 0$ so that

(a) $P(\|X\| > t) \leq K c(X)^2 t^{-2} \quad \forall\, t \geq 0 \quad \forall\, X \in \mathrm{CLT}$

(b) $\mathrm{CLT} \subseteq L_E^p \quad \forall\, 0 \leq p < 2$

(c) If X is in the domain of normal attraction then $X \in \mathrm{CLT}$.

Proof. If $X \in \mathrm{CLT}$, then $\{Z_n\} = \{S_n/\sqrt{n}\}$ is stochastic bounded. So by Lemma 4.9 (with $a_n = \sqrt{n}$ and $\varphi(t) = t^2$) we have

$$P(\|X\| > t) \leq K(X)\ t^{-2} \qquad \forall\, t \geq 0$$

for some finite constant $K(X)$. Now let us consider the Laurent space $L_E^{(2)} = L^{(2)}(\Omega, F, P, E)$ (see Chapter I §2). Then $\mathrm{CLT} \subseteq L_E^{(2)}$ and since the topologies on CLT and on $L_E^{(2)}$ both are stronger than the L_E^1-topology we have that the injection $\mathrm{CLT} \to L_E^{(2)}$ has a closed graph, and so it is continuous by the closed graph principle. So we have

$$\lambda_2(X) \leq K c(X)^{2/3} \qquad \forall\, x \in \mathrm{CLT}$$

since λ_2 is $\frac{2}{3}$-homogeneous. Now (a) is a simple consequence of the definition of $\lambda_2(\cdot)$

(b) is an immediate consequence of (a)

(c): If X belongs to the domain of normal attraction, then $(S_n/\sqrt{n})$ is stochastic bounded and by Lemma 4.9 with $a_n = \sqrt{n}$ and $\varphi(t) = t^2$ we have

$$P(\|S_n\| > \sqrt{n}\ t) \leq K\ t^{-2} \qquad \forall\ t \geq 0$$

for some constant K. But then

$$\mathbb{E}\|Z_n\| = \int_0^\infty P(\|S_n\| > \sqrt{n}\ t)dt$$

$$\leq 1 + \int_1^\infty K\ t^{-2}\ dt = K+1$$

and so $c(X) \subseteq K+1 < \infty$. □

5. Series in a Banach space

In this section we shall apply the result of §2 and §4 to Banach spaces.

So let $(E, \|\cdot\|)$ be a Banach space and $X = (X_n)$ a random sequence in E. Recall that by convention any Banach space-valued random vector is separably valued, so we may assume E to be separable, and $(E, \mathcal{B}(E))$ is then a measurable linear space

In Chap. I §4 we defined weak convergence of Radon measures on a completely regular Hausdorff space. Applying Lemma I.4.1 to $\mathcal{H}$ equal to the set of functions:

$$f(x) = g(\langle x_1', x\rangle \ldots, \langle x_n', x\rangle)$$

with $n \geq 1$, $x_1' \ldots x_n' \in E'$ and $g \in C(\mathbb{R}^n)$, gives:

<u>Lemma 5.2</u>. <u>If</u> μ_n <u>and</u> μ <u>are Radon measures on the Banach space</u> E, <u>such that</u> $\hat{\mu}(x') = \lim_{n\to\infty} \hat{\mu}_n(x')$ <u>for all</u> $x' \in E'$, <u>then</u> $\mu_n \to \mu$ $\sigma(E,E')$<u>-weakly</u>. □

<u>Theorem 5.3</u>. <u>Let</u> (X_n) <u>be independent</u> E<u>-valued random vectors, and put</u> $S_n = \sum_1^n X_j$. <u>Now suppose that there exists a Radon probability</u> μ <u>on</u> E, <u>such that</u>

(a) $$\hat{\mu}(x') = \lim_{n\to\infty} \mathbb{E}\exp(i \langle x', S_n\rangle) \quad \forall\, x' \in E'$$

<u>Then there exists a sequence</u> $\{a_n\} \subseteq E$, <u>with the properties</u>

(b) $\{S_n - a_n\}$ <u>converges a.s.</u>

(c) $\{a_n\}$ <u>is weakly convergent in</u> E.

<u>Moreover in each of the following 6 cases we can conclude that</u> $\{S_n\}$ <u>converges a.s</u>.

(1) X_n <u>is symmetric</u> $\forall\, n\geq 1$

(2) $\sup_n \mathbb{E}\|S_n\| < \infty$, <u>and</u> $\mathrm{cl}\{\mathbb{E}S_n \mid n\geq 1\}$ <u>is compact</u>

(3) $\mathbb{E}(\sup_n \|X_n\|) < \infty$, <u>and</u> $\mathrm{cl}\{\mathbb{E}S_n \mid n\geq 1\}$ <u>is compact</u>

(4) $\int_E \|x\|\,\mu(dx) < \infty$, <u>and</u> $\mathrm{cl}\{\mathbb{E}S_n \mid n\geq 1\}$ <u>is compact</u>

(5) $\exists K$ <u>compact</u> $\subseteq E$, <u>so that</u> $\inf_n P(S_n \in K) > 0$.

(6) $\{S_n\}$ converges in probability.

<u>Proof</u>. We may without loss of generality assume that E is separable.

<u>Case (1)</u>: Suppose that X_n is symmetric for all $n\geq 1$. Since μ is a Radon probability we can find compact convex, symmetric sets $K_1, K_2, \ldots,$ so that $\mu(K_n) \geq 1-2^{-n}$. Let $a_n = \sup_{x\in K_n} \|x\|$ and put

$$K = \{ \sum_{j=1}^{\infty} 2^{-j}(a_j+1)^{-1} x_j \mid x_j \in K_j \quad \forall j\}$$

Then K is a continuous linear image of $\prod_1^{\infty} K_j$, so K is symmetric, convex and compact. Moreover $K_n \subseteq 2^n a_n K$, so if q is the seminorm:

$$q(x) = \inf\{\lambda\geq 0 \mid x \in \lambda K\} \qquad \text{for } x \in E,$$

then $E_q = \{q < \infty\}$ has μ-measure 1 and $K = \{q\leq 1\}$.

From the real-valued version of the equivalence theorem (see e.g. [81] Theorem B p.251), we have that $s(x') = \lim_{n\to\infty} \langle x', S_n\rangle$ exists a.s. for all $x' \in E'$. Now by separability of E there exists a countable set $\{x_j'\} \subseteq K^{o}$ (= the polar of K), such that

$$\text{(i)} \qquad q(x) = \sup_j |\langle x_j', x\rangle| \qquad \forall\, x \in E$$

Now let $q_k: E^\infty \to \overline{\mathbb{R}}_+$ be given by

$$q_k(x) = \begin{cases} \max\limits_{1\le i\le k} |\langle x_i', \sum\limits_1^\infty x_j\rangle| & \text{if } \sum\limits_1^\infty x_j \text{ converges} \\ \infty & \text{otherwise} \end{cases}$$

Then $q_k(\pi_n X) \xrightarrow[n\to\infty]{\text{a.s.}} \max\limits_{1\le i\le k} s(x_i')$, so by Theorem 4.2 (d) we have

$$P(\sup_n \max_{1\le i\le k} |\langle x_i', S_n\rangle| > t) \le 2P(\max_{1\le i\le k} |s(x_i')| > t)$$

$$= 2\mu(x \in E \mid \max_{1\le i\le k} |\langle x_i', x\rangle| > t) \le 2\mu(q>t)$$

Since $L(s(x_1'),\dots,s(x_n')) = \mu_{x_1',\dots,x_n'}$ by assumption (a). Letting $k \to \infty$ it follows from (i) that we have

$$P(\sup_n q(S_n) > t) \le 2\mu(q>t) \qquad \forall\ t\ge 0$$

and since $\mu(q<\infty) = 1$, we have that

(ii) $$M = \sup_n q\ (S_n) < \infty \qquad \text{a.s.}$$

Now let F' be a countable subset of E', which separates points of E, then set

$$\Omega_o = \{M<\infty\} \cap \{\lim_{n\to\infty} \langle x', S_n\rangle \text{ exist } \forall\ x' \in F'\}$$

has probability 1. Moreover, if $\omega \in \Omega_o$, then

$$S_n(\omega) \in M(\omega)K$$

$$\lim_{n\to\infty} \langle x', S_n(\omega)\rangle \qquad \text{exists for all } x' \in F'$$

Hence $\{S_n(\omega)\}$ is $\|\cdot\|$-compact, and since F separates points of E, we have that $\{S_n(\omega)\}$ converges for all $\omega \in \Omega_o$, and case (1) is proved.

<u>The general case</u>: Let $X^s = (X_n^s)$ be a symmetrization of X, and let $S_n^s = \sum_1^n X_j^s$, then

$$\mathbb{E}\exp(i\langle x', S_n^s\rangle) = |\mathbb{E}\exp(i\langle x', S_n\rangle)|^2$$

$$\xrightarrow[n\to\infty]{} |\hat{\mu}(x')|^2 = \hat{\rho}(x')$$

where ρ is the Radon probability given by

$$\rho(A) = \int_E \mu(A+x)\mu(dx) \quad \forall\, A \in \mathcal{B}(E)$$

So by (i) we have that

$$1 = P(\{S_n^s\} \text{ converges})$$

$$= \int_{E^\infty} P(\{S_n - \sum_1^n x_j\} \text{ converges})\, Q(dx)$$

where $Q = L(X)$. So there exists $\{a_n\} \subseteq E$, such that $T = \lim_{n\to\infty} (S_n - a_n)$ exists a.s., and by the real-valued equivalence theorem (see e.g. [81] Theorem B p. 251) we have $\lim_{n\to\infty}\langle x', S_n\rangle$ eixst a.s. Hence we can find a linear map $f: E' \to \mathbb{R}$ with

$$f'(x) = \lim_{n\to\infty} \langle x', a_n\rangle \qquad \forall\, x' \in E'$$

$$\hat{\nu}(x') = \exp(-i\, f(x'))\,\hat{\mu}(x') \qquad \forall\, x' \in E'$$

where $\nu = L(T)$. Now since ν and μ are Radon measures, we have that $\hat{\mu}$ and $\hat{\nu}$ are Makey-continuous, and $\hat{\mu}(0) = \hat{\nu}(0) = 1$. Hence

$$\lim_\alpha \exp(i\, t\, f(x'_\alpha)) = 1 \quad \text{uniformly for } |t| \le 1$$

whenever $x'_\alpha \to 0$ in the Makey-topology. But it is easily checked that $e^{itu_\alpha} \xrightarrow[\alpha]{} 1$ for $t \in [-1,1]$ implies $u_\alpha \xrightarrow[\alpha]{} 0$. So f is Makey-continuous, and hence $f(x') = \langle x', a\rangle$ for some $a \in E$. That is $\{a_n\}$

is weakly convergent in E. This concludes the proof of the general case.

Now assume that $a_n \to a$ weakly, and $S_n - a_n \to T$ a.s. Then $S_n \to T+a$ a.s. in the weak topology so we have

(iii) $$M = \sup_n \|S_n\| < \infty \text{ a.s.}$$

(iv) $$\alpha = \sup \|a_n\| < \infty$$

Now let us look at the 4 remaining cases:

Case (2): In this case it follows from Theorem 4.6 that $\mathbb{E}M < \infty$, so by (iv) and Lebesgue's dominated convergence theorem we have

$$\mathbb{E}(S_n - a_n) \to \mathbb{E}T \quad (\text{in } \|\cdot\|)$$

and since $a_n = \mathbb{E}S_n - \mathbb{E}(S_n - a_n)$, it follows that $\{a_n\}$ is norm compact and weakly convergent. So $\{a_n\}$ is norm convergent, and $\{S_n\}$ converges a.s.

Case (3): Again in this case it follows from Theorem 4.6 that $\mathbb{E}N < \infty$, and we conclude that $\{S_n\}$ converges exactly as in Case (1).

Case (4): Since $\mathcal{L}(T+a) = \mu$ we have

$$\mathbb{E}\|T\| = \int_E \|x-a\| \mu(dx) \leq \|a\| + \int_E \|x\| \mu(dx) < \infty$$

So from Theorem 4.6 we find

$$\mathbb{E}(\sup_n \|S_n - a_n\|) < \infty$$

But $M \leq \alpha + \sup_n \|S_n - a_n\|$, and hence $\mathbb{E}M < \infty$. And we conclude that $\{S_n\}$ converges a.s. exactly as in case (2).

Case (5): Let $\delta = \inf_n P(S_n \in K)$, where K is the compact set in condition (5). We may of course assume that K is taken so large that

$$P(T \in K) > 1 - \tfrac{1}{2}\delta$$

Let $\varepsilon > 0$ be given, then we can cover $K-K$ by balls, $B(x_j, \frac{1}{2}\varepsilon)$ $j = 1,\dots,k$, with centers in x_j and radius $\frac{1}{2}\varepsilon$. And we may determine an $N \geq 1$ so that

$$P(\|S_p - a_p - T\| \leq \tfrac{1}{2}\varepsilon \ \forall\, p \geq N) > 1 - \tfrac{1}{2}\delta$$

Hence the set

$$\{S_n \in K\} \cap \{T \in K\} \cap \{\|S_p - a_p - T\| \leq \tfrac{1}{2}\varepsilon \ \forall\, p \geq N\}$$

has probability > 0, and we can find $\omega_n \in \Omega$ with $S_n(\omega_n) \in K$, $T(\omega_n) \in K$ and

$$\|S_p(\omega_n) - T(\omega_n) - a_p\| \leq \tfrac{1}{2}\varepsilon \ \forall\, p \geq N$$

Let $z_p = S_p(\omega_p) - T(\omega_p)$, then $z_p \in K-K$ and $\|a_p - z_p\| \leq \frac{1}{2}\varepsilon$ for $p \geq N$. That is

$$a_p \in \bigcup_{j=1}^{k} B(x_j, \varepsilon) \quad \forall\, p \geq N$$

$$a_p \in B(a_p, \varepsilon) \quad \forall\, p < N$$

And we see that $\{a_p\}$ is precompact, and weakly convergent. But this implies that $\{a_p\}$ is norm convergent and $\{S_n\}$ converges a.s.

<u>Case (6)</u>: In this case $a_n = S_n - (S_n - a_n)$, so $\{a_n\}$ converges in probability, but a_n being non-random this implies that $\{a_n\}$ converges and so $\{S_n\}$ converges a.s. □

<u>Remarks</u> (1): Of course some condition is necessary to assure that $a_n \to a$ in $\|\cdot\|$, e.g. if $X_j = a_j - a_{j-1}$ $(a_o = 0)$, where $a_n \to a$ weakly but not in $\|\cdot\|$, then $S_n = a_n$ and $\{S_n\}$ does not converge in $\|\cdot\|$. □

(2): Note that conditions (5) or (6) are also necessary for a.s. convergence of $\{S_n\}$. And that conditions (2), (3) or (4) are necessary for L_E^1-convergence of $\{S_n\}$. □

(3): If F' is a norming linear subspace of E', that is

$$\|x\| = \sup\{|\langle x',x\rangle| : x' \in F', \|x'\| \leq 1\}$$

then $\hat{\mu}(x') = \lim_{n\to\infty} \mathbb{E}\exp(i\langle x',S_n\rangle)$ for $x' \in F'$, implies (b) and (c), with $\{a_n\}$ converging in $\sigma(E,F')$ instead of weakly converging. And $\{S_n\}$ converges a.s. in each of the 6 listed cases in Theorem 5.3. □

(4): From the proof it follows that in case (1) we need not to assume the X_n's to be independent. □

Theorem 5.4. *Let (X_n) be a symmetric E-valued random sequence, so that $S = \sum_1^\infty X_j$ exists a.s., and let $\varphi: \mathbb{R}_+ \to \mathbb{R}_+$ be a continuous increasing function. Then the following 5 statements are equivalent*

(a) $\sup_n \mathbb{E}\varphi(\|S_n\|) < \infty$

(b) $\mathbb{E}\varphi(\sup_n \|S_n\|) < \infty$

(c) $\mathbb{E}\varphi(\|S\|) < \infty$

(d) $\mathbb{E}\varphi(\|S\|) < \infty, \quad \lim_{n\to\infty} \mathbb{E}\varphi(\|S-S_n\|) = \varphi(0)$

Proof. The first 3 are equivalent by Theorem 4.2. Clearly (d) implies (c), and (b) implies (d) by Lebesgue's theorem on dominated convergence. □

Theorem 5.5. *Let (X_n) be independent E-valued random vectors, so that $S = \sum_1^\infty X_j$ exists a.s., and let $\varphi: \mathbb{R}_+ \to \mathbb{R}_+$ be increasing* continuous, and satisfying: $\varphi(2t) \leq K\varphi(t)$, then the following 5 statements are equivalent

(a) $\mathbb{E}\varphi(\sup_n \|X_n\|) < \infty$

(b) $\mathbb{E}\varphi(\sup_n \|S_n\|) < \infty$

(c) $\sup_n \mathbb{E}\varphi(\|S_n\|) < \infty$

(d) $\mathbb{E}\varphi(\|S\|) < \infty$

(e) $\mathbb{E}\varphi(\|S\|) < \infty$ and $\lim_{n\to\infty} \mathbb{E}\varphi(\|S-S_n\|) = \varphi(0)$.

Proof. The first 4 statements are equivalent by Theorem 4.6 and evidently (e) implies (d) and (b) implies (e). □

Theorem 5.6. Let (ξ_n) and (η_n) be sequences of real random variables, and (X_n) a symmetric random sequence in E, so that

(a) $(\xi_1,\ldots,\xi_n) \vdash (\eta_1,\ldots,\eta_n) \quad \forall\ n\geq 1$

(b) X and ξ, and X and η are independent.

Then we have

(c) $\{\sum_1^n \eta_j X_j\}$ bounded a.s. $\Rightarrow \{\sum_1^n \xi_j X_j\}$ bounded a.s.

(d) $\sum_1^\infty \eta_j X_j$ converges a.s. $\Rightarrow \sum_1^\infty \xi_j X_j$ converges a.s.

Proof. Immediate consequence of Theorem 2.15. □

Theorem 5.7. Let (η_n) be a sequence of real random variables, and (X_n) a symmetric random sequence in E, so that

(a) $\lim_{n\to\infty} \eta_n = 0$ a.s.

(b) X and η are independent

(c) $\{\sum_1^n X_j\}$ is bounded in probability

Then we have that $\sum_1^\infty \eta_j X_j$ converges a.s.

Proof. Let $N_p = \sup_{j\geq p} |\eta_j|$, then by Theorem 2.15 (with $\varphi = 1_{]t,\infty[}$) we have

$$P(\|\sum_{j=p}^n \eta_j X_j\| > 5t) \leq 6P(N_p \|\sum_{j=p}^n X_j\| > t)$$

for $n \geq p$. Now let

$$\beta(t) = \sup_{1\leq p\leq n<\infty} 6P(\|\sum_{j=p}^n X_j\| > t)$$

Then $0 \leq \beta(t) \leq 1$, and $\beta(t) \to 0$ as $t \to \infty$. If $\mu_p = L(N_p)$, then we have

$$P(\|\sum_{j=p}^n \eta_j X_j\| > 5t) \leq \int_0^\infty 6\, P(\|\sum_{j=p}^n X_j\| > t/x)\mu_p(dx)$$

$$\leq \int_0^\infty \beta(t/x)\mu_p(dx)$$

$$= \mathbb{E}\,\beta(t/N_p)$$

And since $t/N_p \to \infty$ a.s. for $t > 0$, we find by the dominated convergence theorem that

$$\lim_{p\to\infty} \sup_{n\geq p} P(\|\sum_{j=p}^n \eta_j X_j\| > 5t) = 0 \quad \forall\, t>0$$

Hence $\sum_1^\infty \eta_j X_j$ converges a.s. □

Theorem 5.8. Let (ξ_n) be a sequence of independent real random variables, which is bounded away from 0 in probability, (see (2.19)), and let (X_n) be a symmetric random sequence, which is independent of (ξ_n). Then we have

(a) $\{\sum_1^n \xi_j X_j\}$ bounded a.s. $\Rightarrow \{\sum_1^n X_j\}$ bounded a.s.

(b) $\{\sum_1^\infty \xi_j X_j\}$ converges a.s. $\Rightarrow \sum_1^\infty X_j$ converges a.s.

Proof. (a): Let $a,p>0$ be chosen so that (2.23) holds, and let $\beta = (\beta_n)$ be a sequence of independent random variables with distributions given by:

$$P(\beta_n = a) = p \qquad P(\beta_n = 0) = 1-p$$

and such that β, ξ and X are independent.

Then $\beta_j \vdash \xi_j$ for all $j = 1,\ldots,n$, hence by Lemma 2.1 we have $(\beta_1,\ldots,\beta_n) \vdash\!\!- (\xi_1,\ldots,\xi_n)$ for all $n \geq 1$. So from Theorem 5.6 we conclude that $\sup_n \|\sum_1^n \beta_j X_j\| < \infty$ a.s., and so we have $\sup_j \|\beta_j X_j\| < \infty$ a.s. That is by (1.6) exists a $t>0$ with

$$\infty > \sum_1^\infty P(\|\beta_j X_j\| > t) = p \sum_1^\infty P(\|X_j\| > t/a)$$

So we have that $\sup_j \|X_j\| = N < \infty$ a.s. (cf. (1.6)). Now let

$$X_j^* = X_j/(N+1)$$

Then (X_j^*) is a symmetric sequence, since N does not depend on the signs of the X_j's. We shall consider

$$q(t) = \begin{cases} \mathbb{E}\|\sum_1 t_j X_j^*\| & \text{if } t \in \mathbb{R}^{(\infty)} \\ \infty & \text{if } t \notin \mathbb{R}^{(\infty)} \end{cases}$$

where $\mathbb{R}^{(\infty)}$ is the set of $t = (t_j) \in \mathbb{R}^\infty$ with $t_j \neq 0$ for atmost finitely many j's. Then

$$N_o = \sup_n q(p_n\beta_n) = \sup_n \|\beta_n X_n^*\| \leq a$$

$$M_o = \sup_n q(\pi_n\beta) = \sup_n \|\sum_1^n \beta_j X_j^*\| < \infty \text{ a.s.}$$

Now q is a seminorm so by Theorem 4.6 (with $\varphi(t) = t$) we have that $\mathbb{E} M_0 < \infty$. Note that $\mathbb{E}\beta_n = ap$, so Jensen's inequality gives:

$$\mathbb{E}\{ap \|\sum_1^n X_j^*\|\} \leq \mathbb{E}\|\sum_1^n \beta_j X_j^*\| \leq \mathbb{E}\, M$$

Hence $\{\sum_1^n X_j^*\}$ is bounded in L_E^1 and so a.s. bounded (cf. Theorem 4.2(c)). But

$$\sum_1^n X_j = (N+1) \sum_1^n X_j^*$$

so $\{\sum_1^n X_j\}$ is a.s. bounded.

(b) is proved similarly. □

The 3 theorems above go under the name the <u>comparison principle</u>. Using Theorem 2.11 it is possible to prove similar comparison principles for boundedness and convergence in L_E^p $(1 \leq p < \infty)$. We shall leave it to the reader to state and prove these.

6. The type and cotype of a Banach space

Let $(E,\|\cdot\|)$ be a Banach space and $\xi = (\xi_n)$ a sequence independent nonzero random variables, that is

(6.1) $$P(\xi_n = 0) < 1 \qquad \forall\ n\geq 1,$$

Then for $0\leq p\leq\infty$ we define

$$B_E^p(\xi) = \{(x_j) \in E^\infty \mid \{\sum_1^n \xi_j x_j\} \text{ is bounded in } L_E^p\}$$

$$C_E^p(\xi) = \{(x_j) \in E^\infty \mid \sum_1^\infty \xi_j x_j \text{ converges in } L_E^p\}$$

$$|||x_p||| = \sup_n \|\sum_1^n \xi_j x_j\|_p \qquad \forall\ x = (x_j) \in B_E^p(\xi)$$

It is then a matter of routine to check that we have:

(6.2) $(B_E^p(\xi), |||\cdot|||_p)$ is a Fréchet space for $0\leq p\leq 1$, and a a Banach space for $1\leq p<\infty$

(6.3) $|||\cdot|||_p$-topology is stronger than the product topology

(6.4) If $(x_j) \in B_E^p(\xi)$, then $x_j = 0$ whenever $\xi_j \notin L^p$

Now let

$$F_E^p(\xi) = \{(x_j) \in E^{(\infty)} \mid x_j = 0 \ \ \forall\ j:\ \xi_j \notin L^p\}$$

where $E^{(\infty)}$ denotes the set of sequences in E with at most finitely many nonzero coordinates. Then we have:

<u>Theorem 6.1</u>. <u>$C_E^p(\xi)$ is the closure of</u> $F_E^p(\xi)$ <u>in</u> $(B_E^p(\xi), |||\cdot|||_p)$. <u>In particular</u> $C_E^p(\xi)$ <u>is a Fréchet space for</u> $0\leq p<1$ <u>and a Banach space for</u> $1\leq p<\infty$.

<u>Proof</u>. Clearly $F_E^p(\xi) \subseteq C_E^p(\xi)$. And if $x = (x_j) \in C_E^p(\xi)$, then $x^n \to x$ and $x^n \in F_E^p(\xi)$, where $x^n = (x_1,\ldots,x_n,0,0,\ldots)$. Hence

$$F_E^p(\xi) \subseteq C_E^p(\xi) \subseteq cl(F_E^p(\xi))$$

Now suppose that $x \in cl(F_E^p(\xi))$, then there exists $y \in F_E^p(\xi)$ with $|||x-y|||_p < \varepsilon$, where $\varepsilon > 0$ is any given number. Now there exist $N \geq 1$, so that $y_j = 0$ for $j \geq N$.

If $m \geq n \geq N$, then we have

$$\begin{aligned} \|\sum_{j=n}^{m} \xi_j x_j\|_p &= \|\sum_{j=n}^{m} \xi_j (x_j - y_j)\|_p \\ &\leq \|\sum_{j=1}^{m} \xi_j (x_j - y_j)\|_p + \|\sum_{j=1}^{n} \xi_j (x_j - y_j)\|_p \\ &\leq 2\varepsilon \end{aligned}$$

So $\sum_1^\infty \xi_j x_j$ converges in L_E^p, and we have

$$cl(F_E^p(\xi)) \subseteq C_E^p(\xi)$$

from which the theorem follows. □

We shall not here pursue the properties of $C_E^p(\xi)$ and $B_E^p(\xi)$ but refer the interested reader to [40].

The simplest non-trivial sequence of independent random variables is <u>the Bernouilli sequence</u> $\varepsilon = (\varepsilon_j)$, i.e. $\varepsilon_1, \varepsilon_2, \ldots$ are independent and their law is given by: $P(\varepsilon_j = \pm 1) = \frac{1}{2}$ for all $j \geq 1$. From Theorem 5.5 and Lemma 1.3 we have that $B_E^p(\varepsilon) = B_E^o(\varepsilon)$ and $C_E^p(\varepsilon) = C_E(\varepsilon)$ for all $0 \leq p < \infty$. Hence we may introduce the notation:

$$B_E = B_E^p(\varepsilon), \quad C_E = C_E^p(\varepsilon) \qquad (0 \leq p < \infty)$$

Since any symmetric sequence (X_n) is a mixture of sequences $(\varepsilon_j x_j)$, with $(x_j) \in E^\infty$, we have if (X_j) is symmetric then

$$(6.5)\quad \begin{cases} P(\sum_{1}^{\infty} X_n \text{ converges}) = P(X \in C_E) \\ P(\{\sum_{1}^{n} X_j\} \text{ bounded}) = P(X \in B_E) \end{cases}$$

So even though the series $\sum_{1}^{\infty}\varepsilon_j x_j$ are the simplest possible non-trivial random series in E, there are enough of them to determine convergence and boundedness of any series, $\sum_{1}^{\infty} X_j$, of symmetric random vectors in E.

<u>Theorem 6.2</u>. <u>Let $\xi = (\xi_n)$ be a sequence of independent random variables, which is totally non-degenerated, then we have</u>

(a) $$B_E^o(\xi) \subseteq B_E \quad\text{and}\quad C_E^o(\xi) \subseteq C_E$$

<u>Proof</u>. Let $\varepsilon = (\varepsilon_j)$ be a Bernouilli sequence, which is independent of $\xi^s = (\xi_n^s)$, where ξ^s is a symmetrization of ξ. Then clearly we have

$$B_E^o(\xi) \subseteq B_E^o(\xi^s) = B_E^o(\eta)$$

where $\eta = (\varepsilon_j \xi_j^s)$. Let $x \in B_E^o(\eta)$ then it follows from Theorem 5.8 (put $X_j = \varepsilon_j x_j$), that $x \in B_E$. The second inclusion is proved similarly. □

From the Khinchine inequalities (see (I.2.23)) we find

$$(6.6)\qquad B_{\mathbb{R}} = C_{\mathbb{R}} = \ell^2$$

If $E = L^p(S,\Sigma,\mu)$ for some measure space (S,Σ,μ) and some $p \in [1,\infty[$, then by integration of the Khinchine inequalities we find

$$(6.7)\qquad B_E = C_E = \{(f_j) \mid \int_S \{\sum_{j=1}^{\infty} f_j(s)^2\}^{p/2}\mu(ds) < \infty\}$$

$$= L^p(S,\Sigma,\mu,\ell^2)$$

For an arbitrary Banach space E, we have $\ell^1_E \subseteq C_E \subseteq B_E \subseteq \ell^\infty_E$, and if $1 \leq p,q \leq \infty$, we shall say that E is of <u>type</u> p, if $\ell^p_E \subseteq B_E$ and of <u>cotype</u> q if $C_E \subseteq \ell^q_E$. So we have

(6.8) Every Banach space is of type 1 and of cotype $+\infty$.

From (6.7) it follows easily, that we have

(6.9) $L^p(S,\Sigma,\upsilon)$ is of type $\min(2,p)$ and of cotype $\max(2,p)$.

From (6.6) it follows that if E is of type p (cotype q) then $1 \leq p \leq 2$ ($2 \leq q \leq \infty$). And if E is of type p (cotype q), then E is of type p' for all $1 \leq p' \leq p$ (cotype q', for all $q' \geq q$).

Since the injections $\ell^p_E \to B_E$ and $C_E \to \ell^q_E$ has closed graphs, it follows from the closed graph theorem that we have

(6.10) E is of type $p \Leftrightarrow \exists\, K>0$: $\mathbb{E}\, \|\sum_1^n \varepsilon_j x_j\|^p \leq K \sum_1^n \|x_j\|^p$

$$\forall\, n \geq 1\ \forall\, x_1,\ldots,x_n \in E$$

(6.11) E is of cotype $q \Leftrightarrow \exists\, k>0$: $\mathbb{E}\, \|\sum_1^n \varepsilon_j x_j\|^q \geq k \sum_1^n \|x_j\|^q$

$$\forall\, n \geq 1\ \forall\, x_1,\ldots,x_n \in E$$

This shows that the property of being of type p or of cotype q is only of property of the finite dimensional subspaces of E. Hence we have

<u>Proposition 6.3</u>. <u>If</u> E <u>is of type</u> p <u>(cotype</u> q) <u>and</u> F <u>is finitely representable in</u> E, <u>then</u> F <u>is of type</u> p <u>(cotype</u> q). □

Now we may define the numbers $p(E)$ and $q(E)$ by

$$p(E) = \sup\{p \in [1,2] \mid E \text{ is of type } p\}$$

$$q(E) = \inf\{q \in [2,\infty] \mid E \text{ is of cotype } q\}$$

Note that Proposition 6.3 and Theorem I.3.5 show that $p(E) = p(E'')$ and $q(E) = q(E'')$.

Proposition 6.4. *If E is of type p, then E' is of cotype q, where $\frac{1}{p} + \frac{1}{q} = 1$. In particular we have*

(a) $$\frac{1}{p(E)} + \frac{1}{q(E')} \geq 1$$

Note that $p(c_o) = 1$, $q(\ell^1) = 2$, $p(\ell^\infty) = 1$, so in the inequality in (a) may be strict, and the converse of Proposition 6.4 is false in general.

Proof. Let $x_1', \ldots, x_n' \in E'$, and choose $x_1, \ldots, x_n \in E$, so that $\|x_j\| = 1$ and

(i) $$\langle x_j', x_j \rangle \geq \tfrac{1}{2} \|x_j'\| \quad \forall\, j = 1, \ldots, n$$

If we put $t_j = \|x_j'\|^{q-1} = \|x_j'\|^{q/p}$, then we have

$$\sum_{j=1}^n \|x_j'\|^q = \sum_{j=1}^n \|x_j'\|\, t_j \leq 2 \sum_{j=1}^n \langle x_j', t_j x_j \rangle$$

$$= 2\,\mathbb{E}\Big(\sum_{j=1}^n \sum_{i=1}^n \varepsilon_j \varepsilon_i \langle x_j', t_i x_i \rangle\Big)$$

$$= 2\,\mathbb{E} \Big\langle \sum_1^n \varepsilon_j x_j', \sum_1^n \varepsilon_i t_i x_i \Big\rangle$$

$$\leq 2\,\mathbb{E}\Big\{\Big\|\sum_1^n \varepsilon_j x_j'\Big\| \; \Big\|\sum_1^n \varepsilon_i t_i x_i\Big\|\Big\}$$

$$\leq 2\Big(\mathbb{E}\Big\|\sum_1^n \varepsilon_j x_j'\Big\|^q\Big)^{1/q} \Big(\mathbb{E}\Big\|\sum_1^n \varepsilon_j t_j x_j\Big\|^p\Big)^{1/p}$$

$$\leq 2K\Big(\mathbb{E}\Big\|\sum_1^n \varepsilon_j x_j'\Big\|^q\Big)^{1/q} \Big(\sum_1^n t_j^p \|x_j\|^p\Big)^{1/p}$$

$$= 2K\Big(\mathbb{E}\Big\|\sum_1^n \varepsilon_j x_j'\Big\|^q\Big)^{1/q} \Big(\sum_1^n \|x_j'\|^q\Big)^{1/p}$$

where K is constant appearing in (6.10). And since $\frac{1}{q} = 1 - \frac{1}{p}$ we find

$$(\sum_1^n \|x_j'\|^q)^{1/q} \leq 2K(\mathbb{E}\|\sum_1^n \varepsilon_j x_j'\|^q)^{1/q}$$

which shows that E' is of cotype q. □

Theorem 6.5. *Let $\xi = (\xi_n)$ be a sequence of independent real random variables satisfying*

(a) (ξ_n) *is totally non-degenerated*

(b) $\mathbb{E}\,\xi_n = 0 \quad \forall\, n \geq 1$

(c) $\sup_n |\xi_n| \in L^s$

for some $s \in [1,\infty[$. *Then we have*

(d) $B_E = B_E^r(\xi) \quad \forall\, 0 \leq r \leq s$

(e) $C_E = C_E^r(\xi) \quad \forall\, 0 \leq r \leq s$

Remark. We shall later show (see Theorem 7.6) that if E is of cotype q, for some $q < s$ then the theorem holds when (c) is substituted by much weaker condition:

(c*) $\sup_n \mathbb{E}|\xi_n|^s < \infty$

Proof. Let $N = \sup_n |\xi_n|$, then we have

$$(\xi_1,\ldots,\xi_n) \vdash\!\!\!- (\varepsilon_1 N,\ldots,\varepsilon_n N) \quad \forall\, n$$

So by Theorem 2.11 we have

$$\mathbb{E}\|\sum_1^n \xi_j x_j\|^s \leq \mathbb{E}\|4N \sum_1^n \varepsilon_j x_j\|^s$$

$$= 4^s\, \mathbb{E}(N^s)\; \mathbb{E}\|\sum_1^n \varepsilon_j x_j\|^s$$

Hence we find

$$B_E \subseteq B_E^s(\xi); \qquad C_E \subseteq C_E^s(\xi)$$

And from Theorem 6.2 we have

$$B_E^s(\xi) \subseteq B_E^r(\xi) \subseteq B_E^o(\xi) \subseteq B_E$$

for $r \in [0,s]$. So (d) holds and (e) follows similarly. □

Theorem 6.6. *Let $\xi = (\xi_n)$ be a sequence of independent real random variables, satisfying*

(a) (ξ_n) *is totally non-degenerated*

(b) $\mathbb{E}\,\xi_n = 0$

(c) $\sup_n \mathbb{E}\,|\xi_n|^p < \infty$

where p *belongs to* $[1,2]$. *Then the following statements are equivalent*

(1) E *is of type* p

(2) $\exists\, C > 0$ so that $\mathbb{E}\|\sum_1^n X_j\|^p \leq C \sum_1^n \mathbb{E}\|X_j\|^p$
for all independent random vectors $X_1,\ldots,X_n \in L_E^p$ *with mean* 0

(3) $C_E^p(\xi) \supseteq \ell_E^p$

(4) $B_E^o(\xi) \supseteq \ell_E^p$

Proof. (1) ⇒ (2): Let $X_1,\ldots,X_n \in L_E^p$ be independent and have mean 0, and let $\varepsilon = (\varepsilon_j)$ be a Bernouilli sequence independent of $(X_1,\ldots,X_n)$, then by Theorem 2.11:

$$\mathbb{E}\|\sum_1^n X_j\|^p \leq 4^p \ \mathbb{E}\|\sum_1^n \varepsilon_j X_j\|^p$$

$$= 4^p \int_{E^n} \mathbb{E}\|\sum_1^n \varepsilon_j x_j\|^p \mu(dx_1,\ldots,dx_n)$$

where $\mu = L(X_1,\ldots,X_n)$. Hence from (6.10) we have

$$\mathbb{E}\|\sum_1^n X_j\|^p \leq 4^p K \sum_{j=1}^n \mathbb{E}\|X_j\|^p$$

So (2) holds with $C = 4^p K$, where K is the constant from (6.10).

(2) ⇒ (3): Let $x_1,\ldots,x_n \in E$. If (2) holds then we have

$$\mathbb{E}\ \|\sum_1^n \xi_j x_j\|^p \leq C \sum_1^n \mathbb{E}|\xi_j|^p \|x_j\|^p$$

$$\leq C \sup_j \mathbb{E}\,|\xi_j|^p \sum_1^n \|x_j\|^p$$

and so $\ell_E^p \subseteq C_E^p(\xi)$.

(3) ⇒ (4): Obvious.

(4) ⇒ (1): Easy consequence of Theorem 6.2. □

<u>Theorem 6.7</u>. <u>Let</u> $\xi = (\xi_n)$ <u>be a sequence of independent real random variables, satisfying</u>

(a) (ξ_n) <u>is totally non-degenerated</u>

(b) $\mathbb{E}\,\xi_n = 0 \quad \forall\ n$

(c) $\sup_n |\xi_n| \in L^q$

<u>where</u> q <u>belongs to</u> $[2,\infty]$. <u>Then the following statements are equivalent</u>

(1) E <u>is of cotype</u> q

(2) $\exists\ C > 0$ <u>so that</u> $\sum_{j=1}^n \mathbb{E}\|X_j\|^q \leq C\,\mathbb{E}\|\sum_1^n X_j\|^q$

<u>for all independent random vectors</u> $X_1,\ldots,X_n \in L^q_E$

<u>with mean</u> 0.

(3) $B^o_E(\xi) \subseteq \ell^q_E$

(4) $C^q_E(\xi) \subseteq \ell^q_E$

<u>Remark</u>. We shall later show (see Theorem 7.7 and the remark to Theorem 6.5) that the theorem holds if we substitute (c) by the condition:

(c*) $\exists\, r > q$ so that $\sup_n \mathbb{E}\,|\xi_n|^r < \infty$

<u>Proof</u>. The proof is exactly as the proof of Theorem 6.6, except that we use Theorem 6.5 to conclude that (4) implies (1). □

7. Geometry and type

We shall now see that there is an intimate connection between the type, the cotype and the geometric notions introduced in §3 of chapter I. We start with the problem of boundedness and convergence:

Theorem 7.1. Let E be a Banach space, then the following 3 statements are equivalent

(a) $B_E \subseteq c_o(E)$

(b) $B_E = C_E$

(c) E does not contain c_o

Proof. (a) ⇒ (b): If $x \in B_E \setminus C_E$, then we can find $1 = n_1 < n_2 < \cdots$, and $a > 0$ so that

$$\mathbb{E} \| \sum_{n_k \le j < n_{k+1}} \varepsilon_j x_j \| \ge a \qquad \forall\ k \ge 1$$

Now let

$$X_k = \sum_{n_k \le j < n_{k+1}} \varepsilon_j x_j$$

Then $X_1, X_2, \dots$ are independent symmetric random variables, so that $M = \sup_n \| \sum_1^n X_k \| < \infty$ a.s. Hence by (6.5) we have

(i) $$P((X_n) \in B_E) = 1$$

Moreover since $\|X_k\| \le 2M$, $\mathbb{E} M < \infty$ and $\mathbb{E}\|X_k\| \ge a$, we have

(ii) $$P(X_k \to 0) < 1$$

So from (i) and (ii) we see that $B_E \not\subseteq c_o(E)$. So we have proved "(a) ⇒ (b)".

(b) ⇒ (c): Let T be a bounded linear map: $c_o \to E$, and let $f_j = Te_j$ where e_j is the j-th unit vector in c_o. Then we have

$$\| \sum_{j=1}^{n} \varepsilon_j(\omega) f_j \| \leq \|T\| \quad \forall\ \omega \quad \forall\ n \geq 1$$

so $(f_j) \in B_E = C_E$. Hence $f_j \to 0$, and T is not an isomorphism.

(c) ⇒ (a): Let (ε_j) be a Bernouilli sequence and let $(x_j) \in B_E$. Then $X_j = \varepsilon_j x_j$ are vectors in L_E^1, and since $(x_j) \in B_E$, we have that (X_j) has unconditionally bounded partial sums (see §3, chapter I). Now by Theorem I.3.12 we know that c_o is not contained in L_E^1, so by Theorem I.3.13 we have

$$\|x_j\| = \mathbb{E}\|X_j\| = \|X_j\|_1 \longrightarrow 0 \quad \square$$

<u>Corollary 7.2</u>. <u>Let</u> (X_j) <u>be a sequence of independent E-valued random vectors, so that the partial sums:</u> $S_n = \sum_1^n X_j$ <u>are bounded. If</u> E <u>does not contain</u> c_o, <u>then there exists a bounded non random sequence</u> (a_n) <u>of vectors in</u> E <u>so that</u>

(a) $\{S_n - a_n\}$ <u>is a.s. convergent</u>

<u>And in each of the following 3 cases we have that</u> $\sum_1^\infty X_j$ <u>converges a.s.</u>:

(1) (X_j) <u>is a symmetric sequence</u>.

(2) $\mathbb{E} X_j = 0$ <u>and</u> $\sup \mathbb{E}\|S_n\| < \infty$

(3) $\mathbb{E} X_j = 0$ <u>and</u> $\mathbb{E}(\sup \|X_n\|) < \infty$

<u>Proof</u>. Case (1) is an immediate consequence of (6.5) and Theorem 7.1. The general case then follows by a standard symmetrization procedure. And the cases (2) and (3) then follows easily from Theorem 5.5. □

<u>Theorem 7.3</u>. <u>Let</u> E <u>be a Banach space. Then</u> $\ell^1 \to \ell^p$ <u>is mimicked through</u> E <u>for all</u> $p \geq p(E)$, <u>but is not mimicked through</u> E <u>for any</u> $1 \leq p < p(E)$.

<u>Remark</u>. In particular we see that $p(E) = 1$ if and only if E mimicks ℓ^1.

<u>Proof</u>. Suppose that $\ell^1 \to \ell^p$ is mimicked through E. Then there exists $x_1^n, \ldots, x_n^n$, so that

$$\|x_j^n\| \leq 1; \quad (\sum_1^n |t_j|^p)^{1/p} \leq 2 \|\sum_1^n t_j x_j^n\|$$

So putting $t_j = \varepsilon_j$ we find

$$n^{1/p} \leq 2\, \mathbb{E} \|\sum_1^n \varepsilon_j x_j^n\| \leq 2K_r (\sum_1^n \|x_j^n\|^r)^{1/r}$$

$$\leq 2K_r\, n^{1/r}$$

for any $r < p(E)$. Hence $p \geq r$ for all $r < p(E)$ and so $p \geq p(E)$. That is $\ell^1 \to \ell^p$ is not mimicked through E for any $1 \leq p < p(E)$.

Let us make a small digression in order to study two important functions $\lambda(n)$ and $\nu(n)$, which also will be of use in our study of the law of large numbers.

(7.1) $$\lambda(n) = n^{-1} \sup\{\min_{\pm} \|\sum_1^n \pm x_j\| : \|x_j\| \leq 1 \ \forall j = 1, \ldots, n\}$$

(7.2) $$\nu(n) = n^{-\frac{1}{2}} \sup\{(\mathbb{E}\|\sum_{j=1}^n \varepsilon_j x_j\|^2)^{\frac{1}{2}} : \sum_1^n \|x_j\|^2 \leq 1\}$$

where (ε_j) as usual is a Bernouilli sequence. Then clearly $\lambda(n)$ and $\nu(n)^2$ are the smallest numbers for which we have

(7.3) $$\min_{\pm} \|\sum_1^n \pm x_j\| \leq n\lambda(n) \max_{1 \leq j \leq n} \|x_j\|$$

(7.4) $$\mathbb{E}\,\|\sum_1^n \varepsilon_j x_j\|^2 \le n\,\nu(n)^2 \sum_1^n \|x_j\|^2$$

for all $x_1,\ldots,x_n$. Hence we have

(7.5) $\sqrt{n}\,\nu(n)$ and $n\,\lambda(n)$ are increasing in n and $\nu(1) = \lambda(1) = 1$.

By integrating (7.4) we find

(7.6) $$\mathbb{E}\,\|\sum_1^n X_j\|^2 \le n\,\nu(n)^2 \sum_1^n \mathbb{E}\,\|X_j\|^2$$

whenever $(X_1,\ldots,X_n)$ is symmetric. From (7.6) it then easily follows that we have

(7.7) $$\nu(nk) \le \nu(n)\,\nu(k) \qquad \forall\, n,k \ge 1$$

And since

$$\min_{\pm} \|\sum_1^n \pm x_j\| \le (\mathbb{E}\,\|\sum_1^n \varepsilon_j x_j\|^2)^{\frac12} \le \sqrt{n}\,\nu(n)\,(\sum_1^n \|x_j\|^2)^{\frac12}$$

$$\le n\,\nu(n) \max_{1\le j\le n} \|x_j\|$$

we have

(7.8) $$\lambda(n) \le \nu(n) \qquad \forall\, n\ge 1$$

The next two propositions are of more nontrivial nature:

(7.9) If $\nu(m) = m^{-\alpha}$ for some $m\ge 2$ and some $\alpha>0$, then $\alpha\le\frac12$ and $p(E) \ge (1-\alpha)^{-1}$.

(7.10) $$\frac{\nu(n)}{1+\sqrt{2n(1-\nu(n)^2)}} \le \sqrt{2^{-n}\lambda(n)^2 + 1-2^{-n}} \qquad \forall\, n$$

<u>Proof of (7.9)</u>. By (7.5) we have $\nu(m) \geq m^{-\frac{1}{2}}$ so $\alpha \leq \frac{1}{2}$. If $m^{k-1} \leq n \leq m^k$ for some $k \geq 1$, then we have by (7.5) and (7.7):

$$\nu(n) \leq \{m^k/n\}^{\frac{1}{2}} \nu(m^k) \leq \sqrt{m}\, \nu(m)^k = \sqrt{m}\, m^{-\alpha k}$$

Hence

(i) $$\nu(n) \leq \sqrt{m}\, n^{-\alpha} \qquad \forall\ n \geq 1$$

Now let $x_1, \ldots, x_n \in E$ and $1 \leq p < (1-\alpha)^{-1}$. Then we split the x_j's in groups of almost the same norm:

$$A_k = \{1 \leq j \leq n \mid \beta 2^{-k-1} < \|x_j\| \leq \beta 2^{-k}\} \qquad k \geq 0$$

where $\beta = (\sum_1^n \|x_j\|^p)^{1/p}$. Then $A_o, A_1, \ldots$ is a disjoint partition of $\{1, \ldots, n\}$. If $\sigma(k) = \# A_k$ then we have

$$\|\sum_1^n \varepsilon_j x_j\|_2 \leq \sum_{k=o}^{\infty} \|\sum_{j \in A_k} \varepsilon_j x_j\|_2$$

$$\leq \sum_{k=o}^{\infty} \sigma(k)^{\frac{1}{2}} \nu(\sigma(k)) \left(\sum_{j \in A_k} \|x_j\|^2\right)^{\frac{1}{2}}$$

$$\leq \sum_{k=o}^{\infty} \sigma(k)^{1-\alpha} 2^{-k} \beta \sqrt{m}$$

Now $\sigma(k)$ may be estimated as follows:

$$\beta^p = \sum_{j=1}^n \|x_j\|^p \geq \sum_{j \in A_k} \|x_j\|^p \geq \beta^p 2^{-p(k+1)} \sigma(k)$$

So $\sigma(k) \leq 2^{p(k+1)}$. Hence

$$\|\sum_1^n \varepsilon_j x_j\|_2 \leq C \beta \sum_{k=o}^{\infty} 2^{-k(1-p(1-\alpha))} = K\{\sum_{j=1}^n \|x_j\|^p\}^{1/p}$$

and since $1-p(1-\alpha) > 0$, K is finite and independent of $x_1, \ldots, x_n \in E$. Hence E is of type p for any $p < (1-\alpha)^{-1}$ that is $p(E) \geq (1-\alpha)^{-1}$. □

Proof of (7.10). Let $\alpha < \nu(n)$, then there exists $x_1,\dots,x_n \in E$, so that

(ii) $$\sum_1^n \|x_j\|^2 = n$$

(iii) $$\alpha n \le (\mathbb{E}\|\sum_1^n \varepsilon_j x_j\|^2)^{\frac12}$$

Let $1 \le \nu \le n$, then we have

$$\sum_{j=1}^n (\|x_\nu\| - \|x_j\|)^2 \le \sum_{i=1}^n \sum_{j=1}^n (\|x_i\| - \|x_j\|)^2$$

$$= 2n \sum_{j=1}^n \|x_j\|^2 - 2(\sum_{j=1}^n \|x_j\|)^2 \le 2n^2 - 2\,\mathbb{E}\|\sum_1^n \varepsilon_j x_j\|^2$$

$$\le 2n^2(1-\alpha^2)$$

So we have

$$\sqrt{n} = (\sum_{j=1}^n \|x_j\|^2)^{\frac12} \ge (\sum_{j=1}^n \|x_\nu\|^2)^{\frac12} - (\sum_{j=1}^n \|x_\nu\| - \|x_j\|)^2)^{\frac12}$$

$$\ge \sqrt{n}\,\|x_\nu\| - n\sqrt{2(1-\alpha^2)}$$

That is

(iv) $$\max_{1\le\nu\le n} \|x_\nu\| \le 1 + \sqrt{2n(1-\alpha^2)}$$

Hence we find

$$\max_{1\le\nu\le n} \|x_\nu\| \{1 + \sqrt{2n(1-\alpha^2)}\}^{-1} \alpha n$$

$$\le \alpha n \le (\mathbb{E}\|\sum_1^n \varepsilon_j x_j\|^2)^{\frac12}$$

$$= \{2^{-n} \sum_{\pm} \|\sum_1^n \pm x_j\|^2\}^{\frac12}$$

$$\le \{2^{-n} \min_{\pm} \|\sum_1^n \pm x_j\|^2 + (1-2^{-n})\, n^2 \max_{1\le j\le n} \|x_j\|^2\}^{\frac12}$$

$$\le \{2^{-n} \lambda(n)^2 + 1 - 2^{-n}\}^{\frac12}\, n \max_{1\le j\le n} \|x_j\|$$

And so

$$\alpha\{1 + \sqrt{2n(1-\alpha^2)}\}^{-1} \leq \{2^{-n}\lambda(n)^2 + 1 - 2^{-n}\}^{\frac{1}{2}}$$

for all $\alpha < \nu(n)$. Letting $\alpha \to \nu(n)$ we have proved (7.10). □

Next we show:

(7.11) E mimicks ℓ^1 if and only if $\lambda(n) = 1$ for all $n \geq 1$.

If E mimicks ℓ^1, then for every $\varepsilon>0$ and every $n\geq 1$, there exists $x_1^n,\ldots,x_n^n$, so that $\|x_j^n\| = 1$, and

$$\|\sum_1^n t_n x_j^n\| \geq (1-\varepsilon) \sum_1^n |t_j| \qquad \forall\ t_1,\ldots,t_n \in \mathbb{R}$$

Hence

$$\min_{\pm} \|\sum_1^n \pm x_j^n\| \geq (1-\varepsilon)n$$

which shows that $\lambda(n) = 1$. If E does not mimick ℓ^1, then for some $m\geq 2$ and some $\gamma < 1$, we have

$$\min\{\|\sum_1^m t_j x_j\| : \sum_1^n |t_j| = 1\} \leq \gamma$$

for all $x_1,\ldots,x_m \in E$ with $\|x_j\| \leq 1$. Now let $x_1,\ldots,x_m$ be vectors in E, with $\|x_j\| \leq 1$, and choose $t_1,\ldots,t_m \in \mathbb{R}$ so that

$$\sum_1^m |t_j| = 1 \quad \text{and} \quad \|\sum_1^m t_j x_j\| \leq \gamma$$

Let $\alpha_j = \text{sign}(t_j)$, then we have

$$\|\sum_1^m \alpha_j x_j\| = \|\sum_1^m (\alpha_j(1-|t_j|)x_j) + t_j x_j\|$$

$$\leq \sum_1^m (1-|t_j|)\ \|x_j\| + \|\sum_1^m t_j x_j\|$$

$$\leq m - 1 + \gamma$$

Hence we have $\lambda(m) \leq 1 - (1-\gamma)/m < 1$. So (7.11) is proved. □

<u>End of the proof of Theorem 7.3</u>. If $p(E) = 2$ then we have by Theorem I.3.7 that $\ell^1 \to \ell^{p(E)}$ factorizes through E. We shall here only prove the theorem in the other extreme case: $p(E) = 1$. So we assume that $p(E) = 1$, and we have to prove that E mimicks ℓ^1.

Suppose to the contrary that E does not mimick ℓ^1. Then by (7.11) we have $\lambda(m) < 1$ for some $m \geq 2$, and by (7.10) we then have $\nu(m) < 1$. But then $p(E) \geq (1-\alpha)^{-1} > 1$, where

$$\alpha = -\log \nu(m)/\log m$$

by (7.9). So are led to do a contradiction, and so E mimicks ℓ^1.

The general case $(p(E) \geq 1)$ is much more complicated, and we refer the interested reader to [89]. □

<u>Theorem 7.4</u>. <u>Let</u> E <u>be a Banach space. Then the injection</u> $\ell^p \to \ell^\infty$ <u>is mimicked through</u> E, <u>if</u> $1 \leq p \leq q(E)$, <u>but not mimicked through</u> E <u>if</u> $q(E) < p \leq \infty$.

<u>Remark</u>. Note that the injection $\ell^p \to \ell^\infty$ is the same as $\ell^p \to c_o$ for $p < \infty$. So we have $q(E) = \infty$, if and only if E mimicks c_o.

<u>Proof</u>. Exactly as in the easy part of the proof of Theorem 7.3 one shows that $\ell^p \to \ell^\infty$ is not mimicked through E for $p > q(E)$.

So let us then assume that $\ell^p \to \ell^\infty$ is not mimicked through E. By Theorem I.3.9 and Theorem I.3.12, there exists a $q < p$, so that $\ell^q \to \ell^\infty$ is not mimicked through L_E^2. Now let $(x_j) \in B_E$ and let (ε_j) be a Bernoulli sequence. Then $(\varepsilon_j x_j)$ has unconditionally bounded partial sums in L^2_E so by Corollary I.3.11 we have $(x_j) \in \ell_E^q$

That is $B_E \subseteq \ell_E^q$, and so we have $q(E) \leq q < p$, and Theorem 7.4 is proved. □

Corollary 7.5. *If* $p(E) > 1$ *or* $q(E) < \infty$ *then we have*

$$B_E = C_E \subseteq c_o(E) \quad \square$$

Theorem 7.6. *Let* $\xi = (\xi_n)$ *be a sequence of independent real random variables satisfying*

(a) $\mathbb{E}\, \xi_n = 0$

(b) $\sup \mathbb{E}\, |\xi_n|^q < \infty$

(c) $\{\xi_n\}$ *are totally non degenerated*

for some $q \in [2,\infty[$. *If* E *is a Banach space with* $q(E) < q$, *then we have*

(d) $B_E = C_E = B_E^r(\xi) = C_E^r(\xi) \qquad \forall\ 0 \leq r \leq q$

Proof. By Theorem 7.4 we have that $\ell^q \to \ell^\infty$ is not mimicked through E. So by Theorem I.3.12 it is not mimicked through L_E^2, and by Theorem I.3.10 there exists a $K > 0$ so that for all $f_1,\ldots,f_n$ there exists $\alpha_1,\ldots,\alpha_n \in \mathbb{R}_+$, with $\sum_1^n \alpha_j = 1$ and:

$$\|\sum_1^n t_j f_j\|_2 \leq K \max_{\pm} \|\sum_1^n \pm f_j\|_2 \ (\sum_1^n \alpha_j |t_j|^q)^{1/q}$$

Applying this to $f_j = \varepsilon_j x_j$, where (ε_j) is a Bernoulli sequence and $x_1,\ldots,x_n \in E$, gives:

$$\mathbb{E}\,\|\sum_1^n t_j\varepsilon_j x_j\|^q \le C\,\|\sum_1^n t_j\varepsilon_j x_j\|_2^q$$

$$\le C\,K^q\ \|\sum_1^n \varepsilon_j x_j\|_2^q\ \sum_1^n |t_j|^q\,\alpha_j$$

Integrating this w.r.t. $L(\xi_1,\ldots,\xi_n)$ gives

$$\mathbb{E}\,\|\sum_1^n \xi_j\varepsilon_j x_j\|^q \le CK^q\,\|\sum_1^n \varepsilon_j x_j\|_2^q\ \sum_1^n \mathbb{E}\,|\xi_j|^q\ \alpha_j$$

$$\le CK^q\ \sup_j\ \mathbb{E}\,|\xi_j|^q\ \|\sum_1^n \varepsilon_j x_j\|_2^q$$

Now $(\xi_1,\ldots,\xi_n) \longmapsto (\varepsilon_1\xi_1,\ldots,\varepsilon_n\xi_n)$, so by Theorem 2.11 we have

$$\mathbb{E}\,\|\sum_1^n \xi_j x_j\|^q \le 4^q\ \mathbb{E}\,\|\sum_1^n\varepsilon_j\xi_j x_j\|^q$$

$$\le C(4K)^q\ \sup_j\ \mathbb{E}\,|\xi_j|^q\,\|\sum_1^n \varepsilon_j x_j\|_2^q$$

And so we have

$$C_E \subseteq C_E^q(\xi) \qquad\qquad B_E \subseteq B_E^q(\xi)$$

and since $q(E) < q < \infty$ we have $B_E = C_E$ by Corollary 7.5. Now the conclusion follows easily from Theorem 6.2. □

<u>Theorem 7.7</u>. <u>Let</u> $\xi = (\xi_n)$ <u>be independent real random variables, satisfying</u>

(a) $\mathbb{E}\,\xi_n = 0 \qquad \forall\ n\geq 1$

(b) $\sup \mathbb{E}\,|\xi_n|^r < \infty$

(c) $C_E^r(\xi) \subseteq \ell_E^q$

<u>where</u> $2\leq q<r$. <u>Then</u> E <u>is of cotype</u> q.

Proof. Let us first show that $q(E) < r$. From the closed graph theorem and (c) it follows that for some constant $C > 0$ we have

(i) $$\left(\sum_{j=1}^{n} \|x_j\|^q\right)^{1/q} \leq C \{ \mathbb{E} \| \sum_{j=1}^{n} x_j \xi_j \|^r \}^{1/r}$$

for all $x_1, \ldots, x_n \in E$, and all $n \geq 1$.

If $q(E) \geq r$, then there exists $x_1^n, \ldots, x_n^n \in E$ with $\|x_j^n\| \geq 1$ and

$$\| \sum_{j=1}^{n} t_j x_j^n \|^r \leq 2 \sum_{1}^{n} |t_j|^r$$

(cf. Theorem 7.4 and Proposition I.3.6). Integrating this with respect to $L(\xi_1, \ldots, \xi_n)$ gives

$$n^{r/q} \leq \{\sum_{1}^{n} \|x_j^n\|^q\}^{r/q} \leq C \, \mathbb{E} \| \sum_{j=1}^{n} x_j \xi_j \|^r$$

$$\leq 2C \sum_{1}^{n} \mathbb{E} |\xi_j|^r \leq 2n \, C \sup_j \mathbb{E} |\xi_j|^r$$

But this is impossible since $r/q > 1$. So we have $q(E) < r$.

But the proof of the previous theorem shows that $C_E \subseteq C_E^r(\xi) \subseteq \ell_E^q$, so E is of cotype q.

CHAPTER III

The two pearls of probability

1. The law of large numbers

The first law of large numbers was stated and proved by James Bernoulli (about 1695; published in 1713 in "Ars Conjectandi" 8 years after the death of James Bernoulli). The study of this theorem has ever since been an important part of probability. In this section we shall study its validity in Banach spaces.

Let (X_n) be a sequence of E-valued random vectors, then we define the average, $\bar{X}_n$, by

$$\bar{X}_n = n^{-1} \sum_{j=1}^{n} X_j$$

And we say that $\{X_n\}$ satisfies the strong law of large numbers if $\{\bar{X}_n\}$ is a.s. convergent.

In the real case $(E = \mathbb{R})$ we have the following 3 well-known versions of the law of large number:

(1.1) If (X_n) are independent integrable and identical distributed, then $\bar{X}_n \to \mathbb{E} X_1$ a.s.

(1.2) If (X_n) are independent, have mean 0, and $\sup_n \mathbb{E}|X_n|^2 < \infty$, then $\bar{X}_n \to 0$ a.s.

(1.3) If (X_n) are independent, have mean 0, and $\sum_1^\infty n^{-p}\, \mathbb{E}|X_n|^p$ for some $p \geq 1$, then $\bar{X}_n \to 0$ a.s.

These results are evidently also valid for $\dim E < \infty$ (use the real versions on each coordinate!).

We shall now generalize these propositions to general Banach spaces. A main tool to do this is Kronecker's lemma:

(1.4) $$\sum_1^\infty n^{-1} x_n \text{ convergent} \Rightarrow \lim_{n\to\infty} n^{-1} \sum_{j=1}^{n} x_j = 0$$

which is wellknown if $\dim E = 1$, and is valid for arbitrary Banach spaces (the same proof as in the real case works!)

Theorem 1.1. Let (X_n) be independent identical distributed random vectors in E. Then (X_n) satisfies the strong law of large numbers if and only if $\mathbb{E}\|X_1\| < \infty$, and if so then $\bar{X}_n \to \mathbb{E}X_1$ a.s.

Proof. The "only if" part is contained in Theorem II.1.5.

So suppose that $X_1 \in L_E^1$. We may without loss of generality assume that $\mathbb{E}X_1 = 0$. If $\mu = L(X_1)$, then we can find a simple Borel function, $f: E \to E$, so that

$$\int_E f(x)\mu(dx) = 0, \qquad \int_E \|x-f(x)\|\ \mu(dx) \leq \varepsilon$$

where $\varepsilon > 0$ is a given number. Now let $Y_n = f(X_n)$ then $Y_1, Y_2, ..$ are independent, have mean 0 and the same distribution, and $Y_n \in E_o = \text{span}\, f(E)$, which is finite dimensional. So by (1.1) we have

$$n^{-1} \sum_{j=1}^{n} Y_j \to 0 \quad \text{a.s.}$$

Now $\xi_j = \|X_j - Y_j\|$ are independent, identical distributed real random variables, so we have

$$n^{-1} \sum_{j=1}^{n} \|X_j - Y_j\| \xrightarrow{a.s.} \mathbb{E}\|X_1 - Y_1\| \leq \varepsilon$$

Hence we find

$$\limsup_{n\to\infty} \|\bar{X}_n\| \leq \limsup_{n\to\infty} \{\|\bar{Y}_n\| + n^{-1} \sum_{j=1}^{n} \|X_j - Y_j\|\}$$

$$\leq \varepsilon$$

a.s., and since $\varepsilon > 0$ is arbitrary, we have $\bar{X}_n \to 0$ a.s. $\square$

So for the identitcal distributed case, there is no difference between $\mathbb{R}$ and general Banach spaces. However, this is not so for (1.2) and (1.3).

Let us recall the definition of $\lambda(n)$ (see (II.7.1)) in the proof of Theorem II.7.3):

$$\lambda(n) = \sup\{\min_{\pm} n^{-1} \|\sum_1^n \pm x_j\| : \max_{1\le j\le n} \|x_j\| \le 1\}$$

A Banach space E is called B-convex, if there exists $n\geq 2$ so that $\lambda(n) < 1$, that is if there exists $n\geq 2$ and $\lambda_o<1$, so that

$$\text{(1.5)} \qquad \min_{\pm} \|\sum_1^n \pm x_j\| \le \lambda_o n \qquad \forall\, x_1,\dots,x_n \in B$$

where B is the unit ball in E. With this definition we have the following version of (1.2):

Theorem 1.2. Let E be a Banach space then the following 7 statements are equivalent

(a) E is B-convex

(b) E does not mimick ℓ^1

(c) E does not parody ℓ^1

(e) $p(E) > 1$

(f) $\lim_{n\to\infty} \lambda(n) = 0$

(g) For all independent mean 0 random vectors in E, with $\sup_n \mathbb{E}\|X_n\|^2 < \infty$, we have $\bar{X}_n \to 0$ a.s.

(h) $n^{-1}\sum_1^n \varepsilon_j x_j \to 0$ in L_E^0 $\quad \forall (x_j) \in \ell_E^\infty$

Proof. (a) - (f) are equivalent by Theorem I.3.8, Theorem II.7.3 and (II.7.7) - (II.7.11).

Suppose that $p(E) > 1$. Then there exists a $1<p\leq 2$ so that E is of type p. Hence for some constant $K>0$ we have

$$\mathbb{E}\|\sum_{j=n}^{m} j^{-1}X_j\|^p \leq K \sum_{j=n}^{m} j^{-p}\,\mathbb{E}\|X_j\|^p$$

$$\leq K \sum_{j=n}^{m} j^{-p}(\mathbb{E}\|X_j\|^2)^{p/2}$$

$$\leq K \sup_j (\mathbb{E}\|X_j\|^2)^{p/2} \sum_{j=n}^{\infty} j^{-p}$$

whenever (X_n) satisfies the hypotheses in (g). Hence we have that $\sum_1^{\infty} j^{-1}X_j$ converges a.s. and in L_E^p (cf. Theorem II.5.3 case (6)). So from (1.4) we have $\bar{X}_n \to 0$ a.s. (and in L_E^p). That is (e) implies (g).

Clearly (g) implies (h).

(h) ⇒ (a): Suppose that (a) does not hold, that is $\lambda(n) = 1$ for all $n\geq 1$. Then we can find $x_j \in E$ with $\|x_j\| \leq 1$ and

$$\min_{\pm} \|\sum_{3^k<j\leq 3^{k+1}} \pm x_j\| \geq \frac{3}{4}(3^{k+1} - 3^k) \quad \forall\ k\geq 0$$

Let (ε_j) be a Bernoulli sequence, and let $\omega \in \Omega$ then $\varepsilon_j(\omega) = \pm 1$, so we have

$$\|\sum_{j=1}^{3^k} \varepsilon_j(\omega)x_j\| \geq \|\sum_{3^{k-1}<j\leq 3^k} \varepsilon_j(\omega)x_j\| - \sum_{j=1}^{3^{k-1}} \|x_j\|$$

$$\geq \frac{3}{4}(3^k - 3^{k-1}) - 3^{k-1}$$

$$= \frac{1}{2}\, 3^{k-1}$$

Hence $(x_j) \in \ell_E^{\infty}$, but

$$\|\frac{1}{n}\sum_{j=1}^{n}\varepsilon_j(\omega)x_j\| \geq 1/6 \qquad \forall\,\omega\in\Omega \qquad \forall\,n = 1,3,9,27,\ldots$$

and so $n^{-1}\sum_1^n \varepsilon_j x_j \not\to 0$ in L_E^0. Hence (h) does hot hold. □

<u>Theorem 1.3</u>. <u>Let</u> E <u>be a Banach space and</u> $1 \leq p \leq 2$. <u>Then the following 3 statements are equivalent:</u>

(a) E <u>is of type</u> p

(b) <u>For all independent mean 0 random vectors</u> (X_n) <u>with</u> $\sum_1^\infty n^{-p}\,\mathbb{E}\|X_n\|^p < \infty$ <u>we have</u> $\bar{X}_n \to 0$ a.s.

(c) $n^{-1}\sum_1^n \varepsilon_j x_j \to 0$ <u>in</u> L_E^0 <u>for all</u> $(x_j)\in E^\infty$ <u>with</u> $\sum_1^\infty n^{-p}\|x_n\|^p < \infty$

<u>Proof</u> (a) ⇒ (b): This can be done exactly as the proof of "(e) ⇒ (g)" in Theorem 1.2.

(b) ⇒ (c): Evident!

(c) ⇒ (a): Let us define

$$G = \{(x_j)\in E^\infty \mid q(x) = \left\{\sum_1^\infty n^{-p}\|x_n\|^p\right\}^{1/p} < \infty\}$$

$$F = \{(x_j)\in E^\infty \mid r(x) = \sup_n \|\frac{1}{n}\sum_1^n \varepsilon_j x_j\|_p < \infty\}$$

Then (G,q) and (F,r) are Banach spaces and since $|||\cdot|||_p$ is equivalent to $|||\cdot|||_0$ on B_E (cf. Theorem II.5.5), (c) implies that $G\subseteq F$. The injection: $G \to F$ clearly has a closed graph so by the closed graph theorem, there exists a constant $K<\infty$ with

$$\|\mathbb{E}\sum_1^n \varepsilon_j x_j\|^p \leq K\, n^p \sum_{j=1}^n j^{-p}\|x_j\|^p$$

for $n \geq 1$ and all $x_1,\ldots,x_n \in E$.

Applying this to the vector $(0,\ldots,0,x_1,\ldots,x_n)$ (m zeros), we find

$$\mathbb{E}\|\sum_1^n \varepsilon_j x_j\|^p \leq K\ (n+m)^p \sum_{j=1}^n (j+m)^{-p}\|x_j\|^p$$

$$\leq K\ (\frac{n+m}{1+m})^p \sum_{j=1}^n \|x_j\|^p$$

for all $n,m\geq 1$ and all $x_1,\ldots,x_n \in E$. Letting $m \to \infty$ gives

$$\mathbb{E}\|\sum_1^n \varepsilon_j x_j\|^p \leq K \sum_{j=1}^n \|x_j\|^p \qquad \forall\ x_1,\ldots,x_n \in E$$

so E is of type p. □

2. The central limit theorem

The first version of the central limit theorem was proved by de Moivre in 1733 (published in his book: "Doctrine of Chance", 3rd ed. 1756). And the study of this theorem has been a major subject of probability theory ever since.

If $(E, \|\cdot\|)$ is a B-space and X is a random vector, then X is said to belong to the domain of normal attraction (see Chap. II, §4) if

$$Z_n = (X_1 + \cdots + X_n)/\sqrt{n} \tag{2.1}$$

converges in law whenever $X_1, X_2, \ldots$ are independent copies of X. The domain of normal attraction is denoted DNA.

In Chap.II, §4 we defined $c(X)$ and CLT by

$$c(X) = \sup_n \mathbb{E}\, \|n^{-\frac{1}{2}} \sum_{j=1}^{n} X_j\|$$

$$\mathrm{CLT} = \{X \in L_E^0 \mid c(X) < \infty\}$$

where $X_1, \ldots, X_n, \ldots$ are independent copies of X. And we saw (see Theorem II.4.10)

(2.2) $\qquad \mathrm{DNA} \subseteq \mathrm{CLT} \subseteq L_E^{(2)} \qquad$ (cf. Chap. I, §2)

(2.3) $\qquad \lambda_2(X) \leq K\, c(X)^{2/3} \qquad$ (cf. Chap. I, (2.11))

(2.4) $\qquad \mathbb{E}\, X = 0 \qquad \forall\, X \in \mathrm{CLT}$

If $X \in \mathrm{DNA}$, then it is easy to determine the limit distribution of Z_n given by (2.1): By the real-valued central limit theorem we have

$$\hat{\nu}(x') = \exp\{-\tfrac{1}{2}\, E\langle X, x'\rangle^2\} \qquad \forall\, x' \in E \tag{2.5}$$

if $\nu = \lim_n L(Z_n)$. That is ν is the Gaussian Radon measure with mean 0 and the same covariance function as X. So the first problem is to determine whether there exists such a measure and then secondly to determine whether $\{Z_n\}$ converges in law to this measure.

Inspired by these considerations we shall say that X is <u>pregaussian</u> if $\langle X,x'\rangle \in L^2$ for all $x' \in E'$ and if there exists a gaussian Radon measure ν satisfying (2.5). The set of all pregaussian random vectors is denoted PG. Clearly we have

(2.6) $$DNA \subseteq PG$$

From Lemma 2.1 in [86] it follows easily that we have:

<u>Preposition 2.1</u>. <u>Let</u> ν <u>be a gaussian Radon measure on</u> E <u>with Fourier transform</u>:

$$\hat{\nu}(x') = \exp(-\tfrac{1}{2} R(x'))$$

<u>If</u> X <u>is any random vector satisfying</u>

$$\mathbb{E}\langle X,x'\rangle^2 \leq R(x') \qquad \forall x' \in E'$$

<u>then</u> X <u>is pregaussian</u>. □

If $X,Y \in PG$, then we have

$$\mathbb{E}\langle X+Y,x'\rangle^2 \leq 2\mathbb{E}\langle X,x'\rangle^2 + 2\mathbb{E}\langle \bar{Y},x'\rangle^2$$

and we deduce from Proposition 2.1:

(2.7) PG is a linear subspace of L^0_E.

We shall later introduce a norm on PG.

<u>Proposition 2.2</u>. DNA <u>is a closed linear subspace of</u> (CLT,c).

<u>Proof</u>. Let X belong to the closure of DNA under the norm $c(\cdot)$. Let $\varepsilon > 0$ be given, then we can find $Y \in DNA$ with

$$c(X-Y) \leq \varepsilon^2$$

Let (X_j) and (Y_j) be independent copies of X respectively Y, and put

$$U_n = n^{-\frac{1}{2}} \sum_{j=1}^{n} X_j \quad ; \qquad \mu_n = L(U_n)$$

$$V_n = n^{-\frac{1}{2}} \sum_{j=1}^{n} Y_j \quad ; \qquad \nu_n = L(V_n)$$

Let ρ be the Prohorov metric on the set of probability measures on E (cf. Chap. I, §4), then by (I.2.16) and (I.4.4) we find

$$\rho(\mu_n,\nu_n) \leq \lambda_1(U_n-V_n) \leq \sqrt{\mathbb{E}\|U_n-V_n\|} \leq \sqrt{c(X-Y)}$$

since $c(X-Y) = \sup_n \mathbb{E}\|U_n-V_n\|$. So for $n,m \geq 1$ we have

$$\rho(\mu_n,\mu_m) \leq \rho(\mu_n,\nu_n) + \rho(\nu_n,\nu_n) + \rho(\nu_m,\mu_m)$$

$$\leq 2\varepsilon + \rho(\nu_n,\nu_m)$$

And since $\{\nu_n\}$ is a ρ-Cauchy sequence, it follows that $\{\mu_n\}$ is a ρ-Cauchy sequence. Hence X belongs to the domain of normal attraction (cf. (I.4.2)), and DNA is closed.

To see that DNA is linear we first note that evidently we have $aX \in DNA$, whenever $X \in DNA$ and $a \in \mathbb{R}$. Now suppose that $X', X'' \in DNA$, and let $X = X' + X''$. Then we put

$$Z_k' = k^{-\frac{1}{2}} \sum_{1}^{k} X_j', \qquad Z_k'' = k^{-\frac{1}{2}} \sum_{1}^{k} X_j'', \quad Z_k = Z_k' + Z_k''$$

where $(X_1', X_1''), (X_2', X_2''), \ldots$ are independent copies of (X', X'').

Since $\{Z_k'\}$ and $\{Z_k''\}$ converges in law it follows from (I.4.9), that for any given $\varepsilon > 0$ we can find K compact $\subseteq E$, such that

$$P(Z_k' \notin K) < \varepsilon, \quad P(Z_k'' \notin K) \leq \varepsilon \qquad \forall\ k \geq 1$$

But then

$$P(Z_k \notin K+K) \leq 2\varepsilon \qquad \forall\ k \geq 1$$

and $K+K$ is compact. Hence $\{L(Z_k)\}$ is uniformly tight, and since $\{X_j' + X_j''\}$ are independent copies of X, and X is pre-gaussian, we have that the only possible limit point of $\{L(Z_k)\}$ is the gaussian Radon measure with mean 0 and covariance $= \mathbb{E}\langle X, x'\rangle^2$. Hence from (I.4.8) we deduce that $\{Z_k\}$ converges in law, and $X = X' + X'' \in \mathrm{DNA}$. □

__Theorem 2.3__. __Let__ (M_n) __be an E-valued martingale, such that__

(a) $\lim_{n\to\infty} M_n = X$ __exists a.s. and in__ L_E^1

(b) M_n __and__ X __belongs to__ DNA __for all__ $n \geq 1$

__Then we have:__ $\lim_{n\to\infty} c(M_n - X) = 0$.

__Proof__. Let $(M_n^1), (M_n^2), \ldots$ be independent copies of (M_n), and let $X^j = \lim_{n\to\infty} M_n^j$, which of course exists a.s. and in L_E^1 by (a). Let

$$Z_k = k^{-\frac{1}{2}} \sum_{j=1}^{k} X^j, \qquad Z_{kn} = k^{-\frac{1}{2}} \sum_{j=1}^{k} M_n^j$$

Since (M_n) is a martingale we have

$$\mathbb{E}\langle M_n, x'\rangle^2 = \sum_{j=1}^{n} \mathbb{E}\langle D_j, x'\rangle^2 \qquad \forall\ x' \in E'$$

where $D_1 = M_1$ and $D_j = M_j - M_{j-1}$ for $j \geq 2$. Now let us choose independent gaussian random vectors, $U_1, U_2, \ldots$, with mean 0 and covariance equal to the covariance of $D_1, D_2, \ldots$

Since $\langle X, x' \rangle \in L^2$, we have that

$$\mathbb{E} \langle X, x' \rangle^2 = \sum_1^\infty \mathbb{E} \langle D_j, x' \rangle^2 \qquad \forall\, x' \in E'$$

And since X is pregaussian it follows from Theorem II.5.3, that

$$U = \sum_{j=1}^\infty U_j$$

converges a.s., and U is a gaussian random vector.

Now if $U^n = \sum_{j=n+1}^\infty U_j$, then U^n is gaussian with mean 0 and covariance given by:

$$\sum_{j=n+1}^\infty \mathbb{E} \langle D_j, x' \rangle^2 = \mathbb{E} \langle X - M_n, x' \rangle^2$$

Since X and $X - M_n$ belongs to DNA (cf. Proposition 2.2), we have

(i) $$L(Z_k) \xrightarrow[n \to \infty]{w} L(U)$$

(ii) $$L(Z_k - Z_{kn}) \xrightarrow[n \to \infty]{w} L(U^n)$$

From Theorem II.3.4, it follows that we have $\mathbb{E}(\sup_n \|U^n\|) < \infty$ and so

(iii) $$\lim_{n \to \infty} \mathbb{E} \|U^n\| = 0$$

From (a) we find

(iv) $$\lim_{n \to \infty} \mathbb{E} \|Z_k - Z_{kn}\| = 0 \qquad \forall\, k \geq 1$$

Since $\{\|Z_k - Z_{kn}\| : k \geq 1\}$ is equiintegrable for any n by Theorem II.4.10, we deduce from (ii) and Proposition I.4.2:

(v) $$\lim_{k \to \infty} \mathbb{E} \|Z_k - Z_{kn}\| = \mathbb{E} \|U^n\| \qquad \forall\, n \geq 1$$

Let $\varepsilon > 0$ be given, then by (iii) there exists $M \geq 1$ so that

$$\mathbb{E}\,\|U^M\| < \varepsilon .$$

By (v) we can then find $m \geq 1$ so that

(vi) $$\mathbb{E}\,\|Z_k - Z_{kM}\| < \varepsilon \quad \forall\ k \geq m$$

Finally by (iv) we find $N \geq M$, so that

(vii) $$\mathbb{E}\,\|Z_k - Z_{kn}\| < \varepsilon \quad \forall\ n \geq N \quad \forall\ 1 \leq k \leq m$$

If $n \geq M$, then

$$Z_k - Z_{kn} = (Z_k - Z_{kM}) - (Z_{kn} - Z_{kM})$$

$$= (Z_k - Z_{kM}) - \mathbb{E}\,(Z_k - Z_{kM} \mid F_{nk})$$

where $F_{nk} = \sigma\{Z_{k1},\dots,Z_{kn}\}$. Since $Z_{k1},\dots,Z_{kn},\dots$ is a martingale which converges a.s. and in L^1_E to Z_k. Hence by (vi) we have

$$\mathbb{E}\,\|Z_k - Z_{kn}\| \leq 2\,\mathbb{E}\,\|Z_k - Z_{kM}\| \leq 2\varepsilon \quad \forall\ n \geq M \ \forall\, k \geq m$$

and combining this with (vii) gives

$$c(X-M_n) = \sup_k \mathbb{E}\|Z_k - Z_{kn}\| \leq 2\varepsilon \quad \forall\ n \geq N . \quad \square$$

Corollary 2.4. DNA *is equal to the closure of* S_o *in* (CLT,c), *where* S_o *is the set of simple random vectors with mean* 0. *I.e.* X *belongs to the domain of normal attraction, if and only if there exists a sequence of simple random vectors,* (X_n) *with mean zero and with* $c(X-X_n) \to 0$.

Proof. Let $F_1 \subseteq F_2 \subseteq \cdots$ be finite σ-algebras so that $\sigma(X) \subseteq \sigma(\bigcup_1^\infty F_n)$. And put $M_n = \mathbb{E}\,(X \mid F_n)$. If $X \in \mathrm{DNA}$, then M_n

exists and is simple, since $DNA \subseteq L^1_E$. Moreover, it is easy to check that (a) and (b) in Theorem 2.3 is satisfied. □

Let us conclude this section with an example of a general nature.

Example 2.5. Let $q > 2$, and let (ξ_j) be independent real random variables, with mean 0 and $\mathbb{E}\,\xi_j^2 = \sigma_j^2 < \infty$, so that

(2.8) $$\sum_{j=1}^{\infty} P(|\xi_j| > a) < \infty$$

(2.9) $$\sum_{j=1}^{\infty} \int_{|\xi_j| \le a} |\xi_j|^q \, dP < \infty$$

Then by the 2-series theorem, we have that $X = (\xi_j)$ is an ℓ^q-valued random vector.

We shall now study this random vector more closely. Put

$$N = \|X\|_\infty = \sup_j |\xi_j|$$

$$S = \|X\|_q = \left(\sum_1^\infty |\xi_j|^q\right)^{1/q}$$

Then (2.8) is equivalent to $N < \infty$ a.s. (cf. (II.1.6)). And if $\varphi: \mathbb{R}_+ \to \mathbb{R}_+$ is continuous increasing, and satisfies

(2.10) $$\varphi(2t) \le K\,\varphi(t) \qquad \forall\ t \ge 0$$

then we have (cf. Theorem II.5.5 and Lemma II.1.2)

(2.11) $$\mathbb{E}\,\varphi(S) < \infty \Leftrightarrow \exists t > 0: \sum_{j=1}^{\infty} \int_{|\xi_j| \ge t} \varphi(|\xi_j|)\, dP < \infty$$

Let $U = (\gamma_j)$ be a sequence of independent gaussian random variables with mean zero, and $\mathbb{E}\,\gamma_j^2 = \sigma_j^2 = \mathbb{E}\,\xi_j^2$. Then U and

X has the same convariance function, and by Theorem II.3.4 we have that $\|U\|_q < \infty$ a.s. if and only if

$$\sum_{j=1}^{\infty} \mathbb{E}\, |\gamma_j|^q < \infty$$

So we find

$$X \in PG \Leftrightarrow \sum_{j=1}^{\infty} (\sigma_j)^q < \infty \tag{2.12}$$

Now suppose that $\sum_j \mathbb{E}\,|\xi_j|^q = K < \infty$, and let $\xi_{j1},\dots,\xi_{jk},\dots$ be independent copies of ξ_j. By the Khintchine's inequality we know that (see (I.2.24), $K = 2^q K(q)$)

$$\mathbb{E}\, |\sum_{i=1}^{k} \xi_{ji}|^q \le K\, k^{q/2}\ \mathbb{E}\,|\xi_j|^q$$

Let

$$X_i^N = (\xi_{1i},\dots,\xi_{Ni},0\ \cdots)$$

$$Z_k^N = k^{-\frac{1}{2}} \sum_{i=1}^{k} X_i^N$$

$$Z_k = Z_k^{\infty}$$

Then we have by Khichine's inequality:

$$\mathbb{E}\, \|Z_k - Z_k^N\|^q = k^{-q/2} \sum_{j=N+1}^{\infty} \mathbb{E}\, |\sum_{i=1}^{k} \xi_{ji}|^q$$

$$\le K \sum_{j=N+1}^{\infty} \mathbb{E}\,|\xi_j|^q$$

So if $X^N = (\xi_1,\dots,\xi_N,0\ \cdots)$, we find

$$c(X - X^N) \le \{K \sum_{j=N+1}^{\infty} \mathbb{E}\,|\xi_j|^q\}^{1/q} \xrightarrow[N\to\infty]{} 0$$

But X_n but has finite dimensional range, and so $X_n \in \mathrm{DNA}$. Hence by Proposition 2.2 we conclude:

(2.13) If $\sum_{j=1}^{\infty} \mathbb{E}|\xi_j|^q < \infty$ then $X \in \mathrm{DNA}$

Case 1. Let us assume that ξ_j only takes the values a_j, 0 and $-a_j$ with probabilities $\frac{1}{2}p_j$, $1-p_j$ and $\frac{1}{2}p_j$, where

$$\sum_{j=1}^{\infty} p_j < \infty$$

Then (2.8) and (2.9) are trivially satisfied.

Now let us choose:

$$p_j = (j+1)^{-1} (\log(j+1))^{-2}$$

$$a_j = (j+1)^{\alpha}$$

where $\alpha = \frac{1}{2} - \frac{1}{q}$ (note that $\alpha > 0$). Then we have

$$\sigma_j = a_j\sqrt{p_j}, \qquad \sigma_j^q = (j+1)^{-1} (\log(j+1))^{-q}$$

So by (2.12) we have $X \in \mathrm{PG}$. If $t>1$, and $k \geq 2$ is chosen so that $(k-1)^{\alpha} \leq t < k^{\alpha}$, then we have

$$\sum_{j=1}^{\infty} \int_{|\xi_j|>t} \log|\xi_j| = \sum_{j=k}^{\infty} \log(a_{j-1})p_{j-1}$$

$$= \sum_{j=k}^{\infty} \alpha(\log j)^{-1} j^{-1} = \infty$$

Hence by (2.11) we have

(2.14) $X \in \mathrm{PG}$ but $\mathbb{E}\{\log^+ \|X\|_q\} = \infty$

Case 2. Let us now assume that $\xi_j = j^{-\frac{1}{2}}\eta_j$, where $\eta_1, \eta_2, \ldots$ are identically distributed random variables with mean 0 and finite variance. Then $\xi_j \to 0$ a.s. so $N<\infty$ a.s. and (2.8) holds.

Let $p = q/2$, and $A_i = \{t\sqrt{i-1} < |\eta| \le t\sqrt{i}\}$, then we have

$$\sum_{j=1}^{\infty} \int_{|\xi_j|\le t} |\xi_j|^q dP = \sum_{j=1}^{\infty} \sum_{i=1}^{j} \int_{A_i} j^{-p}|\eta|^q dP$$

$$= \sum_{i=1}^{\infty} \int_{A_i} |\eta|^q \sum_{j=i}^{\infty} j^{-p}\, dP$$

$$\le \frac{p}{p-1} \sum_{i=1}^{\infty} \int_{A_i} |\eta|^q\, i^{-p+1}\, dP$$

$$\le \frac{q}{q-2} \sum_{i=1}^{\infty} \int_{A_i} |\eta|^q\, t^{q-2}\, |\eta|^{2-q}\, dP$$

$$= \frac{q\, t^{q-2}\, \mathbb{E}|\eta|^2}{q-2}$$

since $i \ge t^{-2}|\eta|^2$ on A_i. Hence (2.9) holds and we see that X is an ℓ^q-valued random vector. And since $\{S\le a\} \subseteq \{|\xi_j|\le a\}$ for all $j\ge 1$, we have actually shown:

$$\int_{\{S\le t\}} S^q\, dP \le K\, t^{q-2}\, \mathbb{E}\,|\eta|^2$$

where $K = q(q-2)^{-1}$. Now we note that

$$1 = q \quad s^q \int_s^{\infty} x^{-q-1}\, dx \qquad \forall\, s>0$$

So for $t>0$ we have

$$P(S\ge t) = \int_{\{S\ge t\}} 1\, dP = \int_{\{S\ge t\}} dP \int_{S(\omega)}^{\infty} q \;\; x^{-q-1}\, S(\omega)^q dx$$

$$\le \int_t^{\infty} q \quad x^{-q-1}\, dx \int_{\{S\le x\}} S^q dP$$

$$\le q \quad K\, \mathbb{E}\,|\eta|^2 \int_t^{\infty} x^{-3}\, dx$$

$$= \tfrac{1}{2}\, q \quad K\, \mathbb{E}\,|\eta|^2\, t^{-2}$$

So by (I.2.11) and (I.2.19) we have

$$\mathbb{E}\, S \le 2\, \lambda_2(S)^{3/2} \le 2(\tfrac{1}{2}q \quad K)^{\frac{1}{2}} \, (\mathbb{E}\,|\eta|^2)^{\frac{1}{2}}$$

Hence we find

$$\mathbb{E}\, \|X\|_q \le C(\mathbb{E}\,|\eta|^2)^{\frac{1}{2}}$$

where $C = \sqrt{2}\; \ell(q-2)^{-\frac{1}{2}}$. And since

$$\mathbb{E}\, |\, k^{-\frac{1}{2}}\, (\eta_1 + \cdots + \eta_k)\,|^2 = \mathbb{E}\,|\eta|^2$$

we conclude that $X \in \mathrm{CLT}$ and

$$(2.15) \qquad c(X) \le C(\mathbb{E}\,|\eta|^2)^{\frac{1}{2}}$$

Now if $\eta \in L^q$, then $\sum_1^\infty \mathbb{E}\,|\xi_j|^q < \infty$, and so $X \in \mathrm{DNA}$ by (2.13). But L^q is dense in L^2 so (2.15) shows that $X \in \mathrm{DNA}$ whenever $\eta \in L^2$, (cf. Proposition 2.2).

By (2.11) we have that $\mathbb{E}\, S^2 < \infty$ if and only if there exists a $t > 0$ with:

$$\infty > \sum_{j=1}^\infty \int_{\xi_j > t} |\xi_j|^2 \, dt = \sum_{j=1}^\infty \int_{\eta > t\sqrt{j}} j^{-1} |\eta|^2 \, dP$$

Now we note that if

$$f(x) = \sum_{j=1}^\infty j^{-1}\, 1_{]t\sqrt{j},\infty[}(x)$$

then $\log(k+1) \le f(x) \le 1 + \log k$, where k is determined by $t\sqrt{k} < x \le t\sqrt{k+1}$. But then we have

$$\log t + \tfrac{1}{2} \log k < \log x \le \log t + \tfrac{1}{2} \log(k+1)$$

so we find

$$2 \log x - 2 \log t \le f(x) \le 1 + 2 \log x - 2 \log t$$

and since

$$\sum_{j=1}^{\infty} \int_{|\eta|>t\sqrt{j}} |\eta|^2 dP = \mathbb{E}\,\{ f(|\eta|)\eta^2\}$$

we have proved:

(2.16) $X \in \mathrm{DNA}$ if $\mathbb{E}\,|\eta|^2 < \infty$, but $\mathbb{E}\,\|X\|_q^2 < \infty$ if and only if $\mathbb{E}\,(\eta^2 \log^+ \eta) < \infty$.

3. Stochastic integration

Recall that a gaussian linear space $G \subseteq L_E^0$ is a linear space $G \subseteq L_E^0$, so that $\mathbb{E}\,X = 0$ and $\mathcal{L}(X)$ is gaussian for all $X \in G$. Clearly we have

(3.1) If G is gaussian, then so is its L_E^0-closure.

And from Theorem II.3.4 it follows that

(3.2) $G \subseteq L_E^{(\Phi)}$, and the L_E^0-topology coincides with the $L_E^{(\Phi)}$-topology

whenever G is a gaussian linear space, and

$$\Phi(t) = e^{-t^2}$$

In particular we have

(3.3) All the L_E^p-topologies $(0 \leq p < \infty)$ coincide on a gaussian linear space $G \subseteq L_E^0$.

We shall now construct an E-valued integral with respect to a white noise:

Let (S,Σ,μ) be a finite positive measure space, then a white noise, W, with variance μ, is a stochastic process $\{W(A) \mid A \in \Sigma\}$, satisfying

(3.4) $\{W(A)\ A \mid A \in \Sigma\}$ is a gaussian process with mean 0.

(3.5) $\mathbb{E}\,W(A)\,W(B) = \mu(A \cap B)\ \forall\ A,B \in \Sigma$

It is wellknown (and easy to verify!) that if $A_1, A_2, \ldots$ are disjoint sets from Σ, then

(3.6) $W(A_1), W(A_2), \dots$ are independent

(3.7) $W(\bigcup_1^\infty A_n) = \sum_1^\infty W(A_n)$ a.s.

Let S_E denote the set of simple measurable functions: $S \to E$. If $f = \sum_1^n x_j 1_{A_j} \in S_E$ then we define

$$\int_S f\, dW = \sum_1^n x_j W(A_j)$$

A standard argument (using (3.7)) shows that $\int f\, dW$ does not depend on the representation $f = \sum_1^n x_j 1_{A_j}$, and that the map

$$f \sim \int_S f\, dW \quad .$$

is a linear map from S_E into L_E^0. The range of this map, $G \subseteq L_E^0$, is clearly a gaussian linear space, and so is its closure which we denote

$$G_E(W) = cl\{ \int_S f\, dW \mid f \in S_E\}$$

where the closure is taken in the L_E^0-topology. Now by (3.2) and (3.3) we know that

$$G_E(W) \subseteq L_E^p \subseteq L_E^{(\Phi)} \ \forall\ 0 \le p < \infty$$

and all the L_E^p-topologies $(0 \le p < \infty)$ coincide and coincide with the $L_E^{(\Phi)}$-topology.

The following 3 propositions are easy and their proofs are left to the reader:

(3.8) $\mathbb{E} \exp(i <x', \int_S f\, dW>) = \exp(-\frac{1}{2} \int_S <x', f>^2 d\mu)$

(3.9) $\mathbb{E} <x', \int_S f\, dW><y', \int_S g\, dW> = \int_S <x', f><y', g> d\mu$

(3.10) $\|f\|_2^* \le \{\mathbb{E} \| \int_S f\, dW\|^2\}^{\frac{1}{2}}$

where $\|f\|_2^*$ is defined by

$$\|f\|_2^* = \sup_{\|x'\|\leq 1} \{\int_S \langle x',f\rangle^2 d\mu\}^{\frac{1}{2}}$$

Let $L_E^p(\mu)$ denote the space $L_E^p(S,\Sigma,\mu)$, and recall that $L_E^p = L_E^p(\Omega,F,P)$, where (Ω,F,P) is our basic probability space. Then we define $L_E^p(W)$ $(0\leq p<\infty)$ by: $f \in L_E^p(W)$ if and only if there exists $\{f_n\} \subseteq S_E$, so that

$$f_n \to f \quad \text{in} \quad L_E^p(\mu)$$

$$\{\int_S f_n \, dW\} \quad \text{converges in} \quad L_E^2$$

It is then a matter of routine (using (3.8)) to see that

$$\lim_{n\to\infty} \int_S f_n \, dW = \int_S f \, dW$$

does not depend on $\{f_n\} \subseteq S_E$ as long as $f_n \to f$ in L_E^o. Hence we may extend the integral with respect to W to all of $L_E^o(W) \supseteq L_E^p(W)$ for $0\leq p<\infty$. An obvious continuity argument shows that (3.8), (3.9) and (3.10) holds for all $f \in L_E^o(W)$.

<u>Proposition 3.1</u>. <u>Let</u> $f_n \in L_E^p(W)$, <u>such that</u>

(a) $\quad f_n \to f$ in $L_E^p(\mu)$, $\quad \{\int_S f_n \, dW\}$ converges in L_E^2

<u>Then</u> $f \in L_E^p(W)$, <u>and we have</u>

(b) $$\int_S f \, dW = \lim_{n\to\infty} \int_S f_n \, dW$$

<u>Proof</u>. We can find $g_n \in S_E$, so that $\|f_n - g_n\|_p \leq 2^{-n}$ and

$$\|\int_S f_n \, dW - \int_S g_n \, dW\|_2 \leq 2^{-n}$$

Then $g_n \to f$ in $L_E^p(\mu)$, and

$$\int_S g_n \, dW \to \lim_{n\to\infty} \int_S f_n \, dW \qquad \text{in } L_E^2$$

So $f \in L_E^p(W)$ and (b) holds. □

<u>Proposition 3.2</u>. <u>If</u> $f \in L_E^p(W)$ <u>and</u> $\varphi \in L^\infty(\mu)$, <u>then</u> $\varphi f \in L_E^p(W)$ <u>and</u>

(a) $$\mathbb{E}\, q(\|\int_S \varphi f \, dW\|) \le 6\mathbb{E}\, q(\|\varphi\|_\infty 5 \|\int_S f \, dW\|)$$

<u>for all increasing continuous functions</u> $q: \mathbb{R}_+ \to \mathbb{R}_+$

<u>If</u> q <u>is convex then</u> (a) <u>holds without the 6 and 5</u>.

<u>Proof</u>. Suppose that $\varphi = \sum_1^n t_j 1_{A_j}$, with $A_1,\dots,A_n$ disjoint. Then

$$\|\varphi\|_\infty = \max t_j$$

$$\varphi f = \sum_1^n t_j 1_{A_j} f$$

A standard "simple function"-argument shows that $1_{A_j} f \in L_E^p(W)$, and that the random variables:

$$X_j = \int_{A_j} f \, dW \qquad j = 1,\dots,n$$

are independent (cf. (3.6)), and

$$\int_S \varphi f \, dW = \sum_1^n t_j X_j$$

So by Theorem II.2.15 we have that (a) holds. The general case follows from the special case by choosing simple functions φ_n so that $\|\varphi_n - \varphi\|_\infty \to 0$, and then applying Proposition 3.1. □

<u>Proposition 3.3</u>. $L_E^p(W) = L_E^0(W) \cap L_E^p(\mu)$.

Proof. The inclusion "$\subseteq$" is evident. Now suppose that $f \in L_E^0(W) \cap L_E^p(\mu)$. Then there exists $f_n \in S_E$ with $f_n \to f$ pointwise and

$$\int_S f_n \, dW \to \int_S f \, dW \quad \text{in } L_E^2$$

Now let

$$A_n = \{s \in S \mid \|f_k(s)\| \le 1 + \|f(s)\| \quad \forall\, k \ge n\}$$

then $A_n \uparrow S$ and

$$g_n = 1_{A_n} f_n \to f \quad \text{in } L_E^p(\mu)$$

by the dominated convergence theorem. For $n \le m$ it follows by Proposition 3.2, that we have

$$\begin{aligned}\|\int_S (g_n - g_m)\,dW\|_2 &\le \|\int_{A_m} (f_n - f_m)\,dW\|_2 + \|\int_{A_m \setminus A_n} f_n \, dW\|_2 \\ &\le \|\int_S (f_n - f_m)\,dW\|_2 + \|\int_{S \setminus A_n} f_n \, dW\|_2\end{aligned}$$

Let

$$\nu_n(A) = \int_A f_n \, dW, \qquad \nu(A) = \int_A f \, dW$$

then ν_n and ν are L_E^2-valued measures on (S, Σ), and $\nu_n(A) \to \nu(A)$ for all $A \in \Sigma$ by Proposition 3.2. So from the Vitali-Saks theorem (see e.g. Theorem IV. . in [19]) we conclude that

$$\lim_{n \to \infty} \|\nu_n(S \setminus A_n)\|_2 = 0$$

since $S \setminus A_n \downarrow \emptyset$. Hence it follows that $\{\int_S g_n \, dW\}$ converges in L_E^2, and so $f \in L_E^p(W)$. □

A standard "simple function"-argument shows that the conditional integral $\mu(f|\Sigma_o)$ exists for all $f \in L^1_E(\mu)$ and all σ-algebras $\Sigma_o \subseteq \Sigma$, and it satisfies (see e.g. [14])

(3.11) $\mu(f|\Sigma_o)$ is Σ_o-measurable: $S \to E$

(3.12) $\int_A f\,d\mu = \int_A \mu(f|\Sigma_o)d\mu \qquad \forall A \in \Sigma_o$

(3.13) $\|\mu(f|\Sigma_o)\| \leq \mu(\|f\| \,|\, \Sigma_o)$

(3.14) If $\Sigma_n \uparrow$ and $\Sigma_o = \sigma(\bigcup_1^\infty \Sigma_n)$, then

$$\lim_{n\to\infty} \mu(f|\Sigma_n) = \mu(f|\Sigma_o)$$

μ - a.e. and in $L^p_E(\mu)$, if $1 \leq p < \infty$ and $f \in L^p_E(\mu)$

<u>Proposition 3.4</u>. <u>Let</u> $\Sigma_o \subseteq \Sigma$ <u>be a</u> σ-<u>algebra and let</u> $f \in L^1_E(W)$. <u>If</u> $F_o = \sigma\{W(A) \,|\, A \in \Sigma_o\}$ <u>then we have</u>

(a) $\mu(f \,|\, \Sigma_o) \in L^1_E(W)$

(b) $\int_S \mu(f|\Sigma_o)dW = \mathbb{E}(\int_S f\,dW \,|\, F_o)$

<u>Proof</u>. <u>Case 1: $E = \mathbb{R}$.</u> Let

$$f_o = \mu(f \,|\, \Sigma_o), \quad X = \int_S f\,dW, \qquad X_o = \mathbb{E}(X|F_o)$$

If π_o is the projection of $L^2(\mu)$ onto

$$H_o = \overline{\text{span}}\,\{W(A) \,|\, A \in \Sigma_o\}$$

then by Proposition I.2.1 we have $X_o = \pi_o X$. Now we note that $L^1_{\mathbb{R}}(W) = L^2(\mu)$ and

$$H_o = \{\int_S f\,dW \,|\, f \in L^2(S,\Sigma_o,\mu)\}$$

Hence

$$Y_o = \int_S f_o \, dW \in H_o$$

and if $Y = \int_S g \, dW \in H_o$ for some $g \in L^2(S, \Sigma_o, \mu)$ we have

$$\mathbb{E}(Y_o Y) = \int_S f_o \, g \, d\mu = \int_S f g \, d\mu = \mathbb{E}(XY)$$

since g is Σ_o-measurable. But this shows that $Y_o = \pi_o X = X_o$.

Case 2: E is general. If $f = \sum_1^n x_j g_j$ for $g_j \in L^2(\mu)$, then $f \in L_E^1(W)$ and

$$\mu(f|\Sigma_o) = \sum_1^n x_j \, \mu(g_j|\Sigma_o)$$

so $\mu(f|\Sigma_o) \in L_E^1(W)$ and from Case 1 we have

$$\int_S \mu(f|\Sigma_o) dW = \sum_1^n x_j \int_S \mu(g_j|\Sigma_o) dW$$

$$= \sum_1^n x_j \, \mathbb{E}(\int_S g_j \, dW \mid F_o)$$

$$= \mathbb{E}(\int_S f \, dW \mid F_o)$$

In general there exist $f_n \in S_E$ with $f_n \to f$ in $L_E^1(\mu)$ and $\int_S f_n \, dW \to \int_S f \, dW$ in L_E^2. Hence (cf. (3.13))

$$\mu(f_n \mid \Sigma_o) \to \mu(f \mid \Sigma_o) \quad \text{in } L_E^1$$

$$\mathbb{E}(\int_S f_n \, dW \mid F_o) \to \mathbb{E}(\int_S f \, dW \mid F_o) \quad \text{in } L_E^2$$

Now by the argument above we have $\mu(f_n|\Sigma_o) \in L_E^1(W)$ and

$$\int_S \mu(f_n \mid \Sigma_o) \, dW = \mathbb{E}(\int_S f_n \, dW \mid F_o)$$

Hence by Proposition 3.1, we have that (a) and (b) holds. □

Corollary 3.5. Let $f \in L_E^1(W)$, and let $H = \sigma\{W(A) \mid A \in \sigma(f)\}$, then we have

(a) $\int_S f\, dW$ is H-measurable

Moreover if $\Sigma_o \subseteq \Sigma$ is a σ-algebra and $F_o = \sigma\{W(A) \mid A \in \Sigma_o\}$, then

(b) $\int_S (f-\mu(f|\Sigma_o))dW$ and F_o are independent. □

Theorem 3.6. Let $f \in L_E^p(\mu)$ for some $p \in [0,\infty[$ then $f \in L_E^p(W)$ if and only if f satisfies:

(a) $\langle x',f(\cdot)\rangle \in L^2(\mu) \quad \forall\, x' \in E'$

(b) $R(x',y') = \int_S \langle x',f\rangle\langle y',f\rangle d\mu$ is the covariance function of a gaussian Radon measure on E.

Proof. If $f \in L_E^p(W)$ then (a) and (b) follows immediately from (3.9) and (3.10). Now suppose that (a) and (b) holds.

Suppose that $f \in L_E^1(\mu)$ and (a) and (b) holds, then we have that $\Sigma_\infty = \sigma(f)$ is countably generated. Hence we can find finite σ-algebras $\Sigma_1 \subseteq \Sigma_2 \subseteq \cdots \subseteq \Sigma_n \subseteq \cdots$ so that

$$\Sigma_\infty = \sigma(\bigcup_1^\infty \Sigma_n).$$

Now put $f_n = \mu(f|\Sigma_n)$. Then $f_n \in S_E$ so we may define

$$S_n = \int_S f_n\, dW \qquad S_o = 0$$

$$X_n = S_n - S_{n-1} = \int_S (f_n - \mu(f_n|\Sigma_{n-1}))dW$$

since $f_{n-1} = \mu(f \mid \Sigma_{n-1}) = \mu(f_n \mid \Sigma_{n-1})$. Let $F_n = \sigma\{W(A) \mid A \in \Sigma_n\}$. Then we have

(i) X_n is F_n-measurable

(ii) X_n is independent of F_{n-1}

(cf. Corollary 3.5). Hence $\{X_n\}$ are independent gaussian random variables, and

$$S_n = \sum_1^n X_j$$

$$\mathbb{E}\exp(i\,\langle x',S_n\rangle) = \exp(-\tfrac{1}{2}\int_S \langle x',f_n\rangle^2 d\mu)$$

$$\xrightarrow[n\to\infty]{} \exp(-\tfrac{1}{2}R(x',x'))$$

So by Theorem II.5.3 (case (1)) and by assumption (b) we have that $\{S_n\}$ converges a.s. Now from (3.14) we deduce:

$$f_n = \mu(f|\Sigma_n) \to \mu(f|\Sigma_\infty) = f \quad \text{in } L_E^1(\mu)$$

and we have just seen that $\{\int_S f_n\, dW\}$ converges in L_E^0 (and so in L_E^2 by (3.3)). Hence $f \in L_E^1(W)$.

Now suppose that $f \in L_E^p(\mu)$ and satisfies (a) and (b). Let $A_n = \{s \mid \|f(s)\| < n\}$, and let $B_n = A_n \setminus A_{n-1}$ for $n \geq 1$. If $f_n = 1_{B_n} f$, then $f_n \in L_E^1(\mu)$, and

$$\int_S \langle x',f_n\rangle^2 d\mu \leq \int_S \langle x',f\rangle^2 d\mu$$

Since f satisfies (a) and (b), then so does f_n by Proposition 2.1. Hence $f_n \in L_E^1(W)$ by the argument above, and we may define

$$U_j = \int_S f_j dW, \qquad S_n = \sum_1^n U_j$$

Since $B_1,\dots,B_n,\dots$ are disjoint we have that $U_1,U_2,\dots$ are independent, and if $g_n = 1_{A_n} f$, then (S_n) has Fourier transform:

$$\hat{\nu}_n(x') = \exp(-\tfrac{1}{2}\int_S \langle x',g_n\rangle^2 d\mu)$$

But by (a) we have

$$\int_S \langle x' g_n\rangle^2 d\mu \to \int_S \langle x', f\rangle^2 d\mu \qquad \forall\, x'$$

So by (b) and Theorem II.5.3 we have that $\{S_n\}$ converges a.s., and by (3.3) and dominated convergence theorem we find

$$\{\int_S g_n \, dW\} \text{ converges in } L_E^2$$

$$g_n \to f \quad \text{in} \quad L_E^p(\mu)$$

So from Proposition 3.1 we deduce that $f \in L_E^p(W)$. □

<u>Theorem 3.7</u>. <u>If</u> E <u>is of type 2, then</u> $L_E^2(\mu) \subseteq L_E^2(W)$. <u>Conversely</u> if $L^2(\mu)$ <u>is of infinite dimension and</u> $L_E^2(\mu) \subseteq L_E^2(W)$, <u>then</u> E <u>is of type 2</u>.

<u>Proof</u>. Suppose that E is of type 2, and that $f = \sum_1^n x_j 1_{A_j} \in S_E$ where $A_1, \dots, A_n \in \Sigma$ are disjoint. Then by Theorem II.6.6 we have

$$\begin{aligned} \mathbb{E}\| \int_S f \, dW\|^2 &= \mathbb{E}\| \sum_1^n W(A_j) x_j\|^2 \\ &\leq C \sum_1^n \mathbb{E}| W(A_j)|^2 \|x_j\|^2 \\ &= C \sum_1^n \mu(A_j)\, \|x_j\|^2 \\ &= C \int_S \|f\|^2 \, d\mu \end{aligned}$$

And since S_E is dense in $L_E^2(\mu)$, we have $L_E^2(\mu) \subseteq L_E^2(W)$ and

(3.15) $$\mathbb{E}\| \int_S f \, dW\|^2 \leq C \int_S \|f\|^2 \, d\mu \qquad \forall\, f \in L_E^2(\mu)$$

Now suppose that $L_E^2(\mu) \subseteq L_E^2(W)$, then by the closed graph theorem we have that (3.15) holds for some constant $C > 0$. Since

$L^2(\mu)$ is infinite dimensional, there exist disjoint sets $A_1, A_2, \ldots$ so that

$$\lambda_j = \mu(A_j) > 0 \qquad \forall\ j \geq 1$$

Now let $f = \sum_{j=1}^{\infty} \lambda_j^{-\frac{1}{2}} x_j 1_{A_j}$ where

$$\sum_{j=1}^{\infty} \|x_j\|^2 < \infty$$

Then we have

$$\int_S \|f\|^2 d\mu = \sum_{j=1}^{\infty} \lambda_j^{-1} \|x_j\|^2 \mu(A_j) < \infty$$

So $f \in L_E^2(W)$, and we have

$$\int_S f\, dW = \sum_{j=1}^{\infty} x_j \gamma_j$$

where $\gamma_j = \lambda_j^{-\frac{1}{2}} W(A_j)$. But then (γ_j) are independent gaussian random variables with mean 0 and variance 1. So we have

$$\sum_{j=1}^{\infty} x_j \gamma_j$$

converges a.s. whenever $x = (x_j) \in \ell_E^2$. That is E is of type 2 (cf. Theorem II.6.6). □

Theorem 3.8. *If E is of cotype 2, then $L_E^0(W) \subseteq L_E^2(\mu)$. Conversely if $L^2(\mu)$ is infinite dimensional, and if $L_E^p(W) \subseteq L_E^2(\mu)$ for some $p < 2$, then F is of cotype 2.*

Proof. The first part is proved exactly as the first part of Theorem 3.7.

So suppose that $L_E^p(W) \subseteq L_E^2(\mu)$ for some $p < 2$. Then we may and shall assume that $p \geq 1$. Let $L_E^p(W)$ be equipped with the norm:

$$|||f|||_p = ||f||_p + ||\int_S f\,dW||_1$$

Then by Proposition 3.1 we have that $(L_E^p(W), |||\cdot|||_p)$ is a Banach space, and the injection: $L_E^p(W) \to L_E^2(\mu)$ clearly has a closed graph. So by the closed graph theorem there exists a constant $C > 0$, such that

$$\text{(i)} \qquad ||f||_2 \le C|||f|||_p \qquad \forall\ f \in L_E^p(W)$$

Now let $A_1, A_2, \ldots, A_j, \ldots$ be disjoint sets from Σ, so that $\mu(A_j) > 0$ for all $j \ge 1$.

Let $\lambda_j = \mu(A_j)^{-\frac{1}{2}}$, and $x_1, \ldots, x_n$ be elements of E. Then we put

$$f_k = \sum_{j=1}^{n} \lambda_{j+k}\, x_j\, 1_{A_{j+k}} \qquad k = 1, 2, \ldots$$

If (γ_j) is a sequence of independent real gaussian random variables with mean 0 and variance 1, we have

$$||\int_S f_k\,dW||_1 = \mathbb{E}\,||\sum_{j=1}^{n} x_j \lambda_{j+k}\, W(A_{j+k})||$$

$$= \mathbb{E}||\sum_{j=1}^{n} \gamma_j\, x_j||$$

$$||f_k||_p = \{\sum_{j=1}^{n} ||x_j||^p\ \mu(A_{j+k})^{2-p}\}^{1/p}$$

$$||f_k||_2 = \{\sum_{j=1}^{n} ||x_j||^2\}^{\frac{1}{2}}$$

Hence by (i) we have

$$\{\sum_1^n ||x_j||^2\}^{\frac{1}{2}} \le C\,||f_k||_p + C\,\mathbb{E}||\sum_{j=1}^{n} \gamma_j x_j||$$

for all $k \ge 1$. Since $p < 2$ and the A_j's are disjoint we have $\mu(A_{j+k})^{2-p} \xrightarrow[k\to\infty]{} 0$, so

$$\lim_{k\to\infty} \|f_k\|_p = 0$$

So we find

$$\{\sum_1^n \|x_j\|^2\}^{\frac{1}{2}} \leq C\ \mathbb{E}\|\sum_1^n \gamma_j x_j\|$$

And from Theorem II.7.7, it then follows that E is of cotype 2. □

__Theorem 3.9__. __Let__ $0\leq r<p\leq 2$ __and suppose that__ μ __satisfies:__

(a) μ __is not atomic__

(b) $L_E^r(W) \subseteq L_E^p(\mu)$

__Then__ E __is of cotype 2__.

__Proof__. Let S_o be the μ-continuous part of S, that is $\mu|S_o$ is non-atomic. Then by (a) we have that $a = \mu(S_o) > 0$. We may of course assume that $r > 0$, then the closed graph theorem shows that there exists a constant K with

(i) $$(\int_S \|f\|^p d\mu)^{1/p} \leq K(\int_S \|f\|^r d\mu)^{1/r} + K\ \mathbb{E}\|\int_S f\, dW\|$$

Now let $x_1 \cdots x_n \in E$ and let $0\leq t\leq a$, then we put

$$\sigma = \sqrt{t}\ (\sum_1^n \|x_j\|^2)^{-\frac{1}{2}}, \quad r_j = \sigma\|x_j\|$$

Since $\sum_1^n r_j^2 = t\leq a = \mu(S_o)$, there exist disjoint sets $A_1,\ldots,A_n \in \Sigma$, so that

$$\mu(A_j) = r_j^2 \qquad \forall\ 1\leq j\leq n$$

Now we put

$$f = \sum_1^n r_j^{-1}\ x_j\ 1_{A_j}$$

and we find

$$\left(\int_S \|f\|^p d\mu\right)^{1/p} = \left(\sum_1^n \|x_j\|^p \mu(A_j) r_j^{-p}\right)^{1/p}$$

$$= \left(\sum_1^n \|x_j\|^p r_j^{2-p}\right)^{1/p}$$

$$= t^{1/p-1/2} \left(\sum_1^n \|x_j\|^2\right)^{\frac{1}{2}}$$

and similarly

$$\left(\int_S \|f\|^r d\mu\right)^{1/r} = t^{1/r-1/2} \left(\sum_1^n \|x_j\|^2\right)^{1/2}$$

More if (γ_j) are independent $N(0,1)$-variables, then

$$\mathbb{E}\left\|\int_S f\,dW\right\| = \mathbb{E}\left\|\sum_{j=1}^n \gamma_j x_j\right\|$$

So from (i) we find

$$\left(\sum_{j=1}^n \|x_j\|^2\right)^{1/2} t^{1/p-1/2}\left(1-K t^{1/r-1/p}\right) \le K\,\mathbb{E}\left\|\sum_1^n \gamma_j x_j\right\|$$

for all $t \in [0,a]$. Since $\frac{1}{r}-\frac{1}{p} > 0$, there exists a $t_o \in [0,a]$ with

$$K\, t_o^{1/r-1/p} \le \frac{1}{2}$$

Hence we have

$$\left(\sum_{j=1}^n \|x_j\|^2\right)^{\frac{1}{2}} \le K_1\, \mathbb{E}\left\|\sum_1^n \gamma_j x_j\right\|$$

and E is of cotype 2 by Theorem II.7.7. □

4. The central limit theorem revisited

Let us return to the central limit theorem, and the problems around this theorem.

Let $PG_p = PG \cap L_E^p$ $(0 \le p < \infty)$, and let $X \in PG_p$. We can then consider the measure space (S, Σ, μ), where $S = E$, $\Sigma = \mathcal{B}(E)$ and $\mu = \mathcal{L}(X)$. If W_X is a white noise with variance μ, then by Theorem 3.6 we have that $f(x) = x$ belongs to $L_E^p(W_X)$, and

$$U = \int_E x\, W_X(dx)$$

is a gaussian random vector with mean 0 and covariance function $\mathbb{E}\langle X, x'\rangle^2$. So we may define

$$u_p(X) = \|X\|_p + \{\mathbb{E}\, \|\int_E x\, W_X(dx)\|^2\}^{\frac{1}{2}}$$

It is then easily verified that we have:

(4.1) $(PG_p, u_p(\cdot))$ is a Banach space for $p \ge 1$ and a Fréchet space for $0 \le p < 1$.

Note that $PG_0 = PG$.

If E and F are Banach spaces, and T is a continuous linear operator: $E \to F$, then T is said to be of type p $(1 \le p \le 2)$ if one of the following 5 equivalent statements are satisfied:

(4.2) $$(Tx_j) \in B_F \quad \forall\ (x_j) \in \ell_E^p$$

(4.3) $$\exists\ K>0:\ \mathbb{E}\|\sum_1^n \varepsilon_j Tx_j\|^p \le K \sum_{j=1}^n \|x_j\|^p$$

(4.4) $$\exists\ K>0:\ \mathbb{E}\|\sum_1^n TX_j\|^p \le K \sum_{j=1}^n \mathbb{E}\,\|X_j\|^p$$ for all independent mean 0 random vectors, $X_1, \dots, X_n \in L_E^p$

(4.5) $(T x_j) \in C_F^p(\xi) \quad \forall\ (x_j) \in \ell_E^p$

(4.6) $(T x_j) \in B_F^o(\xi) \quad \forall\ (x_j) \in \ell_E^p$

where (ε_j) is a Bernoulli sequence, and $\xi = (\xi_j)$ is a sequence of independent real random variables satisfying (a), (b) and (c) in Theorem II.6.6. It is then clear that we have

(4.7) E is of type p if and only if the identity operator: $E \to E$ is of type p.

And the equivalence of (4.2)-(4.6) is proved exactly as if T = identity.

Similarly we say that T is of cotype q $(2 \leq q \leq \infty)$, if T satisfies one of the following 5 equivalent statements

(4.8) $(T x_j) \in \ell_F^q \quad \forall\ (x_j) \in C_E$

(4.9) $\exists K>0: \quad \sum_{j=1}^{n} \|Tx\|^q \leq K\ \mathbb{E}\ \|\sum_1^n \varepsilon_j x_j\|^q$

(4.10) $\exists K>0: \quad \sum_{j=1}^{n} \mathbb{E}\|Tx_j\|^q \leq K\ \mathbb{E}\|\sum_1^n X_j\|^q$ for all independent mean 0 random vectors: $X_1,\dots,X_n \in L_E^q$.

(4.11) $(T x_j) \in \ell_F^q \quad \forall\ (x_j) \in B_E^o(\xi)$

(4.12) $(T x_j) \in \ell_F^q \quad \forall\ (x_j) \in C_E^r(\xi)$ for some $r>q$

where (ε_j) is a Bernoilli sequence, and $\xi = (\xi_j)$ is an independent sequence of real random variable satisfying (a) - (b) of Theorem II.7.7 and $\inf \mathbb{E}\ |\xi_j|^q > 0$. Again we have

(4.13) E is of cotype q if and only if the identity operator: $E \to E$ is of type q.

Theorem 4.1. Let $T: E \to F$ be a bounded linear operator, then the following statements are equivalent

(a) T is of type 2

(b) $TX \in \mathrm{DNA}$ for any random vector with mean 0 and $\mathbb{E}\,\|X\|^2 < \infty$

(c) $TX \in \mathrm{PG}$ for any symmetric, discrete (i.e. countably valued), random vector with $\|X(\omega)\| = 1$ for all ω.

Remark. Putting $T = I_E$ we get two new characterizations of type 2 spaces.

Proof. (a) ⇒ (b): So let T be of type and let $X_1, X_2, \ldots$ be independent identically distributed random vectors, with

$$\mathbb{E}\,X_1 = 0, \qquad \mathbb{E}\,\|X_1\|^2 < \infty$$

Let $Z_n = (X_1 + \cdots + X_n)/\sqrt{n}$, then by (4.4) we have

$$\mathbb{E}\,\|T\,Z_n\| \leq (\mathbb{E}\|T\,Z_n\|^2)^{\frac{1}{2}} \leq \sqrt{K}\,(\mathbb{E}\,\|X_1\|^2)^{\frac{1}{2}}$$

so we find $c(TX) \leq \sqrt{K}\|X\|_2$. And so $TX \in \mathrm{CLT}$, but the simple functions are dense in L^2_E, so by Proposition 2.2 we have $TX \in \mathrm{DNA}$ for $X \in L^2_E$, and $\mathbb{E}\,X = 0$.

(b) ⇒ (c): Obvious.

(c) ⇒ (a): Let $(x_j) \in \ell^2_E$, we shall then show that $(Tx_j) \in C^0_F(\gamma)$, where $\gamma = (\gamma_j)$ are independent $N(0,1)$-variables (cf. (4.5)). We may clearly assume that $x_j \neq 0$ for all j and

$$1 = \sum_{j=1}^{\infty} \|x_j\|^2$$

Now let X be a random vector whose distribution is given by

$$P(X = y_j) = P(X = -y_j) = \tfrac{1}{2} \|x_j\|^2$$

where $y_j = x_j/\|x_j\|$. Then X is symmetric, discrete and $\|X(\omega)\| = 1$ for all ω. Hence $TX \in PG$, and so there exists a gaussian Radon measure with

$$\hat{\nu}(y') = \exp(-\tfrac{1}{2}\ \mathbb{E}\langle y', TX\rangle) \qquad \forall\ y' \in F'$$

Now we note that

$$\mathbb{E}\langle y', TX\rangle^2 = \sum_{j=1}^{\infty} \langle y', Tx_j\rangle^2 = \lim_{n\to\infty} \mathbb{E}\langle y' \sum_{j=1}^{n} \gamma_j T x_j\rangle$$

Hence by Theorem II.5.3 we have that $\sum_1^{\infty} \gamma_j T x_j$ converges a.s., and so $(Tx_j) \in C_E^o(\gamma)$. □

Let PG_p^o denote the set of $X \in PG_p$ satisfying $\mathbb{E}\langle x', X\rangle = 0$ for all $x' \in E'$. Then we have (cf. (2.2), (2.4) and (2.6))

(4.14) $$DNA \subseteq PG_r^o \qquad \forall\ 0 \le r < 2$$

The converse inclusion holds if and only if E is of cotype 2, which is seen from the following theorem

Theorem 4.2. Let $0 \le r < p \le 2$ be given then the following statements are equivalent

(a) $PG_r \subseteq L_E^p$ (i.e. if X is pregaussian, then $\mathbb{E}\|X\|^r < \infty \Rightarrow \mathbb{E}\|X\|^p < \infty$)

(b) E is of cotype 2.

(c) $PG_r^o \subseteq DNA$ (i.e. any pregaussian mean 0 random vector, with finite r-th moment, belongs to the domain of normal attraction).

Remark. (1): Since (b) is independent of p and r we have that all forms of (a) with r and p running over the region: $0 \leq r < p \leq 2$ are equivalent. And all the forms of (c) with r running through $[0,2[$ are equivalent.

(2): A similar theorem (with similar proof) holds for cotype 2 operator. I shall leave the formulation and the proof to the reader.

(3): It can be shown (see [96]) that (a)-(c) is equivalent to

(d) DNA $\subseteq L_E^2$ (i.e. any random vector in the domain of normal attraction has finite second moment).

Proof of Theorem 4.2. (a) $\Rightarrow$ (b): Let X be a random vector with a non-atomic distribution law μ, and let W be a white noise with variance μ.

If $f \in L_E^r(W)$, then by Theorem 3.6 we have that $Y = f(X) \in PG_r$, so by (a) we have

$$L_E^r(W) \subseteq L_E^p(\mu)$$

Hence from Theorem 3.9 it follows that E is of cotype 2.

(b) $\Rightarrow$ (c): Let $X \in PG_r^o$, and put $\mu = L(X)$. If W is a white noise with variance μ, then $f(x) = x \in L_E^r(W)$, by Theorem 3.6. If $g = \sum_1^n x_j 1_{A_j}$ is a Borel measurable simple function: $E \to E$, then since E is of cotype 2 we have

$$\int_E \|g\|^2 d\mu = \sum_1^n \|x_j\|^2 \mu(A_j)$$

$$\leq K\, \mathbb{E}\|\sum_1^n x_j W(A_j)\|^2$$

$$\leq K\, \mathbb{E}\|\int_E g\, dW\|^2$$

where K is some finite constant. Now let (g_n) be simple functions, so that $g_n \to f$ a.s. and

$$\int_E g_n \, dW \to \int_E f \, dW \qquad \text{in } L^1_E$$

Then by Fatou's lemma we have

$$\int_E \|x\|^2 \mu(dx) \leq \liminf_{n\to\infty} \int_E \|g_n(x)\|^2 \mu(dx)$$

$$\leq K \lim_{n\to\infty} \mathbb{E} \left\| \int_E g_n \, dW \right\|^2$$

$$= K \, \mathbb{E} \left\| \int_E x \, dW \right\|^2$$

That is $X \in L^2_E$ and

(i) $$\mathbb{E} \|X\|^2 \leq K \, \mathbb{E} \|U_X\|^2 \qquad \forall \, X \in PG^o_r$$

where U_X is a gaussian random vector with mean 0 and covariance: $\mathbb{E} \langle x', X\rangle^2$.

Now let $X_1, \ldots, X_n, \ldots$ be independent copies of X and put $Z_n = (X_1 + \cdots + X_n)/\sqrt{n}$, then

$$\mathbb{E} \langle x', Z_n\rangle = 0 \qquad \forall \, x'$$
$$\mathbb{E} \langle x', Z_n\rangle^2 = \mathbb{E} \langle x', X\rangle^2 \qquad \forall \, x'$$

So by (i) we have

$$\mathbb{E} \|Z_n\| \leq (\mathbb{E} \|Z_n\|^2)^{\frac{1}{2}} \leq (K \mathbb{E} \|U_X\|^2)^{\frac{1}{2}}$$

That is $X \in CLT$, and

(ii) $$c(X) \leq \sqrt{K \, \mathbb{E} \|U_X\|^2} \leq \sqrt{K} u_r(X) \qquad \forall \, X \in PG^o_r$$

But the simple functions are dense in PG^o_r, by definition of the stochastic integral, so from (ii) and Proposition 2.2 we deduce that $PG_r^o \subseteq DNA$.

(c) ⇒ (a): Immediate consequence of (2.2). □

5. Some operators of type 2

I shall now give some examples of operators of type 2.

Let us begin with the space $C(S)$, where S is a compact metric space. If ρ is a continuous metric on S, then we define

$$|||f|||_\rho = \|f\|_\infty + \sup_{t \neq s} \frac{|f(t) - f(s)|}{\rho(t,s)}$$

$$C^\rho(S) = \{f \in C(S) \mid |||f|||_\rho < \infty\}$$

$$C_0^\rho(S) = \{f \in C^\rho(S) \mid \lim_{(t,s)\to(a,a)} \frac{|f(t)-f(s)|}{\rho(t,s)} = 0 \ \forall a\}$$

Then we have

Theorem 5.1. $(C^\rho(S)\, |||\cdot|||_\rho)$ is a Banach space and $C_0^\rho(S)$ is closed separable subspace of $C^\rho(S)$.

Proof. A standard argument shows that $|||\cdot|||_\rho$ is complete, and that $C_0^\rho(S)$ is closed. Let

$$(Rf)(t,s) = \begin{cases} \dfrac{f(t)-f(s)}{\rho(t,s)} & \text{if } t \neq s \\ 0 & \text{if } t = s \end{cases}$$

Then R is a continuous linear map: $C_0^\rho(S) \to C(S \times S)$, and if we put

$$Sf = (Rf, f)$$

then S is an isometry: $C_0^\rho(S) \to C(S \times S) \times C(S)$ if we equip $C(S \times S) \times C(S)$ with the norm:

$$\|(f,g)\| = \|f\|_\infty + \|g\|_\infty$$

Hence separability of $C_0^\rho(S)$ follows from separability of $C(S \times S)$ and $C(S)$ (S is a compact metric space!) □

A continuous metric ρ on S, is said to be <u>pregaussian</u> if and only if any gaussian process $\{X(t) \mid t \in S\}$ with

(5.1) $\mathbb{E}\, X(t) = 0 \quad \forall t$

(5.2) $\sqrt{\mathbb{E}\,(X(t) - X(s))^2} \leq \rho(t,s) \quad \forall\, t,s$

has a version with continuous sample path. That is, ρ is pregaussian, if any nonnegative definite function R on $S \times S$, satisfying

(5.3) $\sqrt{R(t,t) + R(s,s) - 2R(t,s)} \leq \rho(t,s) \quad \forall\, t,s$

is the covariance function of a gaussian Radon measure on $C(S)$.

From Lemma 2.1 in [86] we have

(5.4) If X is a gaussian proces on S with mean 0 and continuous sample paths, then

$$\rho(t,s) = \sqrt{\mathbb{E}\,(X(t) - X(s))^2}$$

is pregaussian.

From Theorem 3.1 in [21], we have

(5.5) If $\int_0^1 \sqrt{H_\rho(t)}\, dt < \infty$, then ρ is pregaussian

where H_ρ is the <u>entropy</u> of ρ:

$$H_\rho(t) = \log\left\{\begin{matrix}\text{minimal number of balls of diameter} \\ \leq 2t, \text{ which covers } S\end{matrix}\right\}$$

If (f_j) is a sequence in $C(S)$, and $\gamma = (\gamma_j)$ is a sequence of independent normalized real gaussian variables, then by Theorem II.5.3 we have

(5.6) $(f_j) \in C(\gamma)$ if and only if the metric:

$$\rho(t,s) = \sqrt{\sum_{j=1}^{\infty} |f_j(t) - f_j(s)|^2}$$

is pregaussian.

Theorem 5.2. Let E be a Banach space and T a continuous linear operator: $E \to C(S)$. If $T(E) \subseteq C^{\rho}(S)$ for some pregaussian metric ρ, then T is of type 2.

Proof. By the closed graph we have that $T: E \to (C^{\rho}(S), |||\cdot|||_{\rho})$ is continuous. Now let $(x_j) \in \ell_E^2$ and put $f_j = Tx_j$, then

$$\sum_{j=1}^{\infty} |f_j(t) - f_j(s)|^2 \leq \sum_{j=1}^{\infty} \rho(t,s)^2 |||f_j|||^2$$

$$\leq K \left(\sum_{j=1}^{\infty} \|x_j\|^2 \right) \rho(t,s)^2$$

where K is the norm of T as an operator from E into $(C^{\rho}(S), |||\cdot|||_{\rho})$. So the theorem follows from (5.6). □

Proposition 5.3. Let $R: E \to F$, $S: F \to G$ and $T: G \to H$ be continuous linear operators. If S is of type 2, then so is TSR.

Proof. Obvious. □

Corollary 5.4. If the domain or the range of a continuous linear operator T is of type 2, then so is T. □

Corollary 5.5. Let $\{X(t) \mid t \in S\}$, be a stochastic process, and M a random variable satisfying

(a) $P(|X(t) - X(s)| \le M\rho(t,s)) = 1 \quad \forall\, t,s \in S$

(b) $P(|X(t)| \le M) = 1 \quad \forall\, t \in S$

(c) $P(X(\cdot) \in C_o^{\rho}(S)) = 1$

(d) $\mathbb{E}\, M^2 < \infty$

for some pregaussian metric ρ. Then X belongs to the domain of normal attraction on $C(S)$.

Proof. We know that X is a $C(S)$-valued random vector, with $X \in C_o^{\rho}(S)$ a.s. By separability of $C_o^{\rho}(S)$ (cf. Theorem 5.1) we have that X is a $C_o^{\rho}(S)$-valued random vector. But the injection: $C_o^{\rho}(S) \to C(S)$ is of type 2 by Theorem 5.2, so $X \in \mathrm{DNA}$ by Theorem 4.1, since $|||X|||_{\rho} \le M$, and $\mathbb{E}\, M^2 < \infty$.

Remark. $C^{\rho}(S)$ is separable only if S is finite. So (c) cannot be substituted by $X \in C^{\rho}(S)$ a.s.

However, if $X \in C^{\rho}(S)$, and if there exists a pregaussian metric δ, satisfying

(5.7) $$\lim_{(t,s)\to(a,a)} \frac{\rho(t,s)}{\delta(t,s)} = 0 \qquad \forall\, a \in S$$

(5.8) $$\rho(t,s) \le K\delta(t,s) \qquad \forall\, t,s$$

Then clearly $X \in C_o^{\delta}(S)$, and

$$|||X|||_{\delta} \le (K+1)\, |||X|||_{\rho}$$

It is an open problem whether any pregaussian metric admits a pregaussian metric δ satisfying (5.7) and (5.8).

If ρ satisfies (5.5), it is easy to construct a δ satisfying (5.5), (5.7) and (5.8): Let d be the ρ-diameter of S. Then we choose a decreasing continuous function, $g:]0,d] \to]0,\infty]$, satisfying

$$g(0) = \infty, \quad \int_0^d g(t)\sqrt{H_\rho(t)}\,dt < \infty$$

If

$$h(t) = \int_0^t g(s)ds$$

then

$$h(t+s) = \int_0^t g(u)du + \int_0^s g(u+t)du \le h(t) + h(s)$$

So $\delta(t,s) = h(\rho(t,s))$ is a metric. And since any ρ-ball of diameter r is a δ-ball of diameter $h(r)$, we have

$$H_\delta(h(t)) = H_\rho(t) \qquad \forall\, t > 0$$

Hence

$$\int_0^{h(d)} \sqrt{H_\delta(t)}\,dt = \int_0^d h'(t)\sqrt{H_\delta(h(t))}\,dt$$

$$= \int_0^d g(t)\sqrt{H_\rho(t)}\,dt$$

$$< \infty$$

and

$$\frac{\rho(t,s)}{\delta(t,s)} \xrightarrow[\rho(t,s)\to 0]{} \frac{1}{g(0)} = 0$$

$$\rho(t,s) \le \delta(t,s)/g(d)$$

So (5.5), (5.7) and (5.8) holds.

J. Zinn has shown in [117], that if ρ satisfies (5.4), then there exists a pregaussian metric satisfying (5.7) and (5.8).

As corollaries to these remarks we have

Corollary 5.6. Let $\{X(t) \mid t \in S\}$ be a stochastic process with continuous sample path, and let ρ be a continuous metric on S, such that

(a) $$\mathbb{E}\, |||X|||_\rho^2 < \infty$$

(b) $$\int_0^d \sqrt{H_\rho(t)}\, dt < \infty$$

Then X belongs to the domain of normal attraction on $C(S)$. □

Corollary 5.7. Let $\{X(t) \mid t \in S\}$ be a stochastic process with continuous sample paths, such that

(a) $$\mathbb{E}\, |||X|||_\rho^2 < \infty$$

for some metric ρ given by

$$\rho(t,s) = \sqrt{\mathbb{E}\,(Z(t)-Z(s))^2}$$

where $Z(t)$ is a mean 0 gaussian process with continuous sample paths. Then X belongs to the domain of normal attraction. □

Theorem 5.8. Let (S,Σ,μ) be a finite measure space, and let $1 \le p \le 2 \le q \le \infty$. Then the injection: $L^q(\mu) \to L^p(\mu)$ is of type 2.

Proof. Follows from Proposition 5.3 and the factorization:

$$L^q(\mu) \to L^2(\mu) \to L^p(\mu) \qquad \square$$

Corollary 5.9. Let (S,Σ,μ) be a finite measure space and $\{X(t) \mid t \in S\}$ a measurable stochastic process. Let $q \in [2,\infty]$, and suppose that

(a) $X(\cdot) \in L^q(\mu)$ a.s.

(b) $\{X(\cdot,\omega) \mid \omega \in \Omega\}$ <u>is separable in</u> $L^q(\mu)$

(c) $\mathbb{E}\ \|X\|_q^2 < \infty$

<u>Then</u> X <u>belongs to the domain of normal attraction on any</u> $L^p(\mu)$ <u>for</u> $1 \leq p \leq q$. □

BIBLIOGRAPHIE

[1] T.W. Anderson, *The integral of a symmetric unimodal function over a symmetric convex set and some probability inequalities*, Proc.Amer.Math.Soc.6(1955), 170-176.

[2] A. Arajao, *On the central limit theorem for* $C(I^k)$*-valued random variable*, Preprint Stat.Dept.,Univ.of Calif.,Berkeley, (1973).

[3] A. Badrikan, *Prolegomene au calcul des probabilités dans les Banach*, Springer Lecture Notes in Math.,539(1976), 1-165.

[4] A. Beck, *On the strong law of large numbers, Ergodic Theory,* Proc.Int.Symp. New Orleans 1961 (ed.: F.B. Wright), Academic Press.

[5] A. Beck, *A convexity condition in normed linear spaces and the strong law of large numbers*, Proc. Amer. Math.Soc., 13(1962), 329-334.

[6] A. Beck, *Conditional independence*, Bull. Amer.Math.Soc., 80(1974), 1169-1172.

[7] A. Beck and D.P. Giesy, *P-uniform convergence and vector-valued strong law of large numbers*, Trans.Amer.Math.Soc., 147(1970), 541-559.

[8] C. Bessaga and A. Pelczynski, *On basis and unconditional convergence of series in Banach spaces*, Stud.Math., 17(1958), 151-164.

[9] P. Billingsley, *Convergence of probability measures*, John Wiley and Sons, New York (1968).

[10] C. Borell, *Convex set functions in* d*-space*, Period.Math.Hung. 6(1975), 111-136.

[11] C. Borell, *Convex measures on locally conves spaces*, Ark.Math., 120(1974), 390-408.

[12] C. Borell, *The Brunn-Minkovski inequality in Gauss space*, Invent.Math., 30(1975), 207-216.

[13] C. Borell, *Gaussian measures on locally convex spaces*, Math. Scand. (to appear).

[14] S.D. Chatterji, *Martingale convergence and the Radon-Nikodym theorem in Banach spaces*, Math.Scand. 22(1968), 21-41.

[15] S.D. Chatterji, *Martingales of Banach-value random variables*, Bull.Amer.Math.Soc., 66(1960), 395-398.

[16] S.D. Chatterji, *A note on the convergence of Banach space valued martingales*, Math.Ann., 153(1964), 142-149.

[17] K.L. Chung, *Note on some strong laws of large numbers*, Amer.J.Math., 69(1947), 189-192.

[18] J. Delporte, *Fonction aléatoires présque surement continues sur une intervalle fermé*, Ann. Inst. H. Poincaré, 1(B) (1964), 111-215.

[19] N. Dunford and J.T. Schwartz, *Linear operators, vol. I*, Interscience, New York (1967).

[20] R.M. Dudley, *Metric entropy and the central limit theorem in* C(S), Ann.Inst.Fourier 24(1974), 49-60.

[21] R.M. Dudley, *The sizes of compact subsets of Hilbert space and continuity of gaussian Processes*, J.Funct.Anal., 1(1967), 290-330.

[22] R.M. Dudley, *Sample functions of the gaussian process*, Ann.Prob., 1(1973), 66-103.

[23] R.M. Dudley and M. Kanter, *Zero-one laws for stable measure,* Amer.Math.Soc., 45(1974), 245-252.

[24] R.M. Dudley and V. Strassen, *The central limit theorem and ε-entropy,* Springer Lecture notes in Math. 89(1969), 224-231.

[25] T. Figiel and G. Pisier, *Séries aléatoires dans les espaces uniformémenet convexes ou uniformémenet lisses,* C.R. Acad.Sci.Paris, 279(1974), 611-614.

[26] X. Fernique, *Integrabilité des vecteurs gaussiens,* C.R. Acad.Sci. Paris, 270(1970), 1698-1699.

[27] X. Fernique, *Régularité de trajéctoires des fonctions aléatoires gaussiennes,* Springer Lecture notes 480(1975), 2-95.

[28] X. Fernique, *Une démonstration simple du théorème de R.M. Dudley et M. Kanter sur les lois zero-un pour les mesures stables,* Springer Lecture Notes Math., 381(1974).

[29] M.R. Fortet and E. Mourier, *Les fonctions aléatoires commes éléments aléatoires dans les espaces de Banach,* Stud.Math., 15(1955), 62-79.

[30] M.R. Fortet and E. Mourier, *Resultats complementaires sur les élémenets aléatoires prenant leurs valeur dans un espace de Banach,* Bull. Sci. Math., 78(1965), 14-30.

[31] D.J.H. Garling, *Functional limit theorems in Banach spaces,* Ann. Prob. 4(1976), 600-611.

[32] E. Giné, *On the central limit theorem for sample continuous processes,* Ann.Prob. 2(1974), 629-641.

[33] E. Giné, *A note on the central limit theorem in* C(S), preprint.

[34] D.P. Giesy, *On a convexity condition in normed linear spaces*, Trans.Amer.Math.Soc., 125(1966), 114-146.

[35] M. Hahn, *Central limit theorem for* D[0,1]*-valued random variable*, Ph.D. thesis, M.I.T. (1975).

[36] M. Hahn, *Conditions for sample continyity and the central limit theorem*, (to appear).

[37] M. Hahn and M.J. Klass, *Sample continuity of square integrable processes*, (to appear).

[38] B. Heinkel, *Théorème central-limite et loi du logarithme iteré dans* C(S), C.R. Acad.Sci.Paris, 282(1976), 711-713.

[39] B. Heinkel, *Mesures majorantes et théor`eme de la limite centrale dans* C(S), (preprint).

[40] J. Hoffmann-Jørgensen, *Sums of independent Banach space valued random variables*, Aarhus Universitet, preprint (1972-73), No. 15.

[41] J. Hoffmann-Jørgensen, *Sums of independent Banach space valued random variables*, Studia. Math., 52(1974), 159-186.

[42] J. Hoffmann-Jørgensen, *Integrability of seminorms, the 0-1 law and the affine kernel for product measures*, Stud.Math. (to appear).

[43] J. Hoffmann-Jørgensen, *On the modules of smoothness and the* G_a*-condition in* B*-spaces*, (preprint).

[44] J. Hoffmann-Jørgensen and G. Pisier, *The law of large numbers and the central limit theorem in Banach spaces*, Ann. Prob. 4(1976), 587-599.

[45] K. Ito and M. Nisio , *On the convergence of sums of independent Banach space valued random variables*, Osaka Math. J., 5(1968), 35-48.

[46] N.C. Jain, *Central limit theorem in a Banach space*, Springer Lecture Notes in Math., 526 1976.

[47] N.C. Jain, *An example concerning* CLT *and* LIL *in Banach spaces*, Ann. Prob. 4(1976), 690-694.

[48] N.C. Jain, *Central limit theorem and related questions in Banach space*, Proc. AMS symp. Urbana 1976 (to appear).

[49] N.C. Jain, *A zero-one law for gaussian processes*, Proc.Amer.Math.Soc. 29(1971), 585-587.

[50] N.C. Jain, *Tail probabilities for sums of independent Banach space valued random variables*, (preprint).

[51] N.C. Jain and G. Kallianpur, *Note on uniform convergence of stochastic processes*, Ann. Math. Stat., 41(1970), 1360-1362.

[52] N.C. Jain and G. Kallianpur, *Norm convergent expansions for gaussian processes in Banach spaces*, Proc.Amer.Math.Soc., 25(1970), 890-895.

[53] N.C. Jain and M.B. Marcus, *Integrability of infinite sums of independent vector-valued random variables*, Trans. Amer.Math.Soc., 212(1975), 1-36.

[54] N.C. Jain and M.B. Marcus, *Central limit theorems for* C(S)-*valued random variables*, J. Funct. Anal., 19(1975), 216-231.

[55] N.C. Jain and M.B. Marcus, *Sufficient conditions for the continuity of stationary gaussian processes and applications to random series of functions*, Ann. Inst. Fourier, 24(1974), 117-141.

[56] R.C. James, *A non reflexive Banach space, which is uniformly non octahedral*, Israel J. Math. (to appear).

[57] J.P. Kahane, *Some Random Series*, D.C. Heath, Lexington (1968).

[58] G. Kallianpur, *Abstract Wiener spaces and their reproducing kernel Hilbert spaces*, Z. Wahr Verw.Geb., 17(1971), 113-123.

[59] G. Kallianpur, *Zero-one laws for gaussian processes*, Trans.Amer. Math.Soc., 149(1970), 199-211.

[60] W. Krakowiak, *Comparison theorems for and exponential moments of random series in Banach spaces*, (preprint).

[61] J. Kuelbs, *Strassen law of the iterated logarithm*, Ann. Inst. Fourier, 2(1974), 169-177.

[62] J. Kuelbs, *A strong convergence theorem for Banach space valued random variables*, Ann. Prob. (to appear)

[63] J. Kuelbs, *The law of the iterated logarithm and related strong convergence theorems for Banach space valued random variables*, Springer Lecture Notes in Math. 539(1976).

[64] J. Kuelbs, *Kolmogorovs law of the iterated logarithm for Banach space valued random variables*, Ill.J.Math. (to appear).

[65] J. Kuelbs, *The law of the iterated logarithm in* C[0,1], Z. Wahr. Verw. Geb., 33(1976), 221-235.

[66] J. Kuelbs, *The law of the iterated logarithm in* C(S), (preprint).

[67] J. Kuelbs, *A strong convergence theorem for Banach space valued random variables*, Ann. Prob., 4(1976), 744-771.

[68] J. Kuelbs, *The law of iterated logarithm for Banach space valued random variables*, (preprint).

[69] J. Kuelbs, *An inequality for the distribution of a sum of certain Banach space valued random variables*, Stud.Math., 52(1974), 69-87.

[70] J. Kuelbs, *A counter example for Banach space valued random variables*, (preprint)

[71] J. Kuelbs and R. LePage, *The law of the iterated logarithm for Brownian motion in Banach space*, (to appear).

[72] T. Kurtz, *Inequalities for the law of large numbers*, Ann. Math. Stat., 43(1972), 1874-1883.

[73] S. Kwapień, *A theorem on the Rademacher series with vector valued coefficients*, Springer Lecture Notes in Math., 526(1976).

[74] S. Kwapień, *Isomorphic caracterization of inner product spaces by orthogonal series with vector valued coefficients*, Studia. Math. 44(1972) 583-595.

[75] S. Kwapień, *On Banach spaces containing* c_o, Stud.Math. 52(1974), 159-186.

[76] T.L. Lai, *Reproducing Kernel Hilbert spaces and the law of the iterated logarithm for gaussian processes*, Ann. Prob. (to appear).

[77] H.J. Landau and L.A. Shepp, *On the supremum of a Gaussian process*, Sankhya Ser. A, 32(1971), 369-378.

[78] R. LePage, *Loglog law for gaussian processes*, Z. Wahr. Verw. Geb., 25(1973), 103-108.

[79] W. Linde and A. Pietsch, *Mappings of gaussian cylindrical measures in Banach spaces*, Theor.Prob.Appl. 19(1974), 445-460.

[80] J. Lindenstrauss and L. Tzafriri, *Classical Banach spaces*, Springer Lecture Notes in Math., 338(1973).

[81] M. Loève, *Probability theory*, 3rd ed., Van Nostrand, New York, London, Toronto, (1963).

[82] M.B. Marcus, *Continuity of gaussian processes and Random Fourier Series*, Ann.Prob. 1(1975), 968-981.

[83] M.B. Marcus, *Some new results of limit theorems for* C(S)-*valued random variables*, (to appear).

[84] M.B. Marcus, *Uniform convergence of random Fourier series*, (preprint).

[85] M.B. Marcus, *Uniform estimates for certain Rademacher sums*, (preprint).

[86] M.B. Marcus and L.A. Shepp, *Sample behavior of Gaussian processes*, Proc. Sixth Berkeley Symp. on Math.Stat. and Prob. vol.2(1972), 423-441.

[87] B. Maurey, *Espaces de cotype* p, $0<p\leq 2$,
Seminaire Maurey-Schwarz 1972-73, Ecole Polytechnique, Paris.

[88] B. Maurey, *Théorèmes de factorisation pour les opérateurs linéaires à valeur dans un espace* L^p,
Astérique No. 11(1974), Soc.Math.France.

[89] B. Maurey and G. Pisier, *Series de V.A. Vectorielles indépendentes et propriétés géométriques des espaces de Banach*,
Studia Math., 58(1976), 45-90.

[90] B. Maurey and G. Pisier, *Caractérisation d'une classe d'espaces de Banach par de propriétés de séries aléatoires vectorielles*,
C.R. Acad.Sci.Paris, 277(1973), 687-690.

[91] E. Mourier, *Eléments aléatoires dans un espace de Banach*,
Ann.Inst. H. Poincaré, 13(1952), 159-244.

[92] K. Musial and W.A. Woyczynski, *Un principe de contraction pour convergence presque sure de séries aléatoires vectorielles*,
C.R. Acad.Sci. Paris, (to appear).

[93] G. Nordlander, *On sign-independent and almost sign-independent convergence in normed linear spaces*,
Arkiv för Mat. 4, 21(1961), 287-296.

[94] K.R. Parthasarathy, *Probability measures on metric spaces*,
Academic Press, New York.

[95] G. Pisier, *Sur la loi du logarithme itéré dans les espace de Banach*,
Springer Lecture notes in Math., 526(1976).

[96] G. Pisier, *Le theorème de la limite centrale et la loi du logarithme itéré dans les espaces de Banach*,
Seminaire Maurey-Schwarz 1975-76, Ecole Polytechnique, Paris.

[97] G. Pisier, *Sur la loi du logarithme*,
(to appear).

[98] G. Pisier, *Martingales à valeurs dans espaces uniformément convexes*,
C.R. Acad.Sci. Paris, (to appear).

[99] G. Pisier, *Type des espaces normés*,
C.R. Acad.Sci. Paris, 276(1973), 1673-1676.

[100] G. Pisier, *Martingales with values in uniformly convex spaces*,
Israel J.Math. (to appear).

[101] G. Pisier, *Sur les espaces de Banach qui ne contiennent pas uniformémenet de* ℓ_n^1,
C.R. Acad.Sci. 277(1973), 991-994.

[102] G. Pisier, *Sur les espaces qui ne contiennent pas de* ℓ_n^∞ *uniformement*,
Seminaire Maurey-Schwarz, Ecole Polytechnique 1972-73.

[103] Y.V. Prohorov, *Convergence of random processes and limit theorems in probability*,
Teor. Veroy. Prim., 1(1956), 177-238.

[104] P. Révész, *The laws of large numbers*,
Academic Press, New York, 1968.

[105] C. Ryll-Nardzewski and W.A. Woyczynski, *Bounded multiplier convergence in measure of random vector series*,
Proc.Amer.Math. Soc., 52(1975), 96-98.

[106] L. Schwarz, *Les espaces de cotype 2 d'apres B. Maurey, et leurs applications*,
Ann. Inst. Fourier, 24(1974), 179-188.

[107] V. Strassen, *An invariance principle for the law of the iterated logarithm*,
Z. Wahr. Verw. Geb. 3(1964), 211-226.

[108] V. Strassen, *Probability measures with given marginals*, Ann. Math. Stat., 36(1965), 423-439.

[109] S. Swaminathan, *Probabilistic characteirzation of reflexive spaces*, An. Acad.Brasil Cienc.,(1973), 345-347.

[110] J. Szarek, *On the best constant in the Khintchine inequality*, (to appear).

[111] T. Topsøe, *Topology and measure*, Springer Lecture Notes in Math. B3(1970).

[112] S.R.S. Varadhan, *Limit theorem for sums of independent random variables with valued in a Hilbert space*, Sankhya 24(1962, 213-238.

[113] W.A. Woyczynski, *Random series and laws of large numbers in some Banach space*, Teory Prob.Appl. 18(1973), 361-367.

[114] W.A. Woyczynski, *Strong laws of large numbers in certain Banach spaces*, Ann. Inst. Fourier, 24(1974), 205-223.

[115] W.A. Woyczynski, *A central limit theorem for martingales in Banach spaces*, Bull.Acad.Pol.Sci., 23(1975), 917-920.

[116] W.A. Woyczynski, *Geometry and martingales in Banach spaces*, Springer Lecture Notes in Math., 472(1975), 229-275.

[117] J. Zinn, *A note on the central limit theorem in Banach space*, (to appear).

[118] J. Zinn, *Zero-one laws for non-gaussian measures*, Proc.Amer.Math.Soc. 44(1974), 179-185.

THE STOCHASTIC EVOLUTION OF INFINITE SYSTEMS OF INTERACTING PARTICLES

PAR T.M. LIGGETT

The preparation of these notes was supported in part by N.S.F. Grant No. MPS 72-04591, and by an Alfred P. Sloan Founc research fellowship.

INTRODUCTION

Classical statistical mechanics is concerned with the equilibrium theory of certain physical systems. During the past six or eight years, several classes of Markov processes have been proposed as models for the temporal evolution of such systems, and a significant amount of progress has been made in their study. Some of these models, in turn, have been given economic or sociological interpretations. These lectures are intended as an introduction to and survey of some of this recent work. We will say very little about statistical mechanics itself, and will treat our subject as a self-contained branch of probability theory. The reader who would like to know more about the statistical mechanics which lies in the background is referred to [30], [41], [42], and [47].

Even in terms of the temporal theory, our discussion will be far from exhaustive. There are many important and interesting results in the papers listed in the bibliography to which we will refer only briefly, if at all. As might be expected, the choice of material and the emphasis was influenced by my own personal interests. While some of the proofs have been modified, most of the results presented here have appeared elsewhere in the literature. The only material which is essentially new is that in chapter 4 of part II. One of the attractive features of this subject is that there are still a large number of important open problems. We will point some of these out as we proceed.

Two types of processes have received the most attention. The first is the spin-flip process, which is the object of study in part I. While a special case was considered by Glauber [7] several years earlier, the spin-flip process was first formally introduced and studied in some generality by Dobrushin [6]. The state space of the process is $X = \{0,1\}^S$, where S is a countable set of sites. The interpretation is that at each point of S there is a particle (e.g., an iron atom) which has a positive or negative spin (represented by 1 and 0 respectively) at each time. Thus X represents all possible configurations of spins, and the Markov process describes the evolution of the configuration of spins. The process is described in terms of a collection of nonnegative speed functions $c(x,\eta)$ defined for $x \in S$ and $\eta \in X$ which give the rate at which the spin at x flips from $\eta(x)$ to $1-\eta(x)$ when the configuration of the entire system is η. This process has the property that only one coordinate of η changes at a time.

In the second type of process, two coordinates change at once. The state space is X again, and the process is described in terms of nonnegative speed functions $c(x,y,\eta)$ which give the rate at which an interchange of the x and y coordinates of η will occur when the configuration of the system is η. There are two interpretations of this process which suggest two different forms for the functions $c(x,y,\eta)$. In the first, which was proposed by Spitzer [45], the system being modelled is a lattice gas. Here particles are distributed on S with at most one per site, and $\eta(x)$ is the number of particles at x. The function

$c(x,y,\eta)$ then gives the rate at which a particle at x will move to a vacant site y or vice versa. The second interpretation is that of a binary alloy, and appears, for example, in [1] and [2]. In this case there are two types of particles, and each site in S is occupied either by a particle of type 0 or a particle of type 1. Then $c(x,y,\eta)$ gives the rate at which two particles of opposite types at x and y will interchange their positions. These processes will be called exclusion processes, since multiple occupancy is excluded, and will be discussed in part II.

The functions $c(x,\eta)$ or $c(x,y,\eta)$ do not always determine a unique process on X. Further conditions on these functions must generally be imposed in order to guarantee the existence of a unique process with desirable properties. The conditions usually express the requirement that the rates not depend too heavily on distant parts of the configuration. The first problem which requires treatment, then, involves existence and uniqueness questions. This problem is by now rather well understood, although there are some important conjectures relating to it which remain to be settled. These questions will be discussed in detail in chapter 1 of part I, and more briefly in chapter 1 of part II.

Once existence and uniqueness questions are disposed of, the main questions of interest involve the ergodic theory of the process. These are motivated by the statistical mechanical origin of these processes, and can be stated briefly in the following way. If μ is a probability measure on X, let $\mu S(t)$ be the distribution of the process at time t when μ is the initial distribution. Let $\mathcal{I} = \{\mu : \mu S(t) = \mu \text{ for all } t > 0\}$ be the set of invariant measures for the process. Since X is compact and the process will have the Feller property in all the cases we will consider, $\mathcal{I}$ will always be a nonempty compact convex set. The first problem is then to determine the structure of $\mathcal{I}$. In the case of spin-flip processes, $\mathcal{I}$ will often be a singleton, in which case we will say that the process does not exhibit phase transition. Many of the results which we will discuss give sufficient conditions for the absence of phase transition. When phase transition does occur, the situation is not at all well understood. In the case of exclusion processes, there is typically at least a one parameter family of extremal invariant measures, which can often be exhibited explicitly. The problem is then to show that there are no other extremal invariant measures. The results here deal mainly with the simple exclusion process, in which $c(x,y,\eta)$ depends on η only through $\eta(x)$ and $\eta(y)$. Very little is known about the more general case.

After the structure of $\mathcal{I}$ is understood, the next step in the development of the ergodic theory of the process is to determine the domain of attraction of each element of $\mathcal{I}$. Here one wants to find for each $\nu \in \mathcal{I}$ the set of all probability measures μ on X so that $\mu S(t)$ converges weakly to ν as $t \to \infty$. In the case of a spin flip process without phase transition, of course, one wishes to prove that this occurs for all μ, in which case we will say that the process is ergodic. When $\mathcal{I}$ is not a singleton, it is unreasonable to expect to be able to determine the domains of attraction completely. There are a few cases where this is possible (see, for example, sections 3.3 of part I and 2.1 of part II), but normally one should be satisfied by finding the limit of $\mu S(t)$ for sufficiently nice μ.

The primary aim of the subject is to solve the problems described in the previous three paragraphs in the greatest possible generality. There are related problems which are also important, however. Recently, for example, Holley and Stroock [26] have used rates of convergence in $\mu S(t) \to \nu$ for the ergodic spin-flip process to obtain new information about the Gibbs measures of statistical mechanics.

The most direct relationship between the spin-flip and exclusion processes on the one hand and equilibrium statistical mechanics on the other, is that the Gibbs measures are contained in $\mathcal{J}$ when $c(x,\eta)$ or $c(x,y,\eta)$ is chosen to depend in an appropriate way on the potential involved in the definition of the Gibbs measures. This is of course one of the justifications for considering these processes as models for nonequilibrium statistical mechanics. This relationship will be seen in more detail in section 4.1 of part I and sections 1.3 and 4.2 of part II.

We will conclude this introduction by mentioning three techniques which have been used frequently in this subject, and which we will see more of in these lectures. One of their interesting features is that, while the two types of processes we will discuss in parts I and II are quite different in many respects, each of these techniques has proved very useful for both types of processes. Probably the most useful of these techniques is that of coupling two or more processes together. This will be a key tool in chapter 2 of part I and chapters 2, 3, and 4 of part II. A second tool involves obtaining relations between the infinite particle system of interest and an auxiliary finite system, thus reducing many questions to the study of the finite system. This will be used in chapter 3 of part I and chapter 2 of part II. Finally, there is the use of the monotonicity of the free energy of the system, which was exploited by Holley in [19] and [20], and will be referred to briefly in section 4.2 of part I and section 4.2 of part II.

PART I. SPIN-FLIP PROCESSES

Chapter 1. Existence results and first ergodic theorems. Several approaches to the existence problem have been proposed. In [12], Harris used a direct probabilistic construction of the process of interest in the exclusion context. In [6], Dobrushin obtained existence in the spin-flip context via finite approximations. More recently, Holley and Stroock [25] have changed the problem into one involving martingales, in much the same way that Stroock and Varadhan had done earlier in the context of diffusion processes. They then used compactness arguments to obtain a very general existence result. From their point of view, the principal problem then becomes one of proving the uniqueness of the process, since that does not follow directly from the martingale approach. We will concentrate primarily on the semigroup approach, which proceeds via the Hille-Yosida theorem. It requires the imposition of smoothness conditions on the speed functions, but has the advantage of yielding immediately a Feller process which is uniquely determined by them.

Section 1.1. The Hille-Yosida theorem. By a semigroup of operators on the Banach space W, we will mean a collection $\{S(t),\ t \geq 0\}$ of linear operators on W which satisfy the following conditions: $S(0) = I$, $S(t_1 + t_2) = S(t_1)S(t_2)$, $\|S(t)\| \leq 1$, and $S(t)f \to f$ as $t \to 0$ for all $f \in W$. If $W = C(X)$ where X is compact metric, a semigroup of operators will be said to be a Markov semigroup if $S(t)1 = 1$ and $S(t)f \geq 0$ whenever $f \geq 0$. As is well known, a Markov semigroup on $C(X)$ determines a unique strong Markov process η_t on X via $S(t)f(\eta) = E^{\eta}f(\eta_t)$.

A (possibly unbounded) linear operator Ω on W with domain $\mathcal{D}(\Omega)$ is said to be dissipative if $f - \lambda\Omega f = g$ implies $\|f\| \leq \|g\|$ whenever $f \in \mathcal{D}(\Omega)$ and $\lambda \geq 0$. Ω is closed if its graph is a closed subset of $W \times W$.

Theorem 1.1.1. (Hille-Yosida). There is a one-to-one correspondence between semigroups $S(t)$ of operators on W and closed dissipative operators Ω on W with dense domain which satisfy $R(I - \lambda\Omega) = W$ for all sufficiently small $\lambda > 0$. The correspondence is given by:

(a) $$S(t)f = \lim_{n\to\infty} (I - \frac{t}{n}\Omega)^{-n} f \quad \text{for } f \in W \text{ and } t \geq 0, \quad \text{and}$$

(b) $$\Omega f = \lim_{t \downarrow 0} \frac{S(t)f - f}{t} \quad \text{for } f \in \mathcal{D}(\Omega).$$

Furthermore, $\frac{d}{dt} S(t)f = \Omega S(t)f = S(t)\Omega f$ for $f \in \mathcal{D}(\Omega)$. In case $W = C(X)$ for a compact X, $S(t)$ is Markov if and only if $\Omega 1 = 0$ and $(I - \lambda\Omega)^{-1}f \geq 0$ whenever $f \geq 0$ and $\lambda \geq 0$. Ω is called the generator of $S(t)$.

Remarks: If Ω is a semigroup generator, then $\mathcal{R}(I - \lambda\Omega) = W$ for all $\lambda \geq 0$. A bounded dissipative operator is of course automatically a semigroup generator.

Two easily verified facts which are useful in applying this theorem are: (a) if Ω is closed and dissipative, then $\mathcal{R}(I - \lambda\Omega)$ is closed in W for $\lambda > 0$, and (b) if Ω is dissipative and has dense domain, then the closure of Ω exists and is again a dissipative operator. As a consequence, if Ω is dissipative, has dense domain, and satisfies $\overline{\mathcal{R}(I - \lambda\Omega)} = W$ for all sufficiently small $\lambda > 0$, then the closure $\overline{\Omega}$ of Ω is a semigroup generator. A core of the generator Ω is a linear subspace D of $\mathcal{D}(\Omega)$ with the property that Ω is the closure of the restriction of Ω to D. Thus Ω is determined by its values on a core. In order for D to be a core, it is not in general sufficient that D be dense in W.

In case $W = C(X)$ for a compact X and $S(t)$ is a Markov semigroup, define $\mu S(t)$ for probability measures μ on X by

$$\int f \, d[\mu S(t)] = \int S(t)f \, d\mu .$$

Recalling that $\mathcal{I} = \{\mu : \mu S(t) = \mu \text{ for all } t \geq 0\}$, the Hille-Yosida theorem makes it possible to rewrite this definition in terms of the generator Ω of $S(t)$ in the following way:

$$(1.1.2) \qquad \mathcal{I} = \left\{\mu : \int \Omega f \, d\mu = 0 \quad \text{for all } f \in \mathcal{D}(\Omega)\right\} .$$

In order to verify that a given μ is in $\mathcal{I}$, it suffices to check that $\int \Omega f \, d\mu = 0$ for all f in a core for Ω. $\mathcal{I}$ is convex and is compact in the topology of weak convergence of measures, so that $\mathcal{I}$ is the closed convex hull of $\mathcal{I}_e$, its set of extreme points. Note that $\mathcal{I}$ is nonempty, since any weak limit of $\frac{1}{T}\int_0^T \mu S(t)dt$ as $T \to \infty$ is in $\mathcal{I}$ for any probability measure μ on X. Suppose Ω_n and Ω are Markov semigroup generators such that $\Omega_n f \to \Omega f$ for all f in a core for Ω. Then if $\mu_n \in \mathcal{I}_n$ and $\mu_n \to \mu$ weakly, it follows from (1.1.2) that $\mu \in \mathcal{I}$. It is not in general the case that $\mu \in \mathcal{I}$ implies that there exists $\mu_n \in \mathcal{I}_n$ so that $\mu_n \to \mu$ weakly, so that an explicit knowledge of $\mathcal{I}_n$ for each n leads to an identification of only a subset of $\mathcal{I}$.

We will have occasion to use the Trotter-Kurtz convergence theorem for semigroups. The following version [31] will suffice for our purposes.

Theorem 1.1.3. Suppose Ω_n and Ω are generators of semigroups $S_n(t)$ and $S(t)$ respectively. If there is a core V for Ω such that $V \subset \mathcal{D}(\Omega_n)$ for all n and $\Omega_n f \to \Omega f$ for all $f \in V$, then $S_n(t)f \to S(t)f$ for all $f \in W$ uniformly for t in compact sets.

Section 1.2. Speed functions which are Lipschitz continuous. Let S be a countable set and $X = \{0,1\}^S$ with the product topology. $C(X)$ is the space of continuous functions on X with the norm $\|f\| = \sup_\eta |f(\eta)|$. For $\eta \in X$ and $u \in S$, define $\eta_u \in X$ by

$$\eta_u(x) = \begin{cases} \eta(x) & \text{if } x \neq u \\ 1 - \eta(x) & \text{if } x = u . \end{cases}$$

For $u \in S$, define $\Delta_u : C(X) \to C(X)$ by $\Delta_u f(\eta) = f(\eta_u) - f(\eta)$, and put $C^1(X) = \{f \in C(X) : \ |||f||| = \Sigma_u \|\Delta_u f\| < \infty\}$. Suppose that $c(x,\eta)$ is a uniformly bounded, nonnegative, continuous function on $S \times X$, and define Ω on $C^1(X)$ by

$$\Omega f(\eta) = \sum_x c(x,\eta)\Delta_x f(\eta) . \tag{1.2.1}$$

$\Omega f \in C(X)$ for $f \in C^1(X)$ since the convergence is uniform. Ω is densely defined since $C^1(X) \supset \mathcal{F}$, the set of all functions which depend on finitely many coordinates. If $f \in C^1(X)$, $\lambda \geq 0$, and $f - \lambda\Omega f = g$, choose $\eta, \zeta \in X$ so that $f(\eta) = \inf\{f(\gamma) : \gamma \in X\}$ and $f(\zeta) = \sup\{f(\gamma): \gamma \in X\}$. Then $\Omega f(\eta) \geq 0$ and $\Omega f(\zeta) \leq 0$, so $f(\eta) \geq g(\eta)$ and $f(\zeta) \leq g(\zeta)$. Therefore Ω is dissipative and $g \geq 0$ implies $f \geq 0$. In order to conclude that the closure of Ω generates a unique Markov semigroup, it then suffices to verify that $\mathcal{R}(I - \lambda\Omega)$ is dense in $C(X)$ for all sufficiently small $\lambda \geq 0$.

We will carry out this verification under the assumption that $\{c(x,\eta), x \in S\}$ is a bounded subset of $C^1(X)$, and at the same time, we will obtain sufficient conditions for the ergodicity of the resulting process. The existence proof is based on [32], and the modification required to obtain the ergodicity result is based on [26]. For further generalizations, see [52] and [10].

The following lemma contains the essential a priori bound which is required for the proof. In order to state it, let $c(u) = \inf_\eta[c(u,\eta) + c(u,\eta_u)]$ and $\gamma(x,u) = \|\Delta_u c(x,\eta)\|$ for $x \neq u$.

<u>Lemma 1.2.2</u>. Suppose $f \in C^1(X)$, $\lambda \geq 0$, and $f - \lambda\Omega f = g$. Then

$$\|\Delta_u f\|[1 + \lambda c(u)] \leq \|\Delta_u g\| + \lambda \sum_{x \neq u} \gamma(x,u) \|\Delta_x f\| .$$

<u>Proof</u>. For $\eta \in X$ and $u \in S$,

$$\Delta_u f(\eta) = \Delta_u g(\eta) + \lambda \sum_{x \neq u} c(x,\eta_u)\Delta_u\Delta_x f(\eta)$$

$$+ \lambda \sum_{x \neq u} [\Delta_u c(x,\eta)]\Delta_x f(\eta) - \lambda[c(u,\eta_u) + c(u,\eta)]\Delta_u f(\eta) .$$

Choose $\zeta \in X$ so that $\Delta_u f(\zeta) = \|\Delta_u f\|$, which is possible since $\Delta_u f(\eta_u) = -\Delta_u f(\eta)$ for all η. Then $\Delta_u\Delta_x f(\zeta) \leq 0$, from which the result follows.

Now put $c = \inf_u c(u) \geq 0$, and assume that

$$M = \sup_x \sum_{u \neq x} \gamma(x,u) \leq \sup_x |||c(x,\cdot)||| < \infty . \tag{1.2.3}$$

Then $(\Gamma\beta)(u) = \Sigma_{x \neq u} \beta(x)\gamma(x,u)$ defines a bounded operator on $\ell_1(S)$ with norm M.

<u>Corollary 1.2.4</u>. Suppose $f, g \in C^1(X)$, $\lambda \geq 0$, and $f - \lambda\Omega f = g$. Define $\beta \in \ell_1(S)$ by $\beta(u) = \|\Delta_u g\|$. Then

$$\|\Delta_u f\| \le [(1+\lambda c)I - \lambda\Gamma]^{-1}\,\beta(u)$$

provided that $\lambda M/(1+\lambda c) < 1$.

Theorem 1.2.5. Under assumption (1.2.3), the closure of Ω generates a unique Markov semigroup $S(t)$. For $g \in C^1(X)$, put $\beta(u) = \|\Delta_u g\|$. Then

$$\|\Delta_u S(t)g\| \le e^{t[\Gamma - cI]}\,\beta(u)\ , \quad \text{so} \tag{1.2.6}$$

$$|||S(t)g||| \le e^{(M-c)t}\,|||g|||\,. \tag{1.2.7}$$

In particular, if $M < c$, the corresponding process is ergodic with unique invariant measure μ on X and

$$\left\|S(t)g - \int g\,d\mu\right\| \le \sup_x \|c(x,\cdot)\|\,\frac{e^{(M-c)t}}{c-M}\,|||g|||\,.$$

Proof. Let S_n be a sequence of finite sets such that $S_n \uparrow S$. Define $c_n(x,\eta) = c(x,\eta)$ for $x \in S_n$ and $c_n(x,\eta) = 0$ if $x \notin S_n$, and Ω_n and γ_n as before in terms of $c_n(x,\eta)$ instead of $c(x,\eta)$. Then Ω_n is a bounded dissipative operator, and therefore it is the generator of a Markov semigroup. Take $\lambda > 0$ so that $\lambda M < 1 + \lambda c$. We will show that $\mathcal{R}(I - \lambda\Omega)$ is dense in $C(X)$ for such a λ. Fix $g \in C^1(X)$ and define f_n by $f_n - \lambda\Omega_n f_n = g$. By the argument of lemma 1.2.2, $\|\Delta_u f_n\| \le \|\Delta_u g\| + \lambda\,\Sigma_{x\in S_n}\,\gamma(x,u)\|\Delta_x f\|$ for $u \notin S_n$, so $f_n \in C^1(X)$. Therefore by Corollary 1.2.4 and the fact that $\gamma_n(x,y) \le \gamma(x,y)$ for all $x,y \in S$, it follows that

$$\|\Delta_u f_n\| \le [(1+\lambda c)I - \lambda\Gamma]^{-1}\,\beta(u)\ , \tag{1.2.8}$$

where $\beta(u) = \|\Delta_u g\|$. Put $g_n = f_n - \lambda\Omega f_n$. Then

$$\|g_n - g\| \le \lambda \sup_u \|c(u,\cdot)\| \sum_{x\notin S_n} \|\Delta_x f_n\|\ ,$$

which tends to zero by (1.2.8). Since $g_n \in \mathcal{R}(I - \lambda\Omega)$, it follows that $g \in \overline{\mathcal{R}(I - \lambda\Omega)}$. Therefore $\overline{\mathcal{R}(I - \lambda\Omega)}$ contains $C^1(X)$, which is dense in $C(X)$, so $\overline{\Omega}$ generates a unique Markov semigroup $S(t)$ by the Hille-Yosida theorem. Furthermore, since $g_n \to g$ and $\|f_n - f_m\| \le \|g_n - g_m\|$, it follows that $f = \lim_n f_n$ exists, $\lim_n \Omega f_n$ exists, $f \in \mathcal{D}(\overline{\Omega})$, $\overline{\Omega} f = \lim_n \Omega f_n$, and $f - \lambda\overline{\Omega} f = g$. Taking the limit in (1.2.8) yields

$$\|\Delta_u (I - \lambda\overline{\Omega})^{-1}\,g\| \le [(1+\lambda c)I - \lambda\Gamma]^{-1}\,\beta(u)\,.$$

This can then be iterated to obtain

$$\|\Delta_u(I - \lambda\overline{\Omega})^{-k} g\| \le [(1+\lambda c)I - \lambda\Gamma]^{-k} \beta(u) ,$$

from which (1.2.6) follows by using part (a) of theorem 1.1.1. Now suppose that $M < c$, and take $\tau < t$. Then for $g \in C^1(X)$,

$$(1.2.9) \qquad \|S(t)g - S(\tau)g\| \le \int_\tau^t \|\Omega S(s)g\| \, ds$$

$$\le \sup_x \|c(x,\cdot)\| \int_\tau^t |||S(s)g||| \, ds$$

$$\le \sup_x \|c(x,\cdot)\| \frac{e^{(M-c)\tau}}{c-M} |||g||| .$$

Therefore $\lim_{t\to\infty} S(t)g$ exists, and it is constant by (1.2.7). This constant must be $\int g\,d\mu$ for any $\mu \in \mathcal{J}$, since $\int S(t)g\,d\mu = \int g\,d\mu$ for all t. The final conclusion of the theorem is obtained by letting $t \to \infty$ in (1.2.9).

Remark: It is easy to see that $\mathfrak{F}$ is a core for $\overline{\Omega}$. Since we have proved that $C^1(X)$ is a core, it suffices to show that given $f \in C^1(X)$, there are $f_n \in \mathfrak{F}$ such that $f_n \to f$ and $\Omega f_n \to \Omega f$. To do this, it suffices to choose f_n so that $f_n \to f$ and $\|\Delta_u f_n\| \le \|\Delta_u f\|$, which is always possible.

Section 1.3. Speed functions with absolutely convergent Fourier series.

This section is based on the work of Holley and Stroock [25], although the proofs are different. Let ν be the product measure on X with $\nu\{\eta : \eta(x) = 1\} = 1/2$ for all $x \in S$, and define $\chi_\emptyset(\eta) \equiv 1$ and $\chi_F(\eta) = \prod_{x\in F}[2\eta(x) - 1]$ for nonempty finite subsets F of S. Then $\{\chi_F\}$ is a complete orthonormal family in $L_2(\nu)$, and we may consider speed functions which have the representation

$$c(x,\eta) = \sum_F \hat{c}(x,F)\chi_F(\eta) ,$$

where $\sum_F |\hat{c}(x,F)| < \infty$ for each x. The convergence is then uniform, so $c(x,\eta)$ is continuous for each x. The basic assumption which will be made throughout this section is that there is an $0 < \alpha < 1$ so that

$$(1.3.1) \qquad \sum_{F\neq\phi} |\hat{c}(x,F)| \le \alpha\hat{c}(x,\emptyset).$$

Note that this automatically makes $c(x,\eta)$ nonnegative.

For finite subsets F and G of S, let $F \triangle G$ denote the symmetric difference of F and G, and $\beta(F) = \sum_{x\in F} \hat{c}(x,\emptyset)$. In the computations to follow, we will use the following easily verified facts: $\chi_F\chi_G = \chi_{F\triangle G}$,

$$\Delta_x \chi_F = \begin{cases} -2\chi_F & \text{if } x \in F \\ 0 & \text{if } x \notin F, \end{cases}$$

and

$$\sum_{G \neq F} \sum_{x \in F} |\hat{c}(x, F \triangle G)| \leq \alpha\beta(F) . \tag{1.3.2}$$

Let D be the class of all functions f which have the representation

$$f(\eta) = \sum_F \hat{f}(F)\, \chi_F(\eta)$$

where $\Sigma_F[1 + \beta(F)]|\hat{f}(F)| < \infty$. For $f \in D$, the series

$$\Omega f(\eta) = \sum_x c(x,\eta)\, \Delta_x f(\eta)$$

$$= -2 \sum_x \sum_G \hat{c}(x,G)\chi_G(\eta) \sum_{F \ni x} \hat{f}(F)\chi_F(\eta) \tag{1.3.3}$$

$$= -2 \sum_H \chi_H(\eta) \sum_F \hat{f}(F) \sum_{x \in F} \hat{c}(x, F \triangle H)$$

converges absolutely and uniformly by (1.3.2). Therefore Ω is a densely defined operator on $C(X)$, which is again dissipative. We will show that its closure is a semigroup generator under assumption (1.3.1). The required a priori bound is given in the following lemma. For finite subsets H and F of S, let $\gamma(H,F) = \Sigma_{x \in F}|\hat{c}(x, F \triangle H)|$ if $H \neq F$ and $\gamma(H,H) = 0$. Note that

$$\sum_H \gamma(H,F) \leq \alpha\beta(F) \tag{1.3.4}$$

by (1.3.2).

<u>Lemma 1.3.5</u>. Suppose $f \in D$, $\lambda \geq 0$ and $f - \lambda\Omega f = g$. Then for any finite subset H of S,

$$|\hat{f}(H)|[1 + 2\lambda\beta(H)] \leq |\hat{g}(H)| + 2\lambda \sum_F \gamma(H,F)|\hat{f}(F)|$$

<u>Proof</u>. Separating out the terms on the right side of (1.3.3) for which H = F, we obtain

$$\sum_H \hat{f}(H)\chi_H(\eta)[1 + 2\lambda\beta(H)] = g(\eta) - 2\lambda \sum_{H \neq F} \left[\sum_{x \in F} \hat{c}(x, F \triangle H)\right] \hat{f}(F)\chi_H(\eta) .$$

Identifying the Fourier coefficients of both sides gives

$$\hat{f}(H)[1 + 2\lambda\beta(H)] = \hat{g}(H) - 2\lambda \sum_{F \neq H} \left[\sum_{x \in F} \hat{c}(x, F \triangle H)\right] \hat{f}(F)$$

from which the result follows by taking absolute values.

Theorem 1.3.6. Under assumption (1.3.1), the closure of Ω generates a unique Markov semigroup $S(t)$.

Proof. Let S_n be a sequence of finite sets so that $S_n \uparrow S$. Define speed functions $c_n(x,\eta)$ by $c_n(x,\eta) = 0$ if $x \notin S_n$ and $c_n(x,\eta) = \Sigma_{F \subset S_n} \hat{c}(x,F)\chi_F(\eta)$ if $x \in S_n$. Let Ω_n be defined as before with $c_n(x,\eta)$ replacing $c(x,\eta)$. Then Ω_n is a bounded dissipative operator on $C(X)$ which maps the class of functions which depend only on coordinates in S_n into itself. Therefore, if $\lambda \geq 0$ and g depends only on coordinates in S_n, there is a function f_n depending on the coordinates in S_n which solves $f_n - \lambda\Omega_n f_n = g$. So, take $g \in \mathfrak{F}$ and $\lambda > 0$ and define $f_n \in \mathfrak{F} \subset D$ for all sufficiently large n by $f_n - \lambda\Omega_n f_n = g$. By Lemma 1.3.5,

$$(1.3.7) \qquad |\hat{f}_n(H)|[1 + 2\lambda\beta(H)] \leq |\hat{g}(H)| + 2\lambda \sum_F \gamma(H,F)|\hat{f}_n(F)|$$

for all finite $H \subset S$. Note that this needs to be verified only for $H \subset S_n$, since $\hat{f}_n(H) = 0$ otherwise. By (1.3.4),

$$\Gamma u(H) = \sum_{F:\beta(F)>0} \frac{\gamma(H,F)}{\beta(F)} u(F)$$

defines an operator of norm $\leq \alpha < 1$ on the ℓ_1 space on the finite subsets of S. Define $u(H) = (1/2\lambda)|\hat{g}(H)|$. Then $u \in \ell_1$, so $v = (I - \Gamma)^{-1}u \in \ell_1$ also, and $\beta(H)|\hat{f}_n(H)| \leq v(H)$ by (1.3.7). Put $g_n = f_n - \lambda\Omega f_n$, and observe from this last estimate and (1.3.3) that $\|g_n - g\| = \lambda\|\Omega_n f_n - \Omega f_n\| \to 0$ by the dominated convergence theorem. Therefore $g \in \overline{\mathcal{R}(I - \lambda\Omega)}$, so the result is a consequence of the Hille-Yosida theorem.

For the ergodic theorem, we will assume that

$$(1.3.8) \qquad c = \inf_x \hat{c}(x,\emptyset) > 0 .$$

From lemma 1.3.5 and the above proof, it follows that if $\Sigma_F|\hat{g}(F)| < \infty$, then $f = (I - \lambda\Omega)^{-1}g \in D$ for $\lambda > 0$. Summing the inequality in the statement of lemma 1.3.5 over $H \neq \emptyset$, and using $\beta(H) \geq c$ for $H \neq \emptyset$, $\gamma(H,\emptyset) = 0$, and (1.3.4), we see that

$$[1 + 2\lambda c(1 - \alpha)] \sum_{H \neq \emptyset} |\hat{f}(H)| \leq \sum_{H \neq \emptyset} |\hat{g}(H)| .$$

Iterating this and using (a) of the Hille-Yosida theorem yields

$$\exp[2ct(1-\alpha)] \sum_{H\neq\emptyset} |\widehat{S(t)g}(H)| \leq \sum_{H\neq\emptyset} |\hat{g}(H)| ,$$

and hence

$$\|S(t)g - \widehat{S(t)g}(\emptyset)\| \leq \sum_{H} |\hat{g}(H)| \exp[-2ct(1-\alpha)] . \tag{1.3.9}$$

Therefore, if μ is any invariant measure for the process,

$$\left|\int g\, d\mu - \widehat{S(t)g}(\emptyset)\right| \leq \sum_{H} |\hat{g}(H)| \exp[-2ct(1-\alpha)] .$$

Using (1.3.9) again gives the following theorem.

Theorem 1.3.10. Under assumptions (1.3.1) and (1.3.8), the process is ergodic with unique invariant measure μ, and

$$\left\|S(t)g - \int g\, d\mu\right\| \leq 2 \sum_{H} |\hat{g}(H)| \exp[-2ct(1-\alpha)]$$

for all g for which $\Sigma_H |\hat{g}(H)| < \infty$.

In order to compare the results in this section with those of the previous one, note that

$$\sum_{u} \sup_{\eta} |c(x,\eta_u) - c(x,\eta)| \leq 2 \sum_{F} |F|\, |\hat{c}(x,F)| , \quad \text{and}$$

$$\inf_{\eta} [c(u,\eta) + c(u,\eta_u)] \geq 2\hat{c}(u,\emptyset) - 2 \sum_{\substack{F\not\ni u\\ F\neq\emptyset}} |\hat{c}(u,F)| .$$

Therefore the existence and uniqueness of the process follows from theorem 1.2.5 if $\sup_x \Sigma_F[|F| + 1]|\hat{c}(x,F)| < \infty$, while for the ergodicity,

$$\sup_{x} \sum_{F} |F|\, |\hat{c}(x,F)| + \sup_{x} \sum_{\substack{F\not\ni x\\ F\neq\emptyset}} |\hat{c}(x,F)| < \inf_{x} \hat{c}(x,\emptyset)$$

would be required.

In addition to the ergodicity result in theorem 1.3.10, Holley and Stroock also prove that the process is ergodic when (1.3.8) is replaced by the assumption that $\hat{c}(x,\emptyset) > 0$ for all $x \in S$, although then, of course, there is no exponential rate of convergence to the invariant measure. In [26], they have extended these results to the case in which ν is replaced by a product measure with density different from 1/2.

Section 1.4. The martingale approach; uniqueness problems. In [25], Holley and Stroock have replaced the problem of showing that the closure of Ω is a semigroup generator by the problem of proving existence and uniqueness for a related martingale problem. We will discuss their results briefly. Consider Λ, the canonical path space of right continuous functions with left limits with values in X, and the σ-algebra on Λ generated by the cylinder sets. Given a nonnegative function $c(x,\eta)$ on $S \times X$, define Ωf for $f \in \mathfrak{F}$ by (1.2.1). A probability measure P^{η} on Λ is said to solve the martingale problem for Ω with initial configuration η if $P^{\eta}[\eta(0) = \eta] = 1$ and $f(\eta(t)) - \int_0^t \Omega f(\eta(s))ds$ is a martingale for all $f \in \mathfrak{F}$ relative to P^{η} and the natural increasing class of σ-algebras.

The relationship between the semigroup and martingale problems in case $c(x,\eta)$ is continuous is the following: (a) If the closure of Ω generates a Markov semigroup, then the measure P^{η} defined by the resulting Feller process is the unique solution to the martingale problem for each $\eta \in X$. (b) If the martingale problem has a unique solution P^{η} for each $\eta \in X$, then P^{η} determines a Feller process, so some extension of Ω generates a Markov semigroup. It is not known whether uniqueness for the martingale problem implies that the closure of Ω itself generates a Markov semigroup. (See the recent paper [63], however.)

Holley and Stroock prove that the martingale problem always has a solution provided that $c(x,\eta)$ is continuous. From this point of view, the real problem is then to determine conditions on $c(x,\eta)$ which guarantee that the solution is unique. Essentially, uniqueness is known only under conditions of the type discussed in sections 1.2 and 1.3. It appears that the estimates required to prove uniqueness in the martingale context are quite similar to those which yield both existence and uniqueness in the semigroup context.

Uniqueness certainly does not always hold. In section 3 of [25], Holley and Stroock give an example of nonuniqueness in which $c(x,\eta)$ is uniformly bounded away from 0 and ∞, and $c(x,\eta) \in \mathfrak{F}$ for each x. In [10], Gray and Griffeath give another example which is related to the processes to be discussed in chapter 3.

Probably the most important open problem along these lines is to prove uniqueness in case $S = Z^d$, the d-dimensional integer lattice, and $c(x,\eta)$ is (a) translation invariant, (b) continuous, and (c) strictly positive for all x and η. The positivity of $c(x,\eta)$ is needed here, since, as was observed in [25], there are examples of nonuniqueness in this context if $c(x,\eta)$ is allowed to be zero for some η.

Chapter 2. The use of coupling techniques. One of the important tools which will be used in this and succeeding chapters is that of coupling two (or more) processes together. This simply means that two Markov processes are defined on the same probability space, and the transition mechanisms of the two processes are linked in a convenient way. In the case of spin-flip processes, the most natural way to do this is to insure that, insofar as possible, two coordinates which agree at a given time will flip together. This technique was introduced in [6] and [57], and has since been used in the spin-flip context in [11], [13], [21], and [23], for example.

Section 2.1. The basic coupling. Suppose $c_1(x,\eta)$ and $c_2(x,\eta)$ are uniformly bounded, nonnegative, continuous functions which satisfy (1.2.3). For functions $f(\eta,\zeta)$ on $X \times X$ which depend on finitely many coordinates, define

$$\tilde{\Omega}f(\eta,\zeta) = \sum_x c_1(x,\eta)[f(\eta_x,\zeta) - f(\eta,\zeta)] + \sum_x c_2(x,\zeta)[f(\eta,\zeta_x) - f(\eta,\zeta)]$$

$$+ \sum_{x:\eta(x)=\zeta(x)} \min[c_1(x,\eta),c_2(x,\zeta)][f(\eta_x,\zeta_x) - f(\eta_x,\zeta) - f(\eta,\zeta_x) + f(\eta,\zeta)] .$$

Using the techniques of section 1.2, one can prove that the closure of $\tilde{\Omega}$ is the generator of a Markov semigroup. The coupled process $\gamma_t = (\eta_t,\zeta_t)$ is the corresponding Markov process on $X \times X$.

For $i = 1$ and 2, let Ω_i be defined as in (1.2.1) in terms of the functions $c_i(x,\eta)$. Among the important properties of the coupled process are the following:

(a) If f depends only on η, then $\tilde{\Omega}f(\eta,\zeta) = \Omega_1 f(\eta,\zeta)$. Therefore η_t is the Feller process whose generator is the closure of Ω_1.

(b) If f depends only on ζ, then $\tilde{\Omega}f(\eta,\zeta) = \Omega_2 f(\eta,\zeta)$. Therefore ζ_t is the Feller process whose generator is the closure of Ω_2.

(c) If c_1 and c_2 satisfy

(2.1.1) $$[c_1(x,\eta) - c_2(x,\zeta)][\eta(x) + \zeta(x) - 1] \geq 0$$

for all $x \in S$ whenever $\eta \leq \zeta$ coordinatewise, then

$$P^{(\eta,\zeta)}[\eta_t \leq \zeta_t] = 1 \quad \text{for} \quad \eta \leq \zeta .$$

The simplest way to prove facts such as (c) is often to prove them first in case S is finite, in which case γ_t is a finite Markov chain, and then to extend them to the case of infinite S via an application of theorem 1.1.3. For the proof in the finite case, note first that if f is constant on $M = \{(\eta,\zeta) : \eta \leq \zeta\}$, then $\tilde{\Omega}f = 0$ on M, and therefore $\tilde{\Omega}^n f = 0$ on M for $n \geq 1$. The result then follows from $\tilde{S}(t)1_M = \Sigma_{n=0}^{\infty} (t^n/n!)\tilde{\Omega}^n 1_M = 1$ on M.

Property (c) above suggests defining the following order relation on probability measures on X: We will say that $\mu_1 \leq \mu_2$ if there is a probability measure ν on $X \times X$ which has marginals μ_1 and μ_2 respectively and such that $\nu\{(\eta,\zeta) : \eta \leq \zeta\} = 1$. Note that $\mu_1 \leq \mu_2$ implies that $\mu_1\{\eta : \eta(x) = 1$ for $x \in T\} \leq \mu_2\{\eta : \eta(x) = 1$ for $x \in T\}$ for $T \subset S$. Property (c) then yields the following result.

Lemma 2.1.2. Suppose c_1 and c_2 satisfy (2.1.1), and let $S_1(t)$ and $S_2(t)$ be the corresponding semigroups. Then $\mu_1 \leq \mu_2$ implies $\mu_1 S_1(t) \leq \mu_2 S_2(t)$ for all $t \geq 0$

Section 2.2. Applications to attractive processes. This section is based on Holley's work in [21]. We will consider the Feller process η_t on X whose generator is the closure of the operator defined in (1.2.1). In addition to assuming the smoothness conditions on $c(x,\eta)$ from section 1.2 which guarantee the existence and uniqueness of the process, we will suppose throughout this section that the process is attractive in the sense that

$$[c(x,\eta) - c(x,\zeta)][\eta(x) + \zeta(x) - 1] \geq 0 \qquad (2.2.1)$$

for all $x \in S$ and all $\eta, \zeta \in X$ such that $\eta \leq \zeta$ coordinatewise. The interpretation of this condition is that there is a tendency for coordinates to try to line up with their neighbors.

Theorem 2.2.2. If $\mathcal{I} = \{\nu\}$ is a singleton, then the process is ergodic.

Proof. Let ν_0 and ν_1 be the pointmasses on $\eta \equiv 0$ and $\eta \equiv 1$ respectively. Then $\nu_0 \leq \mu \leq \nu_1$ for all probability measures μ on X, so

$$\nu_0 S(t) \leq \mu S(t) \leq \nu_1 S(t) \qquad (2.2.3)$$

for all such μ and all $t \geq 0$ by lemma 2.1.2. Since $\nu_0 \leq \nu_0 S(t)$ and $\nu_1 \geq \nu_1 S(t)$, it follows from the semigroup property that $\nu_0 S(t)$ increases and $\nu_1 S(t)$ decreases in t. Therefore the weak limits of $\nu_0 S(t)$ and $\nu_1 S(t)$ exist as $t \to \infty$, and thus both limits are ν by assumption, since they must be invariant. The result then follows by letting $t \to \infty$ in (2.2.3).

By the translation invariant case, we will mean the case in which $S = Z^d$ for some $d \geq 1$ and $c(x,\eta)$ is translation invariant in the sense that $c(x,\eta) = c(y,\zeta)$ if $\eta(x + u) = \zeta(y + u)$ for all $u \in S$. In the translation invariant case, let $\mathcal{S}$ be those probability measures on X which are invariant under translations in S. In some cases (see the comment at the end of section 4.2, for example) it is easier to show that $\mathcal{I} \cap \mathcal{S}$ is a singleton than to show that $\mathcal{I}$ is a singleton. The following strengthened form of theorem 2.2.2 is therefore sometimes useful.

Theorem 2.2.4. In the translation invariant case, if $\mathcal{S} \cap \mathcal{I}$ is a singleton, then so is $\mathcal{I}$, and the process is therefore ergodic.

Proof. It suffices to note that in the proof of theorem 2.2.2, $\nu_0 S(t)$ and $\nu_1 S(t)$ are in $\mathcal{S}$ for all $t \geq 0$, and therefore their limits are in $\mathcal{S} \cap \mathcal{I}$.

In order to obtain sufficient conditions for $\mathcal{I}$ to be a singleton, we will compare the process η_t with finite systems. Let S_n be finite sets which increase to S, and define for $i = 0$ and 1

$$c_i^n(x,\eta) = \begin{cases} c(x,\eta^i) & \text{if } x \in S_n \\ 0 & \text{if } x \notin S_n \text{ and } \eta(x) = i \\ M(x) & \text{if } x \notin S_n \text{ and } \eta(x) = 1 - i \end{cases}$$

where $M(x) = \sup_\zeta c(x,\zeta)$, $\eta^i(u) = \eta(u)$ for $u \in S_n$, and $\eta^i(u) = i$ for $u \notin S_n$. Then c_i^n satisfies (2.2.1), $[c_0^n(x,\eta) - c(x,\zeta)][\eta(x) + \zeta(x) - 1] \geq 0$, and $[c(x,\eta) - c_1^n(x,\zeta)][\eta(x) + \zeta(x) - 1] \geq 0$ whenever $\eta \leq \zeta$. Let $S_i^n(t)$ be the semi-group corresponding to c_i^n, and note that $\nu_i^n = \lim_{t\to\infty} \nu_i S_i^n(t)$ exists as in the proof of theorem 2.2.3. By lemma 2.1.2, ν_0^n increases in n and ν_1^n decreases in n, so that $\nu^i = \lim_{n\to\infty} \nu_i^n$ exists for $i = 0$ and 1. By the comments immediately preceeding theorem 1.1.3, ν^0 and ν^1 are in $\mathcal{I}$. In fact, by lemma 2.1.2, $\nu^0 \leq \nu \leq \nu^1$ for all $\nu \in \mathcal{I}$. Thus we have proved

Theorem 2.2.5. $\mathcal{I}$ is a singleton if and only if $\nu^0 = \nu^1$.

The usefulness of this theorem is especially evident in reversible cases (see section 4.1), where ν_i^n can be computed explicitly. One of the fundamental problems in classical statistical mechanics is to determine when $\nu^0 = \nu^1$ in certain reversible cases. Thus one can often use results from that field to determine whether particular attractive reversible processes are ergodic.

Section 2.3. Reduction to attractive processes. In this section, we will describe a technique which can be used to reduce the problem of proving ergodicity for a process with general $c(x,\eta)$ to a corresponding one for an associated attractive process which has a very simple invariant measure. There is in general a loss of information inherent in this technique, in the sense that the original process may in fact be ergodic, even though the associated attractive process is not. However, it is useful in many cases. We will assume throughout this section that $c(x,\eta)$ satisfies the smoothness assumptions of section 1.2.

The idea is to couple three processes η_t, ζ_t and ξ_t together, where η_t and ζ_t are two copies of the process with speed function $c(x,\eta)$ with different initial configurations, and ξ_t is an attractive process with $\xi \equiv 0$ as an absorbing point. The coupling between η_t and ζ_t will be as in section 2.1, and the process ξ_t will give a bound on the discrepancies between η_t and ζ_t. The speed function for the process ξ_t will be written in the form $\xi(x)\delta(x,\xi) + [1 - \xi(x)]\beta(x,\xi)$, where $\delta(x,\xi)$ and $\beta(x,\xi)$ are defined by

$$\beta(x,\xi) = \sup\{|c(x,\eta) - c(x,\zeta)| : |\eta(u) - \zeta(u)| \leq \xi(u) \text{ for all } u \in S\}$$

and

$$\delta(x,\xi) = \inf\{[c(x,\eta) + c(x,\zeta)] : \eta(x) \neq \zeta(x) \text{ and } |\eta(u) - \zeta(u)| \leq \xi(u) \text{ for all } u \in S\}.$$

It is easy to check that this speed function also satisfies the smoothness assumptions of section 1.2, that the process is attractive, and that $\beta(x,\xi) \equiv 0$ for $\xi \equiv 0$, so that the pointmass on $\xi \equiv 0$ is invariant for ξ_t. Therefore by theorem 2.2.2, ξ_t is ergodic if and only if this is the only invariant measure for ξ_t. The coupling will have the property that $|\eta_t(u) - \zeta_t(u)| \leq \xi_t(u)$ for all $t \geq 0$ and $u \in S$ if it has this property at $t = 0$. Therefore we obtain the following result.

Theorem 2.3.1. If ξ_t is ergodic, then η_t is ergodic.

Proof. If ξ_t is ergodic, then $P^{\xi}(\xi_t(u) = 1) \to 0$ as $t \to \infty$ for all $\xi \in X$ and $u \in S$. Therefore $P^{(\eta,\zeta)}(\eta_t(u) \neq \zeta_t(u)) \to 0$ for all $\eta, \zeta \in X$, when η_t and ζ_t are coupled via the basic coupling of section 2.1. If μ_1 and μ_2 are two probability measures on X and $S(t)$ is the semigroup corresponding to the speed function $c(x,\eta)$, it then follows that $\int f\, d\mu_1 S(t) - \int f\, d\mu_2 S(t) \to 0$ as $t \to \infty$ for any $f \in \mathfrak{Z}$. To complete the proof, let μ_1 be invariant for η_t, so that we have $\mu_2 S(t) \to \mu_1$.

The coupling which has the properties required above is described in terms of the following list of transition rates at x as functions of the configuration (η,ζ,ξ). Here $a = 0$ or 1, and $\varepsilon = \varepsilon(x,\zeta,\eta,\xi)$ is defined by $\varepsilon = \delta(x,\xi)[c(x,\eta) + c(x,\zeta)]^{-1}$ if $c(x,\eta) + c(x,\zeta) > 0$, and $\varepsilon = 0$ if $c(x,\eta) + c(x,\zeta) = 0$.

$(a,a,0) \to$	$(1-a,\ 1-a,\ 0)$	$\min[c(x,\eta),\ c(x,\zeta)]$
	$(a,\ 1-a,\ 1)$	$c(x,\zeta) - \min[c(x,\eta),\ c(x,\zeta)]$
	$(1-a,\ a,\ 1)$	$c(x,\eta) - \min[c(x,\eta),\ c(x,\zeta)]$
	$(a,\ a,\ 1)$	$\beta(x,\xi) - |c(x,\eta) - c(x,\zeta)|$

$$
\begin{array}{lll}
(a,a,1) \rightarrow & (1-a,\ 1-a,\ 1) & \min[c(x,\eta),\ c(x,\zeta)] \\
& (a,\ 1-a,\ 1) & c(x,\zeta) - \min[c(x,\eta),\ c(x,\zeta)] \\
& (1-a,\ a,\ 1) & c(x,\eta) - \min[c(x,\eta),\ c(x,\zeta)] \\
& (a,\ a,\ 0) & \delta(x,\xi)
\end{array}
$$

$$
\begin{array}{lll}
(a,1-a,1) \rightarrow & (a,\ a,\ 1) & c(x,\zeta)(1-\varepsilon) \\
& (a,\ a,\ 0) & c(x,\zeta)\varepsilon \\
& (1-a,\ 1-a,\ 1) & c(x,\eta)(1-\varepsilon) \\
& (1-a,\ 1-a,\ 0) & c(x,\eta)\varepsilon\ .
\end{array}
$$

These rates apply only to configurations (η,ζ,ξ) for which $|\eta(u)-\zeta(u)| \leq \xi(u)$ for all $u \in S$. Since this set of configurations is closed for the coupled process, there is no need to define the process on its complement. It is clear that all of the above rates are nonnegative, and it is in fact this nonnegativity requirement which led to our definition of $\beta(x,\xi)$ and $\delta(x,\xi)$. It is not hard to verify that the rates are such as to yield the right marginal processes.

It should be noted that if $\beta_i(x,\xi)$ and $\delta_i(x,\xi)$ for $i = 1,2$ satisfy $\beta_2(x,\xi) \geq \beta_1(x,\xi)$ and $\delta_2(x,\xi) \leq \delta_1(x,\xi)$, $\beta_2(x,\xi) = 0$ for $\xi = 0$, and both processes are attractive, then a simple coupling argument shows that the ergodicity of the process with speed function $\xi(x)\delta_2(x,\xi) + [1-\xi(x)]\beta_2(x,\xi)$ implies the ergodicity of the process with speed function $\xi(x)\delta_1(x,\xi) + [1-\xi(x)]\beta_1(x,\xi)$. Thus, in applying theorem 2.3.1., it is often sufficient to consider functions β and δ which have a certain structure (such as that of the proximity processes of section 3.1) and which majorize the given functions in the above sense.

<u>Chapter 3</u>. <u>The use of duality techniques</u>. Suppose η_t and ζ_t are Markov processes with state spaces U and V respectively, and let $H(\eta,\zeta)$ be a bounded measurable function on $U \times V$. Then η_t and ζ_t are said to be dual to one another with respect to H if

$$E^{\eta}H(\eta_t,\zeta) = E^{\zeta}H(\eta,\zeta_t)$$

for all $\eta \in U$ and $\zeta \in V$. Given a process η_t of interest, it is often useful to find a suitable function H and a process ζ_t which is dual to η_t with respect to H. Many problems involving η_t can then be recast in terms of ζ_t, and often solved more easily in this new context. Various forms of duality (i.e., using various functions H) and various resulting dual processes have been applied in [14],[24],[27],[40], and [58] to obtain ergodic theorems for certain spin-flip processes. In this chapter, we will concentrate primarily on the material in [24]. The processes which we will consider are among those which can arise when the coupling in section 2.3 is carried out.

Section 3.1. The duality theorem for proximity processes. By a proximity process, we will mean a spin-flip process η_t with speed function of the form

$$c(x,\eta) = c(x)\left\{[1 - \eta(x)] + [2\eta(x) - 1] \sum_F p(x,F)\chi_F(\eta)\right\},$$

where the sum is over finite subsets F of S, and $\chi_F(\eta) = \Pi_{x\in F}[1 - \eta(x)]$ for $F \neq \emptyset$ and $\chi_\emptyset(\eta) \equiv 1$. We will assume that $c(x)$ and $p(x,F)$ satisfy $c(x) > 0$, $p(x,F) \geq 0$, and $\Sigma_F\, p(x,F) = 1$ for all $x \in S$, and

(3.1.1) $$\sup_x c(x) \sum_F p(x,F)[|F| + 1] < \infty,$$

where $|F|$ denotes the cardinality of F. Note that for $u \neq x$,

(3.1.2) $$\sup_\eta |c(x,\eta_u) - c(x,\eta)| = c(x) \sum_{F\ni u} p(x,F),$$

so (3.1.1) guarantees that $c(x,\eta)$ is uniformly bounded and satisfies (1.2.3).

For each finite $G \subset S$, $\chi_G(\eta)$ is in the domain of the generator Ω of η_t, and

$$\Omega\chi_G(\eta) = \sum_x c(x,\eta)\,\Delta_x\chi_G(\eta)$$

(3.1.3) $$= \sum_{x\in G} c(x,\eta)[2\eta(x) - 1]\chi_{G\backslash x}(\eta)$$

$$= \sum_{x\in G} c(x) \sum_F p(x,F)[\chi_{(G\backslash x)\cup F}(\eta) - \chi_G(\eta)].$$

This suggests the possibility that η_t is dual with respect to $H(\eta,F) = \chi_F(\eta)$ to the Markov chain whose state space is the set of all finite subsets of S and which has the following description: if the chain is in state G, then (a) each $x \in G$ has an exponential lifetime with rate $c(x)$, (b) when the particle at x dies, it is replaced with probability $p(x,F)$ by all the points in F, and (c) only one particle, rather than two, remains at points in $F \cap (G\backslash x)$. This Markov chain will be called a branching process with interference (BPI). The interference comes from property (c) above.

Consider the Q matrix on the finite subsets of S given by

$$q(A,B) = \sum_{x\in A} c(x) \sum_{F:(A\backslash x)\cup F=B} p(x,F)$$

for $A \neq B$. We want to show that there is a unique nonexplosive Markov chain A_t corresponding to this Q matrix, and that it satisfies $E^A|A_t| \leq e^{\omega t}|A|$, where

$$\omega = \sup_x c(x) \sum_F p(x,F)[|F| - 1] .$$

In order to do this, observe first that

$$(3.1.4) \qquad \sum_B q(A,B) = \sum_{x \in A} c(x)$$

and

$$(3.1.5) \qquad \sum_B q(A,B)[|B| - |A|] \leq \sum_{x \in A} c(x) \sum_F p(x,F)[|F| - 1] \leq \omega |A| .$$

Define

$$q_n(A,B) = \begin{cases} q(A,B) & \text{if } |A| \leq n \\ 0 & \text{if } |A| > n . \end{cases}$$

By (3.1.4), $\sup_A \Sigma_B q_n(A,B) < \infty$, so this Q matrix determines a unique Markov chain A_t^n. By (3.1.5), this chain satisfies

$$(3.1.6) \qquad E^A|A_t^n|e^{-t\omega^+} \leq E^A\left[|A_t^n|e^{-\omega(t \wedge \tau_n)}\right] \leq |A|$$

where $\tau_n = \inf\{t > 0 : |A_t^n| > n\}$ and $\omega^+ = \omega \vee 0$. Therefore

$$P^A(\tau_n \leq t) \leq \frac{1}{n} E^A|A_t^n| \leq \frac{|A|}{n} e^{t\omega^+} .$$

Since $\lim_{n\to\infty} P^A(\tau_n \leq t) = 0$, the minimal solution to the backward equation corresponding to $q(A,B)$ is stochastic, and therefore the chain A_t is nonexplosive. Taking a limit in (3.1.6) yields

$$(3.1.7) \qquad E^A|A_t| \leq |A|e^{\omega t} .$$

Theorem 3.1.8. Let η_t be a proximity process whose parameters satisfy (3.1.1), and let A_t be the corresponding BPI. Then for any $\eta \in X$ and finite $A \subset S$,

$$E^\eta \chi_A(\eta_t) = E^A \chi_{A_t}(\eta) .$$

Proof. Let $u_\eta(t,A) = E^\eta \chi_A(\eta_t) = S(t)\chi_A(\eta)$. Since χ_A is in the domain of Ω for each finite $A \subset S$, the Hille-Yosida theorem and (3.1.3) yield

$$\frac{d}{dt} u_\eta(t,A) = S(t)\Omega\chi_A(\eta) = \sum_B q(A,B)[u_\eta(t,B) - u_\eta(t,A)] .$$

The result then follows from the fact that $E^A \chi_{A_t}(\eta)$ is the unique bounded solution to this differential equation with initial condition $u_\eta(0,A) = \chi_A(\eta)$. (The proof of this uniqueness is similar to the proof that whenever the minimal solution to the backward equation is stochastic, it is the only stochastic solution.)

Corollary 3.1.9. If μ is a probability measure on X, then

$$\mu S(t)\{\eta : \eta(x) = 1 \text{ on } A\} = \sum_B P^A[A_t = B]\, \mu\{\eta : \eta(x) = 0 \text{ on } B\}$$

for all finite $A \subset S$.

Section 3.2. Some applications. Since a proximity process is attractive and $\eta \equiv 0$ is absorbing for it, the process is ergodic if and only if $P^\eta[\eta_t(x)=1] \to 0$ as $t \to \infty$ for $\eta \equiv 1$ and all $x \in S$. Thus the following result is an immediate consequence of theorem 3.1.8.

Theorem 3.2.1. Let η_t be a proximity process and A_t the corresponding BPI. Then η_t is ergodic if and only if $P^A[A_t \neq \emptyset] \to 0$ as $t \to \infty$ for all finite $A \subset S$.

Corollary 3.2.2. Suppose $\omega = \sup_x [c(x) \Sigma_F\, p(x,F)[|F| - 1]] < 0$. Then η_t is ergodic.

Proof. $P^A[A_t \neq \emptyset] \leq E^A|A_t|$, so the result follows from the previous theorem and (3.1.7).

Even if $\omega = 0$ above, one can often conclude that η_t is ergodic. If $\omega = 0$, (3.1.7) says that $|A_t|$ is a (nonnegative) supermartingale, so $\lim_{t\to\infty}|A_t|$ exists with probability one. Therefore if $\inf_x c(x)p(x,\emptyset) > 0$, for example, it follows that $|A_t| = \emptyset$ eventually.

It is of interest to compare the result in corollary 3.2.2 with the ergodicity part of theorem 1.2.5. By 3.1.2, the assumption $M < c$ in that theorem becomes in the present context

$$\sup_x c(x)\left\{\sum_F p(x,F)[|F| - 1] + \sum_{F \not\ni x} p(x,F)\right\} < \inf_x c(x) \sum_{F \not\ni x} p(x,F),$$

which is in general a stronger assumption than $\omega < 0$, although the two conditions coincide in the translation invariant case.

As an example of the application of corollary 3.2.2, consider Harris' contact processes ([13]; see also [11]) in which $S = Z^d$, $c(x,\eta) = \lambda$ if $\eta(x) = 1$ and $c(x,\eta) = \Sigma_{|y-x|=1}\, \eta(y)$ if $\eta(x) = 0$. This speed function can be put in the form of a proximity process by letting $c(x) = 2d + \lambda$, $p(x,\emptyset) = \lambda/(2d+\lambda)$, and

$p(x,F) = 1/(2d+\lambda)$ for $F = \{x,y\}$ with $|x - y| = 1$. Then corollary 3.2.2 and the remark which follows it imply that the process is ergodic for $\lambda \geq 2d$. Harris has proved that the process is not ergodic for sufficiently small values of λ. A simple coupling argument shows that there is a $\lambda_0 > 0$ so that the process is ergodic for $\lambda > \lambda_0$ and is not ergodic for $\lambda < \lambda_0$. The above result of course implies that $\lambda_0 \leq 2d$.

In corollary 3.2.2, no advantage was taken of the interference in the BPI. In other words, the estimate (3.1.7) which was used in it ignored property (c) in the description of the BPI. While it is usually difficult to take advantage of the interference effectively, one can sometimes do better by applying theorem 3.2.1 directly. In the case of Harris' process above, for example, one can prove ergodicity for $\lambda > 2d - 1$ in the following way: for finite $A,B \subset S$,

$$\begin{aligned} P^{A\cup B}[A_t \neq \emptyset] &= P^1[\eta_t \not\equiv 0 \text{ on } A \cup B] \\ &\leq P^1[\eta_t \not\equiv 0 \text{ on } A] + P^1[\eta_t \not\equiv 0 \text{ on } B] - P^1[\eta_t \not\equiv 0 \text{ on } A \cap B] \\ &= P^A[A_t \neq \emptyset] + P^B[A_t \neq \emptyset] - P^{A\cap B}[A_t \neq \emptyset] . \end{aligned}$$

Here 1 represents the configuration $\eta \equiv 1$. Let $\sigma(A) = \lim_{t\to\infty} P^A[A_t \neq \emptyset]$, which is a harmonic function for A_t. Then

$$\sigma(A \cup B) + \sigma(A \cap B) \leq \sigma(A) + \sigma(B) , \quad \text{and} \tag{3.2.3}$$

$$\sigma(A) = \frac{1}{|A|} \sum_{x\in A} \left[\frac{\lambda}{\lambda + 2d} \sigma(A\backslash x) + \frac{1}{\lambda + 2d} \sum_{|y-x|=1} \sigma(A \cup y)\right] . \tag{3.2.4}$$

In order to apply theorem 3.2.1, we need to show that $\sigma(A) = 0$ for all finite A. By (3.2.3), it suffices to prove this when A is a singleton. Since A_t is invariant under translations and rotations, $\sigma_1 = \sigma(\{x\})$ and $\sigma_2 = \sigma(\{x,y\})$ for $|x - y| = 1$ are independent of x and y. Applying (3.2.4) to a singleton A, we obtain $\sigma_1 = [2d/(\lambda+2d)]\sigma_2$. Applying it to a doubleton A containing two nearest neighbors and using (3.2.3) yields

$$\sigma_2 \leq \frac{\lambda}{\lambda + 2d} \sigma_1 + \frac{1}{\lambda + 2d} \sigma_2 + \frac{(2d - 1)}{\lambda + 2d} (2\sigma_2 - \sigma_1) .$$

Simplification gives $(\lambda - 2d + 1)\lambda\sigma_1 \leq 0$, so $\sigma_1 = 0$ if $\lambda > 2d - 1$.

For further improvements of these results, and other applications of this duality, see [11], [13], and [24]. For other general forms of duality and their applications, see [14] and [27]. Of course, any time one proves an ergodic theorem for a proximity process, one can automatically conclude that more general spin-flip processes are ergodic via the coupling of section 2.3.

Section 3.3. The voter model. One of the unpleasant properties of the applications of duality we have seen so far is that in no case do the results appear to be even close to best possible. There is one class of processes, however, where the exploitation of the duality theorem gives complete results. We will refer to this class as the voter model, and will discuss the results briefly, since very similar techniques will be used to analyze the symmetric simple exclusion process in chapter 2 of part II. The details in the case of the voter model can be found in [24].

The voter model is the proximity process in which $p(x,F) = 0$ whenever $|F| \neq 1$. We will write $p(x,y) = p(x,\{y\})$, and will assume that $\sup_x c(x) < \infty$ in order to guarantee (3.1.1). The interpretation of the process is that individuals (voters) are located at the points of S, and that at each time, each individual holds one of two possible positions (denoted by 0 and 1 respectively) on some issue. At exponential times with rate $c(x)$, the individual at x reassesses his position. He does this by choosing a y according to the probabilities $p(x,y)$, and then changes his position to that of the individual at y. Let ν_0 and ν_1 be the pointmasses on $\eta \equiv 0$ and $\eta \equiv 1$ respectively. Then it is clear that $\nu_0, \nu_1 \in \mathcal{I}$, and the main problem is to determine when $\mathcal{I}_e = \{\nu_0, \nu_1\}$. In other words, we want to determine when all the equilibria represent total consensus.

Suppose $p_t(x,y)$ is the transition function for the Markov chain on S which has Q matrix $c(x)p(x,y)$ for $x \neq y$. The BPI A_t which corresponds to the voter model can then be thought of in the following way: The points in A_t move independently according to the Markov chain on S which has transition probabilities $p_t(x,y)$ until the first time that a point moves to an occupied site. At that time, the two particles coalesce, and the system proceeds as before except that its size has been reduced by one. The important property of A_t which leads to a complete analysis of the voter model is that $|A_t|$ never increases. We will assume throughout that $p_t(x,y) > 0$ for all $t > 0$, and $x,y \in S$.

Let $Y_1(t)$ and $Y_2(t)$ be independent copies of the Markov chain with transition probabilities $p_t(x,y)$. The structure of $\mathcal{I}_e$ depends on whether or not $Y_1(t)$ and $Y_2(t)$ hit in finite time with probability one, since that determines whether or not A_t is eventually of cardinality one.

Theorem 3.3.1. Suppose

$$(3.3.2) \qquad P(Y_1(t) = Y_2(t) \text{ for some } t > 0) = 1$$

for all initial states $Y_1(0)$ and $Y_2(0)$. Then $\mathcal{I}_e = \{\nu_0, \nu_1\}$. Furthermore, if μ is a probability measure on X such that $\mu\{\eta : \eta(x) = 1\} = \lambda$ for all $x \in S$, then $\mu S(t) \to \lambda\nu_1 + (1 - \lambda)\nu_0$ as $t \to \infty$.

Proof. Suppose $\mu \in \mathcal{I}$. Then $\mu\{\eta : \eta(x) = 1\}$ is a (bounded) harmonic function for $Y_1(t)$ by corollary 3.1.9. A simple coupling argument using (3.3.2) shows that $Y_1(t)$ has no nonconstant bounded harmonic functions, so $\mu\{\eta : \eta(x) = 1\} = \lambda$ is independent of x. From (3.3.2), it also follows that $P^A[|A_t| = 1] \to 1$ as $t \to \infty$ for all $A \neq \emptyset$. Corollary 3.1.9 gives

$$\mu\{\eta : \eta(x) = 1 \text{ on } A\} = \lambda P^A[|A_t| = 1] + \sum_{|B| \neq 1} P^A[A_t = B]\, \mu\{\eta : \eta(x) = 1 \text{ on } B\}$$

for $A \neq \emptyset$, so it follows that $\mu\{\eta : \eta(x) = 1 \text{ on } A\} = \lambda$. Therefore $\mu = \lambda\nu_1 + (1 - \lambda)\nu_0$. This proves the first part of the theorem. The proof of the second part is similar.

A similar, but more involved, analysis of corollary 3.1.9 yields the following result.

Theorem 3.3.3. Suppose

$$(3.3.4) \qquad P(Y_1(t) = Y_2(t) \text{ for some } t > 0) < 1$$

for some (and hence all) initial states $Y_1(0) \neq Y_2(0)$, and that $p_t(x,y)$ has no nonconstant bounded harmonic functions. Then

(a) $\mu_\rho = \lim_{t\to\infty} \nu_\rho S(t)$ exists for each $\rho \in [0,1]$, where ν_ρ is the product measure on X with $\nu_\rho\{\eta : \eta(x) = 1\} = \rho$ for all $x \in S$.

(b) $\mathcal{I}_e = \{\mu_\rho,\ 0 \leq \rho \leq 1\}$.

(c) In the translation invariant case, μ_ρ is translation invariant and ergodic for each $\rho \in [0,1]$.

(d) In the translation invariant case, if μ is any translation invariant and ergodic probability measure on X, then $\mu S(t) \to \mu_\rho$, where $\rho = \mu\{\eta : \eta(x) = 1\}$.

In the translation invariant case, if $p(x,y)$ satisfies appropriate moment conditions, theorem 3.3.1 applies when $d = 1$ or 2 and theorem 3.3.3 applies when $d \geq 3$. Thus in one and two dimensions, the system approaches a consensus, while in higher dimensions, there are many dynamic equilibria which do not represent consensus.

In the above theorems, we have given only sufficient conditions for convergence to an element of $\mathcal{I}_e$. Necessary and sufficient conditions are given in [24] for both cases. In [40], Matloff has used similar techniques applied to a different dual process to analyze the antivoter model, in which individuals adopt the opposite position from their neighbor's position, instead of imitating the neighbor. Another approach to the antivoter model is in [27].

Chapter 4. Other results. When it is impossible to determine $\mathcal{I}$ completely, it is sometimes possible to identify certain natural subsets of $\mathcal{I}$. In this chapter, we study briefly two results of this type.

Section 4.1. Reversible measures. Suppose Ω is the generator of a Markov semi-group of operators on $C(X)$. A probability measure μ on X is said to be reversible for the corresponding process η_t if

$$(4.1.1) \qquad \int f\Omega g \, d\mu = \int g\Omega f \, d\mu$$

for all $f,g \in \mathcal{D}(\Omega)$. In verifying that μ is reversible, it of course suffices to verify (4.1.1) for f and g in a core for Ω. We will denote by $\mathcal{R}$ the set of all reversible measures. By taking $g \equiv 1$ in (4.1.1), it is seen that $\mathcal{R} \subset \mathcal{I}$. While $\mathcal{I}$ is always nonempty for a Feller process on a compact set, $\mathcal{R}$ is frequently empty. In this section, we will see that for speed functions of a certain form, $\mathcal{R} \neq \emptyset$ for the corresponding spin-flip process, and in fact, $\mathcal{R}$ can be determined completely in that case. These results have been proved under varying assumptions in [6], [17], [39], [41] and [46].

Consider first the case of finite S, and suppose the speed function is of the form

$$(4.1.2) \qquad c(x,\eta) = \exp\left[\sum_{A \ni x} \Phi(A)\chi_A(\eta)\right]$$

where $\chi_A(\eta) = \Pi_{y\in A} [2\eta(y) - 1]$ and $\Phi(A)$ is a real valued function on the power set of S. Then η_t is an irreducible finite state Markov chain, and a simple computation shows that the probability measure

$$\nu\{\eta\} = K \exp\left[- \sum_{A} \Phi(A)\chi_A(\eta)\right],$$

where K is a normalizing constant, is reversible for η_t. It is of course the unique invariant measure for η_t in this case.

Now take S to be countable, and let Φ be a real valued function on the finite subsets of S which satisfies

$$\sup_x \sum_{A \ni x} |A|\, |\Phi(A)| < \infty .$$

Then the function $c(x,\eta)$ defined as in (4.1.2) satisfies assumption (1.2.3), so we may consider the process η_t whose generator is the closure of the operator defined in (1.2.1).

We define the class $\mathcal{G}$ of Gibbs measures corresponding to Φ in the following way. If T is a finite subset of S and $\zeta \in \{0,1\}^{S\setminus T}$, then $\nu_{T,\zeta}$ is the probability measure on $\{0,1\}^T$ given by

$$\nu_{T,\zeta}\{\eta\} = K(T,\zeta) \exp\left[- \sum_{A\cap T \neq \emptyset} \Phi(A)\chi_A(\eta\zeta)\right],$$

where $K(T,\zeta)$ is again a normalizing constant and $\eta\zeta$ is the configuration in $\{0,1\}^S$ which agrees with η on T and with ζ on $S\setminus T$. This measure can also be thought of as a probability measure on $X = \{0,1\}^S$ by setting

$$\nu_{T,\zeta}\{\eta \in X : \eta(x) = \zeta(x) \text{ for } x \in S\setminus T\} = 1 .$$

Note that $\nu_{T,\zeta}$ is reversible for the process on $\{0,1\}^T$ with speed function $c_{T,\zeta}(x,\eta) = c(x,\eta\zeta)$. Let $\mathcal{G}_T$ be the closed convex hull of $\{\nu_{T,\zeta} : \zeta \in \{0,1\}^{S\setminus T}\}$. Then $\mathcal{G}$ is defined as the set of all possible weak limits of μ_n as $n \to \infty$ where $\mu_n \in \mathcal{G}_{T_n}$ and $T_n \uparrow S$. $\mathcal{G}$ is nonempty since X is compact. Note that if $\Phi(A) = 0$ for $|A| > 1$, then $\mathcal{G}$ is a singleton product measure, as would be expected.

<u>Theorem 4.1.3</u>. $\mathcal{R} = \mathcal{G}$.

<u>Proof</u>. To show that $\mathcal{G} \subset \mathcal{R}$, fix $\mu \in \mathcal{G}$ and $f,g \in \mathcal{F}$. Let $\Omega_{T,\zeta}$ be the generator for the process on $\{0,1\}^T$ with speed function $c_{T,\zeta}(x,\eta)$ for T a finite subset of S and $\zeta \in \{0,1\}^{S\setminus T}$. Then if T contains the coordinates on which f and g depend, we have

$$\int_{\{0,1\}^T} f\Omega_{T,\zeta} g \, d\nu_{T,\zeta} = \int_{\{0,1\}^T} g\Omega_{T,\zeta} f \, d\nu_{T,\zeta} .$$

Since $\mu \in \mathcal{G}$, there are $T_n \uparrow S$ and $\mu_n \in \mathcal{G}_{T_n}$ so that $\mu_n \to \mu$. Using

$$\lim_{n\to\infty} \sup_{\zeta \in \{0,1\}^{S\setminus T_n}} \|\Omega_{T_n,\zeta} h - \Omega h\| = 0$$

for $h = f$ and g, we obtain $\int_X f\Omega g \, d\mu = \int_X g\Omega f \, d\mu$. For the reverse containment, suppose $\mu \in \mathcal{R}$. For a finite subset T of S, and $x \in T$, put $f(\eta) = \Pi_{y\in T} \eta(y)$

and $g(\eta) = f(\eta_x)$. Then

$$g(\eta)\Omega f(\eta) = f(\eta_x) \sum_{y\in T} c(y,\eta)[f(\eta_y) - f(\eta)] = c(x,\eta)f(\eta_x) , \text{ and}$$

$$f(\eta)\Omega g(\eta) = f(\eta) \sum_{y\in T} c(y,\eta)[g(\eta_y) - g(\eta)] = c(x,\eta)f(\eta) .$$

Since $\mu \in \mathfrak{R}$,

$$(4.1.4) \qquad \int_X c(x,\eta)[f(\eta_x) - f(\eta)]d\mu = 0 .$$

Using linearity and the fact that $\mathfrak{F}$ is dense in $C(X)$, we conclude that (4.1.4) holds for all $f \in C(X)$. Suppose now that f depends only on the coordinates in T and that g is a continuous function on X which depends only on the coordinates in $S\setminus T$. Then we may rewrite (4.1.4) as

$$(4.1.5) \qquad \int_X c(x,\eta)[f(\eta_x) - f(\eta)]g(\eta)d\mu = 0 .$$

The conditional measure $\mu(d\eta|\zeta)$ on $\{0,1\}^T$ exists for a.e. $\zeta \in \{0,1\}^{S\setminus T}$ with respect to μ, and we may conclude from (4.1.5) that

$$\int_{\{0,1\}^T} c(x,\eta\zeta)[f(\eta_x) - f(\eta)]\mu(d\eta|\zeta) = 0$$

for all f which depend on the coordinates in T and for a.e. ζ. Therefore $\mu(d\eta|\zeta)$ is invariant for the process with generator $\Omega_{T,\zeta}$, and hence $\mu(d\eta|\zeta) = \nu_{T,\zeta}(d\eta)$ for a.e. ζ. It follows that μ is in $\mathcal{G}_T$ when regarded as a measure on $\{0,1\}^T$, and therefore that $\mu \in \mathcal{G}$.

We conclude this section with several remarks. First, that $\mu \in \mathfrak{R}$ implies (4.1.4), while $\mu \in \mathcal{I}$ implies only $\Sigma_x \int_X c(x,\eta)[f(\eta_x) - f(\eta)]d\mu = 0$, is a good indication of why $\mathfrak{R}$ is much easier to treat than is $\mathcal{I}$. Secondly, even though we considered only speed functions of the form (4.1.2), there are others which are covered by this result. In order to see this, it suffices to note that if $c_1(x,\eta)$ and $c_2(x,\eta)$ are positive and are such that $c_1(x,\eta)/c_2(x,\eta)$ is independent of $\eta(x)$ for each $x \in S$, then the generators with these speed functions have the same class of reversible measures. Finally, if $c(x,\eta)$ is of the form (4.1.2) and is attractive, then the measures ν^0 and ν^1 of theorem 2.2.5 are both in $\mathcal{G}$, so in this case, $\mathcal{I}$ is a singleton if and only if $\mathcal{G}$ is a singleton.

One interesting open problem is to determine conditions under which $\mathcal{I} = \mathcal{G}$ in the context of this section. As mentioned above, this is known in the attractive case when $\mathcal{G}$ is a singleton. Holley has proved that $\mathcal{I} = \mathcal{G}$ in the one-dimensional, translation invariant, finite range case, using the free energy technique which will be mentioned in the next section.

Section 4.2. Translation invariant measures. In this section, we mention briefly some cases in which the set of invariant measures which are also translation invariant can be identified. Let $S = Z^d$, and let $\mathcal{S}$ be the class of translation invariant measures on X.

In section 9 of [14], Harris has used duality arguments to give sufficient conditions for a class of attractive processes to satisfy $|(\mathcal{I} \cap \mathcal{S})_e| \leq 2$. Under similar assumptions, Holley and Stroock [27] have shown that it is possible to have $|\mathcal{I}_e| > 2$.

Now suppose $c(x,\eta)$ is given by (4.1.2) where Φ is translation invariant and satisfies

$$(4.2.1) \qquad \sum_{A \ni 0} |A|\ |\Phi(A)| < \infty .$$

The following theorem was proved by Holley [19] in case Φ has finite range (i.e., for some finite $T \subset S$, $\Phi(A) > 0$ and $A \ni 0$ implies $A \subset T$) and was extended to Φ's satisfying (4.2.1) by Higuchi and Shiga [17]. The technique of proof exploits a functional $\varphi(\mu)$, called the free energy, which is defined on the set of probability measures μ on X. It is proved that $\varphi[\mu S(t)]$ is monotone in t for each μ, and is strictly monotone if $\mu \in \mathcal{S} \setminus \mathcal{G}$.

Theorem 4.2.2. (a) $\mathcal{I} \cap \mathcal{S} = \mathcal{G} \cap \mathcal{S}$,

(b) If $\mu \in \mathcal{S}$, $t_n \to \infty$, and $\mu S(t_n) \to \nu$, then $\nu \in \mathcal{G}$.

In the attractive case, this result can be combined with theorem 2.2.4 to conclude that if $\mathcal{G} \cap \mathcal{S}$ is a singleton, then $\mathcal{I}$ is a singleton (and hence, of course, $\mathcal{G}$ is a singleton).

Section 4.3. The one dimensional case. One of the most important open problems in the subject is to prove the ergodicity of all spin-flip processes in the following class: $S = Z^1$, $c(x,\eta)$ is translation invariant, $c(x,\eta) \in \mathcal{F}$, and $c(x,\eta) > 0$. The techniques we have discussed give partial results in this direction, of course, but they seem to be far from sufficient to prove the complete conjecture. This conjecture is suggested by the fact that the statistical mechanical analogue is correct [43].

There are several special cases of this conjecture which are of interest, and which may be more tractable than the general problem. One class of processes for which the result is not known, for example, is that in which $c(x,\eta)$ is attractive. In considering this case, one should presumably take advantage of the results and techniques of section 2.2. Another class of interest is that in which $c(x,\eta)$ is independent of $\eta(x)$. Here, of course, the product measure with density 1/2 is invariant.

A third special case is that in which $c(x,\eta)$ depends on η only through $\{\eta(y),\ y \geq x\}$. Holley and Stroock [27] initiated the study of this one sided process using duality techniques. A particular case of the one sided process which illustrates some of the difficulties involved is that in which $c(x,\eta) = \alpha \geq 1$ if $\eta(x) = \eta(x+1) = 1$ and $c(x,\eta) = 1$ otherwise. When theorem 1.2.5 is applied here, one obtains ergodicity for $\alpha < 3$. The results of Holley and Stroock yield ergodicity for $\alpha < 4$. Furthermore, it can be shown that ergodicity holds for $\alpha = \infty$ in the following sense: there is a uniquely determined probability measure μ such that whenever $\alpha_n \to \infty$ and μ_n is invariant for the process with $\alpha = \alpha_n$, then μ_n converges to μ as $n \to \infty$.

In the intersection of the second and third cases above, it is not hard to prove ergodicity directly, even when $c(x,\eta)$ is not translation invariant and does not have finite range.

Theorem 4.3.1. Suppose $S = Z^1$, $c(x,\eta)$ depends on η only through $\{\eta(y),\ y > x\}$, $c(x,\eta) > 0$, and $c(x,\eta)$ satisfies the conditions for existence and uniqueness of the process from chapter 1. Then the process is ergodic with invariant measure ν, the product measure with $\nu\{\eta : \eta(x) = 1\} = 1/2$.

Proof. Since $c(x,\eta)$ depends on the coordinates of η which are strictly to the right of x,

$$P^{\eta}\Big[\eta_t(y) = \zeta(y) \quad \text{for} \quad x \leq y \leq x+n\Big] = P^{\eta_x}\Big[\eta_t(y) = \zeta_x(y) \quad \text{for} \quad x \leq y \leq x+n\Big]$$

for all choices of $x \in S$, $\eta,\zeta \in X$, and $n \geq 0$. A simple coupling argument, together with the positivity of $c(x,\eta)$ yields

$$P^{\eta}\Big[\eta_t(y) = \zeta(y) \quad \text{for} \quad x \leq y \leq x+n\Big] - P^{\eta_x}\Big[\eta_t(y) = \zeta(y) \quad \text{for} \quad x \leq y \leq x+n\Big] \to 0$$

as $t \to \infty$ for all choices of x, η, ζ, and $n \geq 0$. Combining these two observations leads to

$$P^{\eta}\Big[\eta_t(y) = \zeta(y) \quad \text{for} \quad x \leq y \leq x+n\Big] - P^{\eta}\Big[\eta_t(y) = \zeta_x(y) \quad \text{for} \quad x \leq y \leq x+n\Big] \to 0$$

as $t \to \infty$. For $n = 0$, this gives $P^{\eta}[\eta_t(x) = 1] \to 1/2$. An induction argument then yields the convergence of $P^{\eta}[\eta_t(y) = \zeta(y)$ for $x \leq y \leq x+n]$ to $(1/2)^{n+1}$, which is the desired result. It is easy to see from the coupling argument that the convergence actually occurs at an exponential rate.

PART II. EXCLUSION PROCESSES

Chapter 1. Existence results and identification of simple invariant measures.

Section 1.1. Existence results.

Again we take S to be a countable set and put $X = \{0,1\}^S$. For $\eta \in X$ and $u \neq v \in S$, define $\eta_{uv} \in X$ by

$$\eta_{uv}(x) = \begin{cases} \eta(x) & \text{if } x \neq u,v \\ \eta(u) & \text{if } x = v \\ \eta(v) & \text{if } x = u \end{cases}$$

so that η_{uv} is obtained from η by interchanging the u and v coordinates. Consider a nonnegative continuous function $c(x,y,\eta)$ on $S \times S \times X$ which satisfies $c(x,x,\eta) \equiv 0$, $c(x,y,\eta) = c(y,x,\eta)$, and $c(x,y,\eta) \leq c(x,y)$ for some symmetric function $c(x,y)$ on $S \times S$ which satisfies

$$\sup_x \sum_y c(x,y) < \infty. \tag{1.1.1}$$

For $f \in \mathfrak{F}$, define Ωf by

$$\Omega f(\eta) = \frac{1}{2} \sum_{x,y} c(x,y,\eta)[f(\eta_{xy}) - f(\eta)] . \tag{1.1.2}$$

By assumption (1.1.1), this series converges uniformly and absolutely, and so defines a continuous function. Note that only the values of $c(x,y,\eta)$ for η such that $\eta(x) \neq \eta(y)$ play any role in this expression. The following existence and uniqueness theorem is proved by the same techniques as those used in section 1.2 of part I. The details may be found in [32].

Theorem 1.1.3. In addition to the above conditions, assume that there exists a constant L so that

$$\sum_u \sup_\eta |c(x,y,\eta_u) - c(x,y,\eta)| \leq L\, c(x,y) .$$

Then the closure of Ω generates a unique Markov semigroup $S(t)$ on $C(X)$.

As in the case of spin-flip processes, some conditions are required in order to get a unique process with reasonable properties, although the above conditions are presumably not the best possible. One of the difficulties which occurs here, but does not occur in the spin-flip context, can be seen in the following example. Let $c(x,y,\eta) = \eta(x)q(x,y) + \eta(y)q(y,x)$, where $q(x,y)$ are the transition probabilities for a discrete time Markov chain. If $\sum_x q(x,y) = \infty$ for some $y \in S$, and the process begins in configuration η where $\eta(u) = 1$ for $u \neq y$ and

$\eta(y) = 0$, there is no reasonable way to define the process for small positive times. This example helps explain why a condition like (1.1.1) is needed.

As mentioned in the introduction, two particular forms of the function $c(x,y,\eta)$ have been proposed to model different physical phenomena. In both cases, $c(x,y,\eta)$ is given in terms of nonnegative functions $q(x,y)$ on $S \times S$ with $q(x,x) = 0$ and $c(x,\eta)$ on $S \times X$. In Spitzer's model of a lattice gas [45], $c(x,y,\eta)$ has the form

$$(1.1.4) \qquad c(x,y,\eta) = \eta(x)c(x,\eta)q(x,y) + \eta(y)c(y,\eta)q(y,x) .$$

In this case, sufficient conditions for theorem 1.1.3 to apply are

$$(1.1.5) \qquad \sup_x \sum_y [q(x,y) + q(y,x)] < \infty$$

$$(1.1.6) \qquad \sup_{x,\eta} c(x,\eta) < \infty, \text{ and}$$

$$(1.1.7) \qquad \sup_x \sum_u \sup_\eta |c(x,\eta_u) - c(x,\eta)| < \infty .$$

In [1] and [2], on the other hand, a special case of the exclusion process with $c(x,y,\eta)$ of the form

$$(1.1.8) \; c(x,y,\eta) = \eta(x) \frac{c(x,\eta)}{c(x,\eta) + c(y,\eta_{xy})} q(x,y) + \eta(y) \frac{c(y,\eta)}{c(y,\eta) + c(x,\eta_{xy})} q(y,x)$$

has been considered to model the evolution of a binary alloy. In this case, sufficient conditions for theorem 1.1.3 to apply are (1.1.5), (1.1.7), and

$$(1.1.9) \qquad \inf_{x,\eta} c(x,\eta) > 0 .$$

It is interesting to note that in both cases (1.1.4) and (1.1.8),

$$(1.1.10) \quad \frac{c(x,y,\eta_{xy})}{c(x,y,\eta)} = \eta(y) \frac{c(x,\eta_{xy})q(x,y)}{c(y,\eta)q(y,x)} + \eta(x) \frac{c(y,\eta_{xy})q(y,x)}{c(x,\eta)q(x,y)}$$

for $\eta(x) \neq \eta(y)$. It is this property, in fact, that leads both processes to have the same reversible measures.

The simple exclusion process is obtained by taking $c(x,y,\eta)$ to depend on only through $\eta(x)$ and $\eta(y)$, so that

$$(1.1.11) \qquad c(x,y,\eta) = \eta(x)q(x,y) + \eta(y)q(y,x) .$$

This is of course the simplest special case of both (1.1.4) and (1.1.8). It is not itself of great physical interest, since it corresponds to the situation in which the temperature is infinite in either the lattice gas or binary alloy context. However it is of considerable mathematical interest, and it provides some insight into the more difficult physically relevant processes. A sufficient condition for theorem 1.1.3 to apply is given by (1.1.5).

Throughout the remainder of part II, we will always assume that the appropriate conditions which guarantee the existence and uniqueness of the process via theorem 1.1.3 are satisfied.

An interesting open problem is to prove existence and uniqueness of the process in either the lattice gas or binary alloy context in the translation invariant case under the assumptions that $c(x,\eta)$ is positive and continuous and q satisfies $\Sigma_y q(0,y) < \infty$. This problem is of course closely related to that mentioned at the end of section 1.4 of part I.

Section 1.2. Invariant measures for the simple exclusion process.

One of the pleasant characteristics of the simple exclusion process is that there are often product measures on X which are invariant for the process. Given a function $\alpha(\cdot)$ on S such that $0 \leq \alpha(x) \leq 1$ for all x, let ν_α be the product measure on X with probabilities given by

$$\nu_\alpha\{\eta : \eta(x) = 1 \text{ for all } x \in T\} = \prod_{x\in T} \alpha(x)$$

for all $T \subset S$. In order to determine those simple exclusion processes for which a given ν_α is invariant, we need the following computation, which will be used again in the proof of the duality theorem in chapter 2. Put $\chi_A(\eta) = \Pi_{u\in A}\, \eta(u)$ for finite subsets A of S.

Lemma 1.2.1. If A is a finite subset of S, then

$$\Omega\chi_A(\eta) = \sum_{\substack{x\in A\\ y\notin A}} \{q(y,x)[1 - \eta(x)]\chi_{(A\cup y)\setminus x}(\eta) - q(x,y)[1 - \eta(y)]\chi_A(\eta)\} .$$

Proof. By (1.1.2) and (1.1.11),

$$\Omega\chi_A(\eta) = \frac{1}{2}\sum_{x,y} [\eta(x)q(x,y) + \eta(y)q(y,x)][\chi_A(\eta_{xy}) - \chi_A(\eta)]$$

$$= \sum_{\substack{x\in A\\ y\notin A}} [\eta(x)q(x,y) + \eta(y)q(y,x)][\eta(y) - \eta(x)]\chi_{A\setminus x}(\eta)$$

$$= \sum_{\substack{x\in A\\ y\notin A}} \{\eta(y)[1 - \eta(x)]q(y,x) - \eta(x)[1 - \eta(y)]q(x,y)\}\chi_{A\setminus x}(\eta)$$

$$= \sum_{\substack{x\in A\\ y\notin A}} \{q(y,x)[1-\eta(x)]\chi_{(A\cup y)\setminus x}(\eta) - q(x,y)[1-\eta(y)]\chi_A(\eta)\} .$$

Theorem 1.2.2. Suppose $0 < \alpha(x) < 1$ for all $x \in S$. Then $\nu_\alpha \in \mathcal{J}$ if and only if

$$(1.2.3) \qquad \frac{\alpha(y)}{1-\alpha(y)} q(y,x) = \frac{\alpha(x)}{1-\alpha(x)} q(x,y) \quad \text{whenever } \alpha(x) \neq \alpha(y)$$

and

$$\sum_{y:\alpha(y)=\alpha(x)} q(y,x) = \sum_{y:\alpha(y)=\alpha(x)} q(x,y) \quad \text{for all } x \in S .$$

Proof. By lemma 1.2.1, $\nu_\alpha \in \mathcal{J}$ if and only if for all finite $A \subset S$,

$$(1.2.4) \qquad \sum_{\substack{x\in A\\ y\notin A}} q(y,x)\alpha(y)\frac{1-\alpha(x)}{\alpha(x)} = \sum_{\substack{x\in A\\ y\notin A}} q(x,y)[1-\alpha(y)] .$$

For $A = \{u\}$, (1.2.4) becomes

$$(1.2.5) \qquad [1-\alpha(u)]\sum_y q(y,u)\alpha(y) = \alpha(u)\sum_y q(u,y)[1-\alpha(y)] .$$

Rewriting (1.2.4) for general A yields

$$\sum_{x\in A} \frac{1-\alpha(x)}{\alpha(x)} \left\{\sum_y q(y,x)\alpha(y) - \sum_{y\in A} q(y,x)\alpha(y)\right\}$$

$$= \sum_{x\in A}\left\{\sum_y q(x,y)[1-\alpha(y)] - \sum_{y\in A} q(x,y)[1-\alpha(y)]\right\} ,$$

so using (1.2.5), we have

$$\sum_{x,y\in A}\left\{\frac{\alpha(y)}{1-\alpha(y)} q(y,x) - \frac{\alpha(x)}{1-\alpha(x)} q(x,y)\right\} \frac{[1-\alpha(x)][1-\alpha(y)]}{\alpha(x)} = 0 .$$

Interchanging the roles of x and y, and then adding yields

$$\sum_{x,y\in A}\left\{\frac{\alpha(y)}{1-\alpha(y)} q(y,x) - \frac{\alpha(x)}{1-\alpha(x)} q(x,y)\right\}[\alpha(y)-\alpha(x)]\frac{[1-\alpha(x)][1-\alpha(y)]}{\alpha(x)\alpha(y)} = 0.$$

The above sum is zero for all finite A if and only if the summands are zero for all $x,y \in S$. Therefore $\nu_\alpha \in \mathcal{J}$ if and only if (1.2.3) and (1.2.5) hold. To complete the proof let $\{\alpha_n\}$ be the distinct values of $\{\alpha(x)\}$, and let $C_n = \{x \in S : \alpha(x) = \alpha_n\}$. If $u \in C_m$, (1.2.5) can be written as

$$(1-\alpha_m)\sum_n \alpha_n \sum_{y\in C_n} q(y,u) = \alpha_m \sum_n (1-\alpha_n)\sum_{y\in C_n} q(u,y) .$$

Using (1.2.3), the right hand side becomes

$$\alpha_m(1-\alpha_m)\sum_{y\in C_m} q(u,y) + (1-\alpha_m)\sum_{n\neq m}\alpha_n \sum_{y\in C_n} q(y,u) .$$

Thus, in the presence of (1.2.3), (1.2.5) becomes

$$\sum_{y\in C_m} q(y,u) = \sum_{y\in C_m} q(u,y)$$

for $u \in C_m$.

This result gives two cases in which there is a simple one parameter family of product measures in $\mathcal{I}$.

Corollary 1.2.6. (a) If $\Sigma_y q(x,y) = \Sigma_y q(y,x)$ for all $x \in S$, then $\{\nu_\rho, 0 \leq \rho \leq 1\} \subset \mathcal{I}$, where $\nu_\rho = \nu_\alpha$ for $\alpha(x) \equiv \rho$.

(b) If there is a positive function $\pi(x)$ on S such that $\pi(x)q(x,y) = \pi(y)q(y,x)$, then $\{\tilde{\nu}_\rho, 0 \leq \rho \leq \infty\} \subset \mathcal{I}$, where $\tilde{\nu}_\rho = \nu_\alpha$ for $\alpha(x) = \rho\pi(x)/(1+\rho\pi(x))$.

It is easy to check that in case (b), the measures $\tilde{\nu}_\rho$ are actually reversible for the process, while in case (a), the measures ν_ρ are reversible if and only if q is symmetric. When q is symmetric, which is the case to be studied in chapter 2, both (a) and (b) of the corollary hold, though the two classes of invariant measures it produces are identical. Another important case to which (a) applies is the one in which $S = Z^d$ and $q(x,y)$ is translation invariant. On the other hand, (b) applies for example when $S = Z^1$ and $q(x,y) = 0$ for $|x-y| \geq 2$. One case in which both apply and produce different classes of invariant measures is the asymmetric one dimensional simple random walk, which will be discussed in section 3.4. This gives an example in which $\mathcal{R} \neq \emptyset$ but $\mathcal{R} \neq \mathcal{I}$.

Section 1.3. Invariant measures for the exclusion process with speed change. In this section, we will take $c(x,y,\eta)$ to be of the form (1.1.4) or (1.1.8) where $q(x,y) = q(y,x)$ and $c(x,\eta)$ is given by

$$c(x,\eta) = \exp\left[2\sum_{A\ni x}\Phi(A)\chi_A(\eta)\right], \tag{1.3.1}$$

where $\chi_A(\eta) = \Pi_{u\in A}[2\eta(u)-1]$ and $\Phi(A)$ is a real valued function on the finite subsets of S which satisfies $\sup_x \Sigma_{A\ni x} |A|\,|\Phi(A)| < \infty$. This last condition guarantees that theorem 1.1.3 applies to $c(x,y,\eta)$, provided that

$$\sup_x \sum_y q(x,y) < \infty , \tag{1.3.2}$$

which we also assume.

In order to construct a class of measures which are invariant (and in fact reversible) for the process, we proceed in a manner similar to that of section 4.1 of part I. The major difference comes from the fact that there is no creation or destruction of particles in the exclusion process. The effect of this can be seen most clearly in case that S is finite. The typical spin-flip system is then an irreducible Markov chain on X, and therefore has a unique invariant measure. The typical exclusion process, on the other hand, has $|S|+1$ closed irreducible classes, corresponding to the fact that the cardinality of η_t does not change in time.

Returning to general S, define probability measures $\nu_{T,\zeta,k}$ on $\{0,1\}^T$ for a finite subset T of S, $\zeta \in \{0,1\}^{S\setminus T}$, and $0 \leq k \leq |T|$ by

$$\nu_{T,\zeta,k}\{\eta\} = K(T,\zeta,k)\exp[-\Sigma_{A\cap T\neq\emptyset}\ \Phi(A)X_A(\eta\zeta)]$$

if $\Sigma_{x\in T}\ \eta(x)=k$ and $\nu_{T,\zeta,k}\{\eta\}= 0$ otherwise, where $K(T,\zeta,k)$ is a normalizing constant and $\eta\zeta$ is the configuration on $\{0,1\}^S$ which agrees with η on T and with ζ on $S\setminus T$. This measure can also be thought of as a probability measure on X by setting $\nu_{T,\zeta,k}\{\eta\in X: \eta=\zeta \text{ on } S\setminus T\} = 1$. Note that $\nu_{T,\zeta,k}$ is reversible for the process on $\{\eta\in\{0,1\}^T: \Sigma_{x\in T}\ \eta(x) = k\}$ with speed function $c_{T,\zeta}(x,y,\eta) = c(x,y,\eta\zeta)$, since

$$c_{T,\zeta}(x,y,\eta)\nu_{T,\zeta,k}\{\eta\} = c_{T,\zeta}(x,y,\eta_{xy})\nu_{T,\zeta,k}\{\eta_{xy}\}$$

for $x,y \in T$ by (1.1.10). Let $\tilde{\mathcal{G}}_T$ be the closed convex hull of $\{\nu_{T,\zeta,k}:$ $\zeta \in \{0,1\}^{S\setminus T},\ 0 \leq k \leq |T|\}$, and let $\tilde{\mathcal{G}}$ be the set of all possible weak limits of μ_n as $n \to \infty$ where $\mu_n \in \tilde{\mathcal{G}}_{T_n}$ and $T_n \uparrow S$. $\tilde{\mathcal{G}}$ is the set of canonical Gibbs measures corresponding to Φ. For further information on $\tilde{\mathcal{G}}$, as well as a discussion of its relationship to the Gibbs measures of section 4.1 of part I, see [8], [9] and [16].

Using the same techniques as in theorem 4.1.3 of part I, one easily proves that $\tilde{\mathcal{G}} \subset \mathcal{R} \subset \mathcal{I}$. In [16] and [39], it is proved that $\mathcal{R} = \tilde{\mathcal{G}}$ under appropriate conditions. If $\Phi(A) \equiv 0$, so $c(x,\eta) \equiv 1$, it is easy to see that $\tilde{\mathcal{G}}$ is the set of all exchangeable probability measures on X. This case should be compared with part (a) of corollary 1.2.6 and the results of chapter 2 in order to get some feeling for what should be expected in the general case in which $c(x,\eta)$ is not constant.

It should be noted that while $\tilde{\mathcal{G}}$ depends on Φ, and therefore on $c(x,\eta)$, it does not depend on $q(x,y)$. In general, it should be expected that $\mathcal{I}$ will depend on $q(x,y)$ as well as on $c(x,\eta)$. As will be seen in chapter 2, this can be the case even when $c(x,\eta) \equiv 1$. It is easily seen, though, that if $\mathcal{I}_q$ denotes the set of invariant measures for the process with speed function given by (1.1.4) or (1.1.8) with $c(x,\eta)$ given by (1.3.1) then $\tilde{\mathcal{G}} = \cap\,\mathcal{I}_q$, where the intersection is over all symmetric functions q which satisfy (1.3.2).

<u>Chapter 2</u>. <u>The symmetric simple exclusion process</u>. Throughout this chapter, we consider the process η_t whose generator is the closure of the operator defined in (1.1.2) with $c(x,y,\eta) = [\eta(x) + \eta(y)]p(x,y)$ for $x \neq y$. We assume that $p(x,y)$ is symmetric, that the Markov chain on S with Q matrix $p(x,y)$ is irreducible,

and that $\sup_x \Sigma_y p(x,y) < \infty$ so that theorem 1.1.3 applies. With this condition, we may assume without loss of generality that $\Sigma_y p(x,y) = 1$ for all x. In order to see this, note that the process obtained by replacing $p(x,y)$ by

$$\overline{p}(x,y) = \begin{cases} \frac{1}{c}\, p(x,y) & \text{for } x \neq y \\ 1 - \frac{1}{c} \sum_{z \neq x} p(x,z) & \text{for } x = y \end{cases}$$

where $c = \sup_x \Sigma_z p(x,z)$ is the same as the original process except for a constant time change. By the translation invariant case, we will mean the case in which $S = Z^d$ for some $d \geq 1$ and $p(x,y) = p(0,y-x)$. The results of this chapter are based on [34], [35] and [48].

<u>Section 2.1</u>. <u>Statement of results</u>. Let $\mathcal{H}$ be defined by

$$\mathcal{H} = \left\{\alpha(\cdot) \text{ on } S : \sum_y p(x,y)\alpha(y) = \alpha(x) \text{ and } 0 \leq \alpha(x) \leq 1\right\} .$$

Then the invariant measures for η_t can be described in the following way. The measures ν_α are those defined in section 1.2.

<u>Theorem 2.1.1</u>. (a) $\mu_\alpha = \lim_{t\to\infty} \nu_\alpha S(t)$ exists for each $\alpha \in \mathcal{H}$.

(b) $\mu_\alpha\{\eta : \eta(x) = 1\} = \alpha(x)$ for all $x \in S$.

(c) The map $\alpha \to \mu_\alpha$ gives a one-to-one and onto correspondence between $\mathcal{H}$ and $\mathcal{J}_e$.

(d) $\mu_\alpha = \nu_\alpha$ if and only if α is constant.

<u>Corollary 2.1.2</u>. If $\mathcal{H}$ consists only of constants, then $\mathcal{J}_e = \{\nu_\rho, 0 \leq \rho \leq 1\}$.

In the translation invariant case, $\mathcal{H}$ consists only of constants by the Choquet-Deny theorem, so it follows that all invariant measures for η_t are exchangeable in that case. One interesting case in which $\mathcal{H}$ has many nonconstant elements and therefore $\mathcal{J}$ contains many nonexchangeable measures, is that of the simple random walk on the connected graph in which there are no loops and each vertex is connected by an edge to exactly $k \geq 3$ other vertices.

In order to state the convergence result, let

$$p_t(x,y) = e^{-t} \sum_{n=0}^{\infty} \frac{t^n}{n!} p^{(n)}(x,y) ,$$

where $p^{(n)}(x,y)$ are the n-step transition probabilities corresponding to $p(x,y)$.

<u>Theorem 2.1.3</u>. Suppose $\alpha \in \mathcal{H}$ and μ is a probability measure on X which satisfies

(2.1.4) $$\sum_y p_t(x,y)[\eta(y) - \alpha(y)] \to 0$$

in probability with respect to μ as $t \to \infty$ for each $x \in S$. Then $\mu S(t) \to \nu_\alpha$ as $t \to \infty$.

Condition (2.1.4) is in fact also necessary for $\mu S(t) \to \nu_\alpha$ in case $p_t(x,y)$ is transient [34]. It is probably necessary in general, though this has not yet been proved. In [48], Spitzer gives another condition which is necessary and sufficient in general, but which is less useful than (2.1.4).

Corollary 2.1.5. Consider the translation invariant case. If μ is translation invariant and ergodic, then $\mu S(t) \to \nu_\rho$ as $t \to \infty$, where $\rho = \mu\{\eta : \eta(x) = 1\}$.

In order to simplify matters, we will present proofs of these results under the assumption that $\mathcal{H}$ consists only of constants. The proofs in case p has nonconstant bounded harmonic functions (in which case, of course, p must be transient) can be found in [34]. The key to the proofs is Spitzer's observation in [45] that the symmetric simple exclusion process is self-dual, in a sense that will be described in section 2.2. This permits the translation of problems concerning the invariant measures for the infinite system into problems involving the bounded harmonic functions for the corresponding finite system. Once this is done the proofs follow different lines in the two cases which must be considered separately. We will say that case I occurs if two independent copies of the chain with transition probabilities $p_t(x,y)$ will hit in finite time with probability one, and that case II occurs otherwise. The proof in case I is based on a coupling argument, which enables one to prove that the finite particle system has no nonconstant bounded harmonic functions. For case II, the finite particle system is compared with the corresponding system without the exclusion interaction to obtain the same conclusion under the assumption that $\mathcal{H}$ consists only of constants. The proof in case I is due to Spitzer [48] and in case II to Liggett [34, 35]. One of the interesting features of the proofs is that the two cases require two essentially different techniques; neither technique works in the other case.

In [44], Schwartz considered the case in which the simple exclusion process is modified by letting particles be created at vacant sites and annihilated at occupied sites. A similar analysis leads to a complete identification of $\mathcal{I}$ in case the creation and annihilation rates are independent of η.

Section 2.2. The duality theorem. This section is devoted to the proof of the following result which is due to Spitzer [45], and to some of its immediate consequences. The symmetry of $p(x,y)$ is essential here.

Theorem 2.2.1. Suppose $\eta, \zeta \in X$ and $\Sigma_x \zeta(x) < \infty$. Then

$$P^{\eta}[\eta_t \geq \zeta] = P^{\zeta}[\zeta_t \leq \eta] , \tag{2.2.2}$$

where the inequalities are interpreted coordinatewise.

Proof. Let $f(\eta,\zeta) = 1_{\{\eta \geq \zeta\}}$. As a function of η, $f(\eta,\zeta)$ is in the domain of Ω for each finite configuration ζ. Therefore if we define $u_\eta(t,\zeta) = E^{\eta}f(\eta_t,\zeta) = S(t)f(\eta,\zeta)$, it follows from lemma 1.2.1 that

$$\frac{d}{dt} u_\eta(t,\zeta) = S(t)\Omega f(\eta,\zeta)$$

$$= S(t) \sum_{\substack{\zeta(x)=1 \\ \zeta(y)=0}} p(x,y)[f(\eta,\zeta_{xy}) - f(\eta,\zeta)]$$

$$= \sum_{\substack{\zeta(x)=1 \\ \zeta(y)=0}} p(x,y)[u_\eta(t,\zeta_{xy}) - u_\eta(t,\zeta)] .$$

Since ζ_t is a nonexplosive Markov chain, the unique bounded solution to this equation with initial condition $u_\eta(0,\zeta) = f(\eta,\zeta)$ is given by $E^{\zeta}f(\eta,\zeta_t)$, which gives the desired result.

By integrating both sides of (2.2.2) with respect to μ, we obtain:

Corollary 2.2.3. If μ is any probability measure on X and ζ is any finite configuration, then

$$\mu S(t)\{\eta : \eta \geq \zeta\} = \sum_{\gamma} P^{\zeta}[\zeta_t = \gamma]\, \mu\{\eta : \eta \geq \gamma\} .$$

Corollary 2.2.4. Suppose that for each $n \geq 1$, the Markov chain ζ_t restricted to $\{\gamma \in X : |\gamma| = \Sigma_x \gamma(x) = n\}$ has only constant bounded harmonic functions. Then $\mathcal{I}$ is the set of exchangeable measures on X, and therefore $\mathcal{I}_e = \{\nu_\rho, 0 \leq \rho \leq 1\}$. Furthermore, if μ is any probability measure on X, $t_n \to \infty$, and $\lim_{n\to\infty} \mu S(t_n)$ exists, then this limit is exchangeable.

The only part of this corollary which is not an immediate consequence of corollary 2.2.3 is the final statement. It is proved by applying the following result, which appears in [40], to the Markov chains corresponding to the finite particle systems.

Lemma 2.2.5. Suppose $q(x,y)$ is the Q matrix for a Markov chain $X(t)$ on S which satisfies $\sup_x \Sigma_{y\neq x} q(x,y) < \infty$. If f is a bounded function on S and $g(x) = \lim_{n\to\infty} E^x f(X(t_n))$ exists for all $x \in S$ and some sequence $t_n \to \infty$, then g is harmonic for $X(t)$.

Proof. Let $c = \sup_x \Sigma_{y\neq x} q(x,y)$, $\tilde{q}(x,y) = (1/c)q(x,y)$ for $x \neq y$, and $\tilde{q}(x,x) = 1 - (1/c)\Sigma_{y\neq x} q(x,y)$. Then $\tilde{q}(x,y)$ is a stochastic transition function, and the transition probabilities $q_t(x,y)$ for the chain $X(t)$ are given by

$$q_t(x,y) = e^{-ct} \sum_{n=0}^{\infty} \frac{(ct)^n}{n!} \tilde{q}^{(n)}(x,y) .$$

It suffices to prove that

$$\lim_{t\to\infty} \sum_y \left| \sum_x q(z,x)q_t(x,y) - q_t(z,y) \right| = 0$$

for all $z \in S$. But this expression is dominated by

$$\lim_{t\to\infty} e^{-ct} \sum_{n=1}^{\infty} \left| \frac{(ct)^{n-1}}{(n-1)!} - \frac{(ct)^n}{n!} \right| ,$$

which is zero.

While the rest of the proofs will be carried out under the assumption that $\mathcal{H}$ consists only of constants, we should note that the existence of the limit in part (a) of theorem 2.1.1 is an immediate consequence of corollary 2.2.3 for all $\alpha \in \mathcal{H}$. The reason is that, in fact, $\nu_\alpha S(t)\{\eta : \eta(x_i) = 1$ for $1 \leq i \leq n\}$ is nonincreasing in t for any choice of $n \geq 1$ and distinct $x_1,\ldots,x_n \in S$. In order to see this, it suffices to show that $Qf \leq 0$, where $f(x_1,\ldots,x_n) = \prod_{i=1}^n \alpha(x_i)$ and Q is the generator for the n-particle simple exclusion process. This is the result of the following computation, which of course uses the harmonicity of α for $p(x,y)$, as well as the symmetry of $p(x,y)$:

$$Qf(\vec{x}) = \sum_{i=1}^{n} \sum_{y\notin\{x_1,\ldots,x_n\}} p(x_i,y)[f(x_1,\ldots,x_{i-1},y,x_{i+1},\ldots,x_n) - f(\vec{x})]$$

$$= \sum_{i=1}^{n} \left[\prod_{k\neq i} \alpha(x_k)\right]\left[\alpha(x_i) - \sum_{j=1}^{n} p(x_i,x_j)\alpha(x_j) - \alpha(x_i)\left\{1 - \sum_{j=1}^{n} p(x_i,x_j)\right\}\right]$$

$$= -\sum_{i=1}^{n} \left[\prod_{k\neq i} \alpha(x_k)\right] \sum_{j=1}^{n} p(x_i,x_j)[\alpha(x_j) - \alpha(x_i)]$$

$$= -\frac{1}{2} \sum_{i,j=1}^{n} \left[\prod_{k\neq i,j} \alpha(x_k)\right] p(x_i,x_j)[\alpha(x_j) - \alpha(x_i)]^2 \leq 0 .$$

Section 2.3. The finite process. Throughout this section, we will assume that $\mathcal{H}$ consists only of constants. Let $X_n(t)$ be the Markov chain on S^n in which the individual coordinates undergo independent Markov chains on S with transition probabilities $p_t(x,y)$, and let $U_n(t)$ be the corresponding transition semigroup:

$$U_n(t)g(\vec{x}) = E^{\vec{x}}g(X_n(t)) = \sum_{\vec{y}} \prod_{i=1}^{n} p_t(x_i,y_i)g(\vec{y}) .$$

Lemma 2.3.1. Suppose g is a bounded function on S^n such that $U_n(t)g = g$ for all $t \geq 0$. Then g is constant.

Proof. Define $W_i g$ for $1 \leq i \leq n$ by

$$W_i g(\vec{x}) = \sum_u p(x_i,u)g(x_1,\ldots,x_{i-1},u,x_{i+1},\ldots,x_n) .$$

Then the operators W_i commute, and any bounded harmonic function g of $X_n(t)$ satisfies

$$g = \frac{1}{n}\sum_{i=1}^{n} W_i g . \tag{2.3.2}$$

Let K be the set of all g which satisty (2.3.2) and $|g(\vec{x})| \leq 1$ for all $\vec{x} \in S^n$. Then K is a weak* compact and convex subset of $\ell_\infty(S^n)$, so K is the closed convex hull of its extreme points by the Krein-Milman theorem. Since W_i maps K into itself for each i, it follows that $W_i g = g$ for all i if g is extreme in K. Since $\mathcal{H}$ consists only of constants, it then follows that every extreme g in K is constant, and therefore that K consists only of constants.

Put $D = \{(x,x),\ x \in S\} \subset S^2$, and define

$$f_2(\vec{x}) = P^{\vec{x}}\,[X_2(t) \in D \text{ for some } t > 0] .$$

It is an immediate consequence of the irreducibility of $p(x,y)$ that if $f_2(\vec{x}) < 1$ for some $\vec{x} \notin D$, then $f_2(\vec{x}) < 1$ for all $\vec{x} \notin D$. Of course $f_2(\vec{x}) = 1$ for $\vec{x} \in D$. We will consider two cases:

Case I: $f_2(\vec{x}) = 1$ for all $\vec{x} \in S^2$.

Case II: $f_2(\vec{x}) < 1$ for all $\vec{x} \in S^2\setminus D$.

The results of this section will be used primarily for the proofs of the main results in case II. Lemma 2.3.4 below, however, will also be used in case I. For $n \geq 2$, let

$$f_n(\vec{x}) = \sum_{1\leq i<j\leq n} f_2(x_i,x_j) .$$

Lemma 2.3.3. In case II, $\lim_{t\to\infty} U_n(t)f_n(\vec{x}) = 0$ for all $n \geq 2$ and $\vec{x} \in S^n$.

Proof. It suffices to prove the result for $n = 2$ since

$$U_n(t)f_n(\vec{x}) = \sum_{\vec{y}} \prod_{k=1}^{n} p_t(x_k,y_k) \sum_{1\leq i<j\leq n} f_2(y_i,y_j)$$

$$= \sum_{1\leq i<j\leq n} p_t(x_i,y_i)p_t(x_j,y_j)f_2(y_i,y_j)$$

$$= \sum_{1\leq i<j\leq n} U_2(t)f_2(x_i,x_j) .$$

But $U_2(t)f_2(x,y) = P^{(x,y)}[X_2(s) \in D$ for some $s > t]$, so

$$U_2(t)f_2(x,y) \to P^{(x,y)}[X_2(s) \in D \text{ for arbitrarily large } s] .$$

By lemma 2.3.1 and proposition 5.19 of [29], this last probability is either zero or one. Since case II holds, it must be zero.

We will say that a bounded symmetric function $g(x,y)$ on S^2 is positive definite if $\sum_{x,y} \beta(x)\beta(y)g(x,y) \geq 0$ whenever $\sum_x |\beta(x)| < \infty$ and $\sum_x \beta(x) = 0$. For $n > 2$, a bounded symmetric function on S^n will be called positive definite if it is a positive definite function of each pair of variables.

Let $T_n = \{\vec{x} \in S^n : x_i \neq x_j$ for all $i \neq j\}$, and let $Y_n(t)$ be the Markov chain on T_n which corresponds to the n particle interacting system. This chain is obtained from $X_n(t)$ by suppressing transitions to points in $S^n\setminus T^n$. Let $V_n(t)$ be the transition semigroup corresponding to $Y_n(t)$.

Lemma 2.3.4. Suppose g is a bounded, symmetric, positive definite function on S^n for $n \geq 2$. Then

$$V_n(t)g(\vec{x}) \leq U_n(t)g(\vec{x}) \quad \text{for } \vec{x} \in T_n .$$

Proof. For $t \geq 0$, $U_n(t)g$ is positive definite also since

$$\sum_{x_1,x_2} \beta(x_1)\beta(x_2)U_n(t)g(\vec{x}) = \sum_{j\neq 1,2} \sum_{y_j} \prod_{i\neq 1,2} p_t(x_i,y_i) \sum_{y_1,y_2} \gamma(y_1)\gamma(y_2)g(\vec{y}) ,$$

where $\gamma(y) = \Sigma_x \beta(x)p_t(x,y)$. Therefore

$$U_n(t)g(x_1,x_1,x_3,\ldots,x_n) + U_n(t)g(x_2,x_2,x_3,\ldots,x_n) - 2U_n(t)g(x_1,x_2,\ldots,x_n) \geq 0 .$$

for $\vec{x} \in T_n$, and hence $(Q_U - Q_V)U_n(t)g(\vec{x}) \geq 0$ for $\vec{x} \in T_n$, where Q_U and Q_V are the generators of $U_n(t)$ and $V_n(t)$ respectively. The desired result then follows from

$$U_n(t) - V_n(t) = \int_0^t V_n(t - s)(Q_U - Q_V)U_n(s)\, ds$$

and the fact that $V_n(t)$ maps functions which are nonnegative on T_n to functions which are nonnegative on T_n.

Lemma 2.3.5. f_n is positive definite for all $n \geq 2$.

Proof. Suppose $\Sigma_x |\beta(x)| < \infty$ and $\Sigma_x \beta(x) = 0$. Then

$$\sum_{x_1,x_2} \beta(x_1)\beta(x_2)f_n(\vec{x}) = \sum_{x_1,x_2} \beta(x_1)\beta(x_2)f_2(x_1,x_2) ,$$

so it suffices to prove the result for $n = 2$. Let $H = \{t_1,\ldots,t_n\}$ be a finite set of times with $0 < t_1 < \cdots < t_n$, and define

$$h(x,y) = P^{(x,y)}[X_2(t) \in D \text{ for some } t \in H] .$$

The decomposition according to the time and place of the last visit to D then yields

$$h(x,y) = \sum_{i=1}^{n} P^{(x,y)}[X_2(t_i) \in D,\ X_2(t_j) \notin D \text{ for all } j > i]$$

$$= \sum_{i=1}^{n} \sum_{u\in S} p_{t_i}(x,u)p_{t_i}(y,u)P^{(u,u)}[X_2(t_j - t_i) \notin D \text{ for all } j > i] .$$

Therefore

$$\sum_{x,y} \beta(x)\beta(y)h(x,y) = \sum_{i=1}^{n} \sum_{u\in S} \left|\sum_x \beta(x)p_{t_i}(x,u)\right|^2 P^{(u,u)}[X_2(t_j - t_i) \notin D \text{ for } j > i]$$

which is nonnegative. Therefore $h(x,y)$ is positive definite. Now let H_k be an increasing sequence of finite sets of times for which $\cup_k H_k$ is dense in $[0,\infty)$, and let $h_k(x,y)$ be defined as before in terms of H_k. Then $h_k \uparrow f_2$, so f_2 is positive definite also.

The previous three lemmas combine to give the following corollary.

Corollary 2.3.6. In case II, $\lim_{t\to\infty} V_n(t)f_n(\vec{x}) = 0$ for all $n \geq 2$ and $\vec{x} \in T_n$.

Section 2.4. Proofs of the main results. We continue to assume that $\mathcal{H}$ consists only of constants. In order to prove corollary 2.1.2 (which is equivalent to theorem 2.1.1 under this assumption), it suffices by corollary 2.2.4 to prove the following result.

Theorem 2.4.1. Suppose g is a bounded symmetric function on T_n for $n \geq 2$ such that $V_n(t)g = g$. Then g is constant.

Proof in case I [48]. In this case, the proof is based on a simple coupling argument. In order to prove that $g(\vec{x}) = g(\vec{y})$ for all $\vec{x},\vec{y} \in T_n$, it suffices to show this when $\vec{x}$ and $\vec{y}$ differ at only one coordinate. To prove it in that case, consider the continuous time Markov chain $(Z_1(t),\ldots,Z_{n+1}(t))$ on $A = \{\vec{z} \in S^{n+1} : z_i \neq z_j \text{ for } i \neq j \text{ and } \{i,j\} \neq \{n,n+1\}\}$ which can be described in the following way: (a) the set $\{\vec{z} \in A : z_n = z_{n+1}\}$ is closed for the process, and on this set, the process $(Z_1(t),\ldots,Z_n(t))$ has the same transition law as the chain $Y_n(t)$, and (b) on the set $\{\vec{z} \in A : z_n \neq z_{n+1}\}$, the process has the following transition intensities:

$$\vec{z} \to (z_1,\ldots,z_{i-1},u,z_{i+1},\ldots,z_n,z_{n+1}) \qquad p(z_i,u) \text{ for } u \neq z_j \ \forall j$$

$$\vec{z} \to (z_1,\ldots,z_{n-1},z_n,z_{n+1}) \qquad p(z_{n+1},z_n)$$

$$\vec{z} \to (z_1,\ldots,z_{n-1},z_{n+1},z_{n+1}) \qquad p(z_n,z_{n+1})$$

$$\vec{z} \to (z_1,\ldots,z_{i-1},z_n,z_{i+1},\ldots,z_{n-1},z_i,z_{n+1}) \qquad p(z_i,z_n) \text{ for } i < n$$

$$\vec{z} \to (z_1,\ldots,z_{i-1},z_{n+1},z_{i+1},\ldots,z_{n-1},z_n,z_i) \qquad p(z_i,z_{n+1}) \text{ for } i < n.$$

Note that $\{Z_1(t),\ldots,Z_{n-1}(t),Z_n(t)\}$ and $\{Z_1(t),\ldots,Z_{n-1}(t),Z_{n+1}(t)\}$ are both Markovian and have the same transition law as the chain $\{Y_n^{(1)}(t),\ldots,Y_n^{(n)}(t)\}$, where these chains are regarded as having as state space the collection of all subsets of S of size n. Furthermore, using the symmetry of $p(x,y)$, it can be seen that the process $(Z_n(t),Z_{n+1}(t))$ is also Markovian and has the same transition law as $X_2(t)$ until the first time that $Z_n(t) = Z_{n+1}(t)$. After that time, of course, $Z_n \equiv Z_{n+1}$. Now suppose $\vec{x},\vec{y} \in T_n$ are such that $x_i = y_i$ for $i < n$ and $x_n \neq y_n$. Then

$$g(\vec{x}) = V_n(t)g(\vec{x}) = Eg(Z_1(t),\ldots,Z_{n-1}(t),Z_n(t)), \text{ and}$$

$$g(\vec{y}) = V_n(t)g(\vec{y}) = Eg(Z_1(t),\ldots,Z_{n-1}(t),Z_{n+1}(t)),$$

where the process $(Z_1(t),\ldots,Z_{n+1}(t))$ has initial point $(x_1,\ldots,x_{n-1},x_n,y_n)$. It follows from $f_2 \equiv 1$ that $P[Z_n(t) = Z_{n+1}(t)$ for all large $t] = 1$. Therefore since g is bounded

$$Eg(Z_1(t),\ldots,Z_{n-1}(t),Z_n(t)) - Eg(Z_1(t),\ldots,Z_{n-1}(t),Z_{n+1}(t))$$

tends to zero as $t \to \infty$, and hence $g(\vec{x}) = g(\vec{y})$.

Proof in case II [34, 35]. We will assume $|g| \leq 1$ for simplicity. Then

$$|V_n(t)g(\vec{x}) - U_n(t)g(\vec{x})| \leq f_n(\vec{x})$$

for $\vec{x} \in T_n$, so

$$|g(\vec{x}) - U_n(t)g(\vec{x})| \leq f_n(\vec{x})$$

for $\vec{x} \in T_n$. By taking a pointwise limit of $U_n(t)g$ along a sequence $t_n \to \infty$, it follows from lemmas 2.2.5 and 2.3.1 that there is a constant c so that $|g(\vec{x}) - c| \leq f_n(\vec{x})$ for $\vec{x} \in T_n$. Applying $V_n(t)$ to this, we see that

$$|g(\vec{x}) - c| = |V_n(t)g(\vec{x}) - c| \leq V_n(t)f_n(\vec{x})$$

for $\vec{x} \in T_n$. Therefore $g(\vec{x}) = c$ for all $\vec{x} \in T_n$ by corollary 2.3.6.

Proof of theorem 2.1.3. Suppose μ is a probability measure on X which satisfies (2.1.4) with $\alpha(x) \equiv \rho$, a constant. The function $h(x,y) = \mu\{\eta: \eta(x)=1, \eta(y)=1\}$ is positive definite, so

$$V_2(t)h(x,y) \leq U_2(t)h(x,y) \tag{2.4.2}$$

for $x \neq y$ by lemma 2.3.4. By (2.1.4),

$$\lim_{t\to\infty} \sum_y p_t(x,y)\mu\{\eta: \eta(y) = 1\} = \rho \tag{2.4.3}$$

for each $x \in S$ and $\lim_{t\to\infty} U_2(t)h(x,y) = \rho^2$ for $x,y \in S$. Therefore by (2.4.2),

$$\overline{\lim_{t\to\infty}}\, V_2(t)h(x,y) \leq \rho^2 \tag{2.4.4}$$

for $x \neq y$. But by corollary 2.2.3, $\Sigma_y\, p_t(x,y)\mu\{\eta: \eta(y) = 1\} = \mu S(t)\{\eta: \eta(x) = 1\}$ for $x \in S$ and $V_2(t)h(x,y) = \mu S(t)\{\eta : \eta(x) = 1, \eta(y) = 1\}$ for $x \neq y$. Therefore all weak limits ν of $\mu S(t)$ as $t \to \infty$ have the following properties:

(a) $\nu\{\eta : \eta(x) = 1\} = \rho$ for $x \in S$, by (2.4.3).

(b) $\nu\{\eta : \eta(x) = 1, \eta(y) = 1\} \leq \rho^2$ for all $x \neq y$, by (2.4.4).

(c) ν is exchangeable by corollary 2.2.4 and theorem 2.4.1.

By deFinetti's theorem, every exchangeable probability measure on X can be written in the form $\int_0^1 \nu_\rho \lambda(d\rho)$ for some probability measure λ on $[0,1]$. Therefore, the only probability measure which satisfies (a), (b) and (c) above is ν_ρ.

<u>Proof of corollary 2.1.5</u>. Consider the translation invariant case, and let μ be a translation invariant and ergodic probability measure on X such that $\mu\{\eta : \eta(x) = 1\} = \rho$ for all $x \in S = Z^d$. By Bochner's theorem, there is a finite measure $\lambda(d\sigma)$ on $[0,1)^d$ so that

$$\mu\{\eta : \eta(u) = 1, \eta(v) = 1\} = \int \exp[2\pi i\langle u - v, \sigma\rangle]\lambda(d\sigma) ,$$

where $\langle\ ,\ \rangle$ is the usual inner product on R^d. Let $Z_t(x,\eta) = \Sigma_y\, p_t(x,y)\eta(y)$ and $h(\sigma) = \Sigma_x\, p(0,x) \exp[2\pi i\langle x,\sigma\rangle]$, and compute

$$\begin{aligned}\int_X Z_t(x,\eta)Z_s(x,\eta)\mu(d\eta) &= \sum_{u,v} p_t(x,u)p_s(x,v)\, \mu\{\eta : \eta(u) = \eta(v) = 1\} \\ &= \int_{[0,1)^d} \left[\sum_u p_t(x,u)e^{2\pi i\langle u,\sigma\rangle}\right]\left[\sum_v p_s(x,v)e^{-2\pi i\langle v,\sigma\rangle}\right] d\lambda \\ &= \int_{[0,1)^d} \exp[-(t + s)(1 - h(\sigma))]d\lambda .\end{aligned}$$

Now $|h(\sigma)| \leq 1$ and $h(\sigma) = 1$ if and only if $\sigma = 0$ since $p(x,y)$ is irreducible, so

$$\lim_{s,t\to\infty} \int Z_t(x,\eta)Z_s(x,\eta)\mu(d\eta) = \lambda(\{0\}) .$$

Therefore $Z_t(x,\eta)$ is Cauchy in $L_2(\mu)$, so $Z(x,\eta) = \lim_{t\to\infty} Z_t(x,\eta)$ exists in $L_2(\mu)$. By definition, $Z_{t+s}(x,\eta) = \Sigma_y\, p_s(x,y)Z_t(y,\eta)$, so $Z(x,\eta) \in \mathcal{H}$ for almost every η with respect to μ. Therefore $Z(x,\eta) = Z(0,\eta)$ for a.e. η, and hence $Z(0,\eta)$ is an invariant random variable with respect to the shifts in Z^d. Since μ is ergodic, it then follows that $Z(0,\eta)$ is constant, and therefore $Z(x,\eta) = \rho$ a.e. since $\int Z(x,\eta)d\mu = \rho$. Therefore $Z_t(x,\eta)$ converges to ρ in probability for each $x \in S$, and thus the result follows from theorem 2.1.3.

Chapter 3. The asymmetric simple exclusion process.

Throughout this chapter, we consider the process η_t whose generator is the closure of the operator defined in (1.1.2) with $c(x,y,\eta) = \eta(x)p(x,y) + \eta(y)p(y,x)$ for $x \neq y$, where $p(x,y)$ are the transition probabilities for an irreducible Markov chain on S. We assume of course that $\sup_y \Sigma_x\, p(x,y) < \infty$, so that theorem 1.1.3 applies. When $p(x,y)$ is not symmetric, theorem 2.2.1 fails, and there is virtually no hope of obtaining results in the general case which are as complete as those obtained in chapter 2 for the symmetric case. Since duality techniques appear to be inapplicable, we will discuss here some of those results which can be obtained via coupling. While the symmetric case is essentially completely understood, there remain many important open problems in the asymmetric case. The results in this chapter are based on [36], [37], and [38].

Section 3.1. The basic coupling.

By the coupled process, we will mean the Markov process $\gamma_t = (\eta_t,\zeta_t)$ with state space $X \times X$ which has the following properties: (a) the marginal processes η_t and ζ_t are Markovian and have the transition law of the simple exclusion process corresponding to $p(x,y)$, and (b) whenever $\eta_t(x) = \zeta_t(x) = 1$ for some $x \in S$, the two marginal processes will use the same random mechanisms to decide when the particle at x will attempt a transition, and where it will attempt to go. Thus the generator $\bar{\Omega}$ of γ_t takes the following form when restricted to functions $f(\eta,\zeta)$ which depend on only finitely many coordinates:

$$\bar{\Omega}f(\eta,\zeta) = \sum_{\eta(x)=1,\eta(y)=0} p(x,y)[f(\eta_{xy},\zeta)-f(\eta,\zeta)] + \sum_{\zeta(x)=1,\zeta(y)=0} p(x,y)[f(\eta,\zeta_{xy})-f(\eta,\zeta)]$$

$$+ \sum_{\substack{\eta(x)=\zeta(x)=1\\ \eta(y)=\zeta(y)=0}} p(x,y)[f(\eta_{xy},\zeta_{xy}) - f(\eta_{xy},\zeta) - f(\eta,\zeta_{xy}) + f(\eta,\zeta)] .$$

One of the important properties of the coupled process is that $\{\eta = \zeta\}$, $\{\eta \leq \zeta\}$ and $\{\eta \geq \zeta\}$ are closed for the motion. This observation leads to the following result, in which we use an upper bar to denote symbols pertaining to the coupled process (so that $\bar{\mathcal{I}}$ is the set of invariant measures for γ_t, and $\bar{\mathcal{S}}$ is the set of translation invariant probability measures on $X \times X$ in the translation invariant case).

Lemma 3.1.1. If $\nu \in \bar{\mathcal{I}}_e$, then $\nu\{\eta = \zeta\}$, $\nu\{\eta \leq \zeta\}$ and $\nu\{\eta \geq \zeta\}$ are each either zero or one. The same is true for $\nu \in (\bar{\mathcal{I}} \cap \bar{\mathcal{S}})_e$ in the translation invariant case.

The approach we will follow is to use information concerning $\bar{\mathcal{I}}$ to draw conclusions about $\mathcal{I}$. In order to do this, we need to obtain relations between $\bar{\mathcal{I}}$ and $\mathcal{I}$. These are given in the next lemma.

Lemma 3.1.2. (a) If $\nu \in \bar{\mathcal{I}}$, then the marginals μ_1 and μ_2 of ν are in $\mathcal{I}$. (b) If $\mu_1, \mu_2 \in \mathcal{I}$, then there is a $\nu \in \bar{\mathcal{I}}$ with marginals μ_1 and μ_2. (c) If $\mu_1, \mu_2 \in \mathcal{I}_e$, then ν can be taken in $\bar{\mathcal{I}}_e$.

Proof. (a) is immediate from the fact that the marginals of γ_t are Markovian with the right transition law. For (b), let $\tilde{\nu}$ be the product measure $\mu_1 \times \mu_2$ on $X \times X$. Then $\tilde{\nu}\bar{S}(t)$ has marginals μ_1 and μ_2 for all $t > 0$, so we can take ν to be any weak limit as $t \to \infty$ of $(1/t)\int_0^t \tilde{\nu}\bar{S}(s)ds$. In order to prove (c), let

$$\mathcal{C} = \{\nu \in \bar{\mathcal{I}} : \nu \text{ has marginals } \mu_1 \text{ and } \mu_2\}.$$

Then $\mathcal{C} \neq \emptyset$ by part (b), and $\mathcal{C}$ is compact and convex, so $\mathcal{C}_e \neq \emptyset$ by the Krein-Milman theorem. The result then follows from $\mathcal{C}_e \subset \bar{\mathcal{I}}_e$, which is a consequence of the assumption that μ_1 and μ_2 are extremal in $\mathcal{I}$.

Lemma 3.1.3. Consider the translation invariant case. (a) If $\nu \in \bar{\mathcal{I}} \cap \bar{\mathcal{S}}$, then the marginals μ_1 and μ_2 of ν are in $\mathcal{I} \cap \mathcal{S}$. (b) If $\mu_1, \mu_2 \in \mathcal{I} \cap \mathcal{S}$, then there is a $\nu \in \bar{\mathcal{I}} \cap \bar{\mathcal{S}}$ with marginals μ_1 and μ_2. (c) If $\mu_1, \mu_2 \in (\mathcal{I} \cap \mathcal{S})_e$, then ν can be taken in $(\bar{\mathcal{I}} \cap \bar{\mathcal{S}})_e$.

The proof of the above lemma is omitted, since it is similar to the proof of the previous one. The fact that $\{\eta \leq \zeta\}$ is closed for the motion of the coupled process leads immediately to the following result.

Lemma 3.1.4. If $\mu_1 \leq \mu_2$, then $\mu_1 S(t) \leq \mu_2 S(t)$ for all $t \geq 0$.

Section 3.2. The reversible positive recurrent case. In addition to the basic assumptions of this chapter, we will assume throughout this section that there is a positive function $\pi(x)$ on S such that $\Sigma_x \pi(x) < \infty$ and

$$\pi(x)p(x,y) = \pi(y)p(y,x) . \tag{3.2.1}$$

By part (b) of corollary 1.2.6, the product measures $\tilde{\nu}_\rho$ on X with $\tilde{\nu}_\rho\{\eta: \eta(x) = 1\} = \rho\pi(x)/(1+\rho\pi(x))$ are invariant for the process for all $\rho \in [0,\infty]$. Since $\Sigma_x \pi(x) < \infty$, $\tilde{\nu}_\rho$ concentrates on $\{\eta: \Sigma_x \eta(x) < \infty\}$ and $\tilde{\nu}_\rho\{\eta: \Sigma_x\eta(x)=n\} > 0$ for $\rho \in (0,\infty)$ and $0 \leq n < \infty$. Therefore the process η_t restricted to $\{\eta: \Sigma_x \eta(x) = n\}$ is an irreducible positive recurrent Markov chain, whose stationary distribution is given by the conditional measure $\nu_n(\cdot) = \tilde{\nu}_\rho(\cdot \mid \Sigma_x\eta(x)=n)$, which is independent of ρ. Let ν_∞ be the point mass on $\eta \equiv 1$. The following result is from [36].

<u>Theorem 3.2.2</u>. (a) $\mathcal{I}_e = \{\nu_n, 0 \leq n \leq \infty\}$. (b) If $\mu\{\eta: \Sigma_x \eta(x) = n\} = 1$, then $\mu S(t) \to \nu_n$ for $0 \leq n \leq \infty$.

<u>Proof</u>. Part (a) follows from part (b), and part (b) for $n < \infty$ is just the ordinary convergence theorem for positive recurrent Markov chains. To prove (b) for $n = \infty$, take μ such that $\mu\{\eta: \Sigma_x \eta(x) = \infty\} = 1$, and choose μ_n so that $\mu_n \leq \mu$ and $\mu_n\{\eta: \Sigma_x \eta(x) = n\} = 1$. By lemma 3.1.4, $\mu_n S(t) \leq \mu S(t)$ for all $t \geq 0$. Since $\mu_n S(t) \to \nu_n$ as $t \to \infty$, it follows that $\nu_n \leq \nu$ for all n if ν is any weak limit of $\mu S(t)$ as $t \to \infty$. But

$$\frac{\rho\pi(x)}{1 + \rho\pi(x)} = \tilde{\nu}_\rho\{\eta : \eta(x) = 1\} = \sum_{n=0}^{\infty} \nu_n\{\eta : \eta(x) = 1\} \tilde{\nu}_\rho\{\eta : \Sigma_x \eta(x) = n\} ,$$

so $\nu\{\eta: \eta(x) = 1\} \geq \rho\pi(x)/(1+\rho\pi(x))$ for all $\rho \in (0,\infty)$. Letting $\rho \to \infty$, it follows that $\nu\{\eta: \eta(x) = 1\} = 1$ for all $x \in S$, and therefore $\nu = \nu_\infty$.

Similar techniques apply if $\Sigma_x \pi(x) < \infty$ is replaced by $\Sigma_x 1/\pi(x) < \infty$. Then the extremal invariant measures concentrate on $\{\eta: \Sigma_x [1 - \eta(x)] < \infty\} \cup \{\eta \equiv 0\}$. This amounts, of course, merely to an interchange in the roles of the zeros and ones, since one can think of the zeros moving according to the transpose of p instead of the ones moving according to p. If $\pi(x)$ satisfies (3.2.1), $\Sigma_x \pi(x) = \infty$ and $\Sigma_x 1/\pi(x) = \infty$, the situation is more complex, and is not fully understood. One such case, which illustrates some of the difficulties involved, will be discussed in section 3.4. Another class of cases (that in which $S = \{0,1,2,\cdots\}$ and $p(x,y) = 0$ for $|x - y| > 1$) is analyzed via coupling techniques in [38].

<u>Section 3.3</u>. <u>The translation invariant case</u>. Throughout this section, we will assume that $S = Z^d$ and $p(x,y) = p(0,y-x)$. In this case, part (a) of corollary 1.2.6 gives $\nu_\rho \in \mathcal{I}$ for all $\rho \in [0,1]$. As will be seen in the next section, it

is not necessarily the case that $\mathcal{I}_e = \{\nu_\rho,\ 0 \le \rho \le 1\}$. It is true, though, that $(\mathcal{I} \cap \mathcal{S})_e = \{\nu_\rho,\ 0 \le \rho \le 1\}$, and the proof is illustrative of the way coupling techniques can be used to determine $\mathcal{I}_e$ completely in some cases. The first step in the proof is given by the following lemma.

Lemma 3.3.1. If $\nu \in \bar{\mathcal{I}} \cap \bar{\mathcal{S}}$, then

$$\nu\{(\eta,\zeta) : \eta \le \zeta \text{ or } \eta \ge \zeta\} = 1. \tag{3.3.2}$$

Proof. Let $f_u(\eta,\zeta) = |\eta(u) - \zeta(u)|$. Then $f_u \in \mathcal{D}(\bar{\Omega})$, so evaluating $\bar{\Omega} f_u$ by using the expression for $\bar{\Omega}$ in section 3.1 yields

$$\bar{\Omega} f_u(\eta,\zeta) = [2\zeta(u) - 1]\left\{\sum_y \eta(u)[1-\eta(y)]p(u,y) - \sum_x \eta(x)[1-\eta(u)]p(x,u)\right\}$$

$$+ [2\eta(u) - 1]\left\{\sum_y \zeta(u)[1-\zeta(y)]p(u,y) - \sum_x \zeta(x)[1-\zeta(u)]p(x,u)\right\}$$

$$- 2\eta(u)\zeta(u)\sum_y p(u,y)[1-\eta(y)][1-\zeta(y)] - 2[1-\eta(u)][1-\zeta(u)]\sum_x p(x,u)\eta(x)\zeta(x)$$

$$= \sum_y p(u,y)\left[1_{\{\eta(u)=\zeta(u)=1,\eta(y)\ne\zeta(y)\}} - 1_{\{\eta(y)=\zeta(y)=0,\eta(u)\ne\zeta(u)\}}\right]$$

$$+ \sum_y p(y,u)\left[1_{\{\eta(u)=\zeta(u)=0,\eta(y)\ne\zeta(y)\}} - 1_{\{\eta(y)=\zeta(y)=1,\eta(u)\ne\zeta(u)\}}\right]$$

$$- \sum_y [p(u,y) + p(y,u)]\, 1_{\{\eta(u)=\zeta(y)\ne\eta(y)=\zeta(u)\}}\,.$$

Since $\nu \in \bar{\mathcal{I}}$, $\int \bar{\Omega} f_u(\eta,\zeta)\,d\nu = 0$, so

$$0 = \sum_y p(u,y)[\nu\{\eta(u)=\zeta(u)=1,\eta(y)\ne\zeta(y)\} - \nu\{\eta(y)=\zeta(y)=0,\eta(u)\ne\zeta(u)\}] \tag{3.3.3}$$

$$+ \sum_y p(y,u)[\nu\{\eta(u)=\zeta(u)=0,\eta(y)\ne\zeta(y)\} - \nu\{\eta(y)=\zeta(y)=1,\eta(u)\ne\zeta(u)\}]$$

$$- \sum_y [p(u,y) + p(y,u)]\,\nu\{\eta(u)=\zeta(y)\ne\eta(y)=\zeta(u)\}\,.$$

Since $\nu \in \bar{\mathcal{S}}$, the first two terms above are zero, and hence

$$\sum_y [p(u,y) + p(y,u)]\,\nu\{\eta(u) = \zeta(y) \ne \eta(y) = \zeta(u)\} = 0\,,$$

so

$$(3.3.4) \qquad \nu\{\eta(u) = \zeta(y) \neq \eta(y) = \zeta(u)\} = 0$$

whenever $p(u,y) + p(y,u) > 0$. Using the irreducibility of $p(x,y)$ and the invariance of ν, it is not hard to deduce from this that (3.3.4) holds for all $u \neq y$, and therefore that (3.3.2) holds.

Corollary 3.3.5. (a) If $\nu \in (\bar{\mathcal{J}} \cap \bar{\mathcal{S}})_e$, then either $\nu\{\eta \leq \zeta\} = 1$ or $\nu\{\eta \geq \zeta\} = 1$. (b) If $\mu_1, \mu_2 \in (\mathcal{J} \cap \mathcal{S})_e$, then either $\mu_1 \leq \mu_2$ or $\mu_2 \leq \mu_1$.

Proof. The first statement follows from lemmas 3.1.1 and 3.3.1, while the second is a consequence of the first and (c) of lemma 3.1.3.

Theorem 3.3.6. $(\mathcal{J} \cap \mathcal{S})_e = \{\nu_\rho, \ 0 \leq \rho \leq 1\}$.

Proof. Since ν_ρ is translation invariant and ergodic, $\nu_\rho \in \mathcal{S}_e$. Also $\nu_\rho \in \mathcal{J}$, so that $\nu_\rho \in (\mathcal{J} \cap \mathcal{S})_e$. By corollary 3.3.5, if $\mu \in (\mathcal{J} \cap \mathcal{S})_e$ and $\rho \in [0,1]$, then either $\mu \leq \nu_\rho$ or $\nu_\rho \leq \mu$. Therefore if $\mu \in (\mathcal{J} \cap \mathcal{S})_e$, there is a $\rho_0 \in [0,1]$ so that $\mu \geq \nu_\rho$ for $\rho < \rho_0$ and $\mu \leq \nu_\rho$ for $\rho > \rho_0$. It then follows that $\int f\, d\mu \geq \int f\, d\nu_\rho$ for $\rho < \rho_0$ and $\int f\, d\mu \leq \int f\, d\nu_\rho$ for $\rho > \rho_0$ for any f of the form $f(\eta) = \Pi_{x\in T}\, \eta(x)$ for finite $T \subset S$. Therefore $\int f\, d\mu = \int f\, d\nu_{\rho_0}$ for all such f, and hence $\mu = \nu_{\rho_0}$.

Using similar techniques, one can prove the following somewhat stronger result.

Theorem 3.3.7. If $\mu \in \mathcal{S}$, then all weak limits of $\mu S(t)$ as $t \to \infty$ are exchangeable.

This suggests the conjecture that if $\mu \in \mathcal{S}_e$, then $\mu S(t) \to \nu_\rho$ as $t \to \infty$, where $\rho = \mu\{\eta: \eta(x) = 1\}$. Of course in the symmetric case, this is just corollary 2.1.5. The conjecture would follow from theorem 3.3.7 if one could prove that all weak limits of $\mu S(t)$ are ergodic when $\mu \in \mathcal{S}_e$, but this appears to be difficult. Clearly a proof of the conjecture would lead to an identification of $\lim_{t\to\infty} \mu S(t)$ for all $\mu \in \mathcal{S}$.

A more careful use of (3.3.3) leads to a proof of the following result [38] in one dimension, which we conjecture to be true for $d > 1$.

Theorem 3.3.8. Suppose $d = 1$, $\Sigma_x |x| p(0,x) < \infty$, and $\Sigma_x x p(0,x) = 0$. Then $\mathcal{J}_e = \{\nu_\rho, \ 0 \leq \rho \leq 1\}$.

As will be seen in the next section, the mean zero assumption is needed for the result to hold. In order to give an indication of the relevance of the mean zero assumption, we will prove a lemma which is required in the proof of theorem 3.3.8. The conclusion of the lemma says that there is no net "flow of discrepancies" (i.e., of x's such that $\eta(x) \neq \zeta(x)$) across zero in equilibrium.

Lemma 3.3.9. Suppose ν is a probability measure on $X \times X$ with exchangeable marginals such that $\nu\{\eta \leq \zeta\} = 1$, and assume $p(x,y)$ satisfies the assumptions of theorem 3.3.8. Then

$$\sum_{x<0\leq y} p(x,y)\, \nu\{\eta(x) \neq \zeta(x), \eta(y) = \zeta(y) = 0\} + \sum_{x<0\leq y} p(y,x)\, \nu\{\eta(x) \neq \zeta(x), \eta(y) = \zeta(y) = 1\}$$

$$= \sum_{x<0\leq y} p(x,y) \nu\{\eta(y) \neq \zeta(y), \eta(x) = \zeta(x) = 1\} + \sum_{x<0\leq y} p(y,x) \nu\{\eta(y) \neq \zeta(y), \eta(x) = \zeta(x) = 0\}$$

Proof. By adding

$$\sum_{x<0\leq y} p(x,y) \nu\{\eta(x) = \zeta(x) = 1, \eta(y) = \zeta(y) = 0\} + \sum_{x<0\leq y} p(y,x) \nu\{\eta(x) = \zeta(x) = 0, \eta(y) = \zeta(y) = 1\}$$

to both sides of the equality, and using $\nu\{\eta \leq \zeta\} = 1$, we can rewrite the conclusion of the lemma in the following way:

$$\sum_{x<0\leq y} p(x,y)\, \nu\{\zeta(x) = 1, \zeta(y) = 0\} + \sum_{x<0\leq y} p(y,x)\, \nu\{\eta(x) = 0, \eta(y) = 1\}$$

$$= \sum_{x<0\leq y} p(x,y)\, \nu\{\eta(x) = 1, \eta(y) = 0\} + \sum_{x<0\leq y} p(y,x)\, \nu\{\zeta(x) = 0, \zeta(y) = 1\}.$$

But this is true since ν has exchangeable marginals and p has mean zero.

Section 3.4. The asymmetric simple random walk on Z^1. In this section, we will describe briefly and without proofs the known results in case $S = Z^1$, $p(x,x+1) = p$ and $p(x,x-1) = q$ for all $x \in S$, where $p+q = 1$. These illustrate some of the results which may be expected in the general translation invariant case, and therefore lead to various natural conjectures. We will assume $p \neq 1/2$, since otherwise $p(x,y)$ is symmetric and the results of chapter 2 apply. By reflection, we may assume without loss of generality that $p > 1/2$. The case $p = 1$ is included, even though the irreducibility assumption is not satisfied.

In order to describe $\mathcal{J}_e$, suppose first that $1/2 < p < 1$. Then the transition probabilities satisfy $\pi(x)p(x,y) = \pi(y)p(y,x)$ for $\pi(x) = (p/q)^x$, so that $\tilde{\nu}_\rho \in \mathcal{J}$ for each $\rho \in [0,\infty]$ by part (b) of corollary 1.2.6. Let $A_n = \{\eta \in X : \Sigma_{x<n}\, \eta(x) = \Sigma_{x\geq n}\, [1 - \eta(x)] < \infty\}$ and $A = \{\eta \in X : \Sigma_{x<0}\, \eta(x) < \infty$ and $\Sigma_{x\geq 0}\, [1 - \eta(x)] < \infty\}$. Then A is countable, $A = \bigcup_{n=-\infty}^{\infty} A_n$, $\tilde{\nu}_\rho(A) = 1$ for all $\rho \in (0,\infty)$, and $\tilde{\nu}_\rho(A_n) > 0$ for all $\rho \in (0,\infty)$ and $-\infty < n < \infty$. Since $\{A_n\}$ are the closed irreducible classes for the Markov chain η_t on A, it then follows that η_t is positive recurrent on A and that its stationary distribution ν_n on

A_n is given by the conditional measure

$$\nu_n(\cdot) = \tilde{\nu}_\rho(\cdot \mid A_n), \tag{3.4.1}$$

which is of course independent of ρ. For $p = 1$, let ν_n be the pointmass on the configuration η_n where $\eta_n(x) = 1$ for $x \geq n$ and $\eta_n(x) = 0$ for $x < n$. Then ν_n is clearly invariant for the process, since no motion takes place if the initial distribution is ν_n. By part (a) of corollary 1.2.6, $\nu_\rho \in \mathcal{I}$ for $0 \leq \rho \leq 1$.

The following theorem is proved in [38] by using coupling techniques similar to those used in section 3.3. The main added ingredient can be described in the following way. The arguments of section 3.3 are based on the idea that the function $f(\eta_t, \zeta_t) = \sum_{u=-n}^{n} |\eta_t(u) - \zeta_t(u)|$ is essentially monotone, in the sense that transitions within $[-n,n]$ can only decrease $f(\eta_t, \zeta_t)$. The only transitions which can increase $f(\eta_t, \zeta_t)$ are those which cross the boundaries $-n$ and n. In the case considered in this section, the function $g(\eta_t, \zeta_t)$ = number of sign changes of $\eta_t(u) - \zeta_t(u)$ in $[-n,n]$ is essentially monotone in the same sense. This monotonicity property of both functions is exploited in the proof.

<u>Theorem 3.4.2.</u> $\mathcal{I}_e = \{\nu_\rho, \ 0 \leq \rho \leq 1\} \cup \{\nu_n, \ -\infty < n < \infty\}$.

We conjecture that a similar result holds whenever $S = Z^1$, $p(x,y) = p(0,y-x)$, $\sum_x |x| p(0,x) < \infty$, and $\sum_x x p(0,x) > 0$. It would be necessary to prove first that the process with initial configuration η_n is a positive recurrent Markov chain, so that ν_n would be identified as the resulting limiting distribution. This is harder than it is in the case considered in this section, because ν_n will not have a representation in the form (3.4.1) relative to a product measure. Then a replacement for the monotonicity of the function $g(\eta_t, \zeta_t)$ would have to be found.

The problem of proving convergence has been resolved only for very simple initial distributions. Even these simple cases, however, illustrate the increased complexity of the process in the asymmetric case. The following theorem is proved in [37] by comparing the process with a related finite process.

<u>Theorem 3.4.3.</u> Suppose that μ is a product measure on X for which the following limits exist:

$$\lambda = \lim_{x \to -\infty} \mu\{\eta: \eta(x) = 1\}, \qquad \rho = \lim_{x \to +\infty} \mu\{\eta: \eta(x) = 1\}.$$

(a) If $\lambda \geq 1/2$ and $\rho \leq 1/2$, then $\lim_{t \to +\infty} \mu S(t) = \nu_{1/2}$.

(b) If $\rho \geq 1/2$ and $\lambda + \rho > 1$, then $\lim_{t \to \infty} \mu S(t) = \nu_\rho$.

(c) If $\lambda \leq 1/2$ and $\lambda + \rho < 1$, then $\lim_{t\to\infty} \mu S(t) = \nu_\lambda$.

(d) If $\lambda < 1/2$ and $\lambda + \rho = 1$, then $\lim_{t\to\infty} \mu S(t)$ may fail to exist.

For purposes of comparison, note that if $p = 1/2$, then theorem 2.1.3 gives $\lim_{t\to\infty} \mu S(t) = \nu_{((\lambda+\rho)/2)}$ under these assumptions for any choice of λ and ρ.

Chapter 4. The exclusion process with speed change. One of the key properties of the coupling used in the previous chapter was that $\{\eta \leq \zeta\}$ was closed for the motion of the coupled process. For the exclusion process with nontrivial speed change, the natural coupling (i.e., the one in which particles move together whenever they can) fails to have this property, and is therefore apparently not very useful. For Spitzer's exclusion process with speed change, however, there is another coupling which does have the desired property, provided that $c(x,\eta)$ and and $q(x,y)$ in (1.1.4) are related in an appropriate manner In this chapter, we will describe this coupling, and will illustrate its use by determining $\mathcal{I}_e$ completely for a class of one-dimensional processes.

Section 4.1. The basic coupling. Throughout this chapter, η_t will be the exclusion process with speed change with $c(x,y,\eta)$ given by (1.1.4), and $S(t)$ will be the corresponding semigroup. We will assume that $q(x,y)$ satisfies (1.1.5), and for technical reasons, that there is an $\varepsilon > 0$ so that for each $x,y \in S$, either $q(x,y) = 0$ or $q(x,y) \geq \varepsilon$. In order that a coupling with the desired properties exist, we must assume that $c(x,\eta)$ satisfies

$$(4.1.1)\quad c(x,\eta) \leq c(x,\zeta) \quad\text{and}\quad c(x,\eta) \sum_{\eta(z)=0} q(x,z) \geq c(x,\zeta) \sum_{\zeta(z)=0} q(x,z)$$

for $\eta \leq \zeta$. Note that this implies that $c(x,\eta)$ can only depend on coordinates $\eta(z)$ for which $q(x,z) > 0$, so condition (1.1.7) is a consequence of (1.1.6), which we assume. Conditions (4.1.1) are quite restrictive, of course, and the results obtained here should be regarded as only a first step toward understanding the general process with nontrivial speed change.

For $\eta,\zeta \in X$, let

$$A = A(\eta,\zeta) = \{x \in S : \eta(x) = \zeta(x) = 1\}$$

$$B = B(\eta,\zeta) = \{x \in S : \eta(x) = 1,\ \zeta(x) = 0\}$$

$$C = C(\eta,\zeta) = \{x \in S : \eta(x) = 0,\ \zeta(x) = 1\}$$

$$D = D(\eta,\zeta) = \{x \in S : \eta(x) = \zeta(x) = 0\}$$

$$G(x,\eta,\zeta) = [c(x,\eta) - c(x,\zeta)]^+ \left\{\sum_{u\in B} q(x,u)\right\}^{-1}$$

$$H(x,\eta,\zeta) = [c(x,\zeta) - c(x,\eta)]^+ \left\{\sum_{u\in C} q(x,u)\right\}^{-1} .$$

The generator $\bar{\Omega}$ of the coupled process then takes the following form on functions f which depend on finitely many coordinates: $\bar{\Omega}f(\eta,\zeta) =$

$$\sum_{x\in A, y\in D} [c(x,\eta) \wedge c(x,\zeta)]q(x,y)[f(\eta_{xy},\zeta_{xy}) - f(\eta,\zeta)]$$

$$+ \sum_{x\in A, y\in D, z\in B} G(x,\eta,\zeta)q(x,y)q(x,z)[f(\eta_{xy},\zeta_{xz}) - f(\eta,\zeta)]$$

$$+ \sum_{x\in A, y\in D, z\in C} H(x,\eta,\zeta)q(x,y)q(x,z)[f(\eta_{xz},\zeta_{xy}) - f(\eta,\zeta)]$$

$$+ \sum_{x\in A, y\in B} \left[c(x,\zeta) - G(x,\eta,\zeta) \sum_{u\in D} q(x,u)\right] q(x,y)[f(\eta,\zeta_{xy}) - f(\eta,\zeta)]$$

$$+ \sum_{x\in A, y\in C} \left[c(x,\eta) - H(x,\eta,\zeta) \sum_{u\in D} q(x,u)\right] q(x,y)[f(\eta_{xy},\zeta) - f(\eta,\zeta)]$$

$$+ \sum_{x\in B, y\in C\cup D} q(x,y)c(x,\eta)[f(\eta_{xy},\zeta) - f(\eta,\zeta)]$$

$$+ \sum_{x\in C, y\in B\cup D} q(x,y)c(x,\zeta)[f(\eta,\zeta_{xy}) - f(\eta,\zeta)]$$

The existence of the coupled process is guaranteed by theorem 2.8 of [32], and the corresponding semigroup will be denoted by $\bar{S}(t)$. The nonnegativity of the coefficients in the fourth and fifth terms follows from (4.1.1), as can be seen, for example, from the following computation for the fourth term:

$$c(x,\zeta) \sum_{u\in B\cup D} q(x,u) \geq c(x,\zeta \vee \eta) \sum_{u\in D} q(x,u) \geq c(x,\eta) \sum_{u\in D} q(x,u).$$

It is not hard to check that if $f(\eta,\zeta) = g(\eta)$, then $\bar{\Omega}f(\eta,\zeta) = \Omega g(\eta)$, while if $f(\eta,\zeta) = g(\zeta)$, then $\bar{\Omega}f(\eta,\zeta) = \Omega g(\zeta)$. Therefore the marginals of the coupled process are Markovian and have the transition law of η_t. Finally, note that the set $\{\eta \leq \zeta\}$ is closed for the motion of the coupled process. Therefore the statements of lemmas 3.1.1, 3.1.2, 3.1.3, and 3.1.4 carry over to the present context.

<u>Section 4.2</u>. <u>The invariant measures.</u> Now we specialize further to the case in which $S = Z^1$, $q(x,y)$ is symmetric, translation invariant and irreducible, and

$c(x,\eta)$ takes the form (1.3.1) where Φ is translation invariant. In this case, the canonical Gibbs states corresponding to Φ which were defined in section 1.3 are invariant for the process η_t, and it is known ([9], [43]) that $\tilde{\mathcal{G}}_e = \{\gamma_\rho, 0 \leq \rho \leq 1\}$ where γ_ρ is translation invariant and ergodic for each $\rho \in [0,1]$, γ_ρ is weakly continuous in ρ, and $\gamma_\rho\{\eta: \eta(x) = 1\} = \rho$. Our aim is to use the coupled process to prove that $\mathcal{J} = \tilde{\mathcal{G}}$. The first step is to prove that $\mathcal{J} \cap \mathcal{S} = \tilde{\mathcal{G}}$. This was proved in [20] under somewhat different assumptions via free energy techniques. Our proof is similar to that of theorem 3.3.6.

Theorem 4.2.1. Suppose $\nu \in \bar{\mathcal{S}}$, and put $\nu_t = \nu\bar{S}(t)$. Then

(4.2.2) $$\lim_{t\to\infty} \nu_t\{(\eta,\zeta) : \eta(x) = \zeta(y) \neq \eta(y) = \zeta(x)\} = 0$$

for all $x,y \in S$. Therefore if ν_∞ is any weak limit of ν_t as $t \to \infty$, $\nu_\infty\{(\eta,\zeta) : \eta \leq \zeta \text{ or } \eta \geq \zeta\} = 1$.

Proof. Define $f_y(\eta,\zeta) = 1 - |\eta(y) - \zeta(y)|$. Then

(4.2.3) $$\bar{\Omega} f_y(\eta,\zeta) = \sum_x q(y,x)\{[c(x,\zeta)+c(y,\eta)]1_C(x)1_B(y)+[c(x,\eta)+c(y,\zeta)]1_C(y)1_B(x)\}$$

$$+ \sum_x q(y,x)[\eta(x)+\eta(y) - 1]\{c(x,\zeta)\zeta(x)[1-\zeta(y)] - c(y,\zeta)\zeta(y)[1-\zeta(x)]\}$$

$$+ \sum_x q(y,x)[\zeta(x)+\zeta(y) - 1]\{c(x,\eta)\eta(x)[1-\eta(y)] - c(y,\zeta)\eta(y)[1-\eta(x)]\}$$

$$- \sum_x q(y,x)\{|c(x,\eta)-c(x,\zeta)|1_A(x)1_D(y) - |c(y,\eta)-c(y,\zeta)|1_A(y)1_D(x)\}.$$

Since f is in the domain of $\bar{\Omega}^n$ for all n,

(4.2.4) $$\frac{d^n}{dt^n}\int f\, d\nu_t = \int \bar{\Omega}^n f\, d\nu_t ,$$

and therefore each derivative of $\int f\, d\nu_t$ is uniformly bounded in t. Since $\nu_t \in \bar{\mathcal{S}}$ for all t, and the integral of each of the last three terms in (4.2.3) with respect to any measure in $\bar{\mathcal{S}}$ is zero, it follows from (4.2.4) with $n = 1$ that $\int f\, d\nu_t$ is nondecreasing in t. Therefore

(4.2.5) $$\lim_{t\to\infty} \int \bar{\Omega}^n f\, d\nu_t = 0$$

for all $n \geq 1$. Using (4.2.5) for $n = 1$ and (4.2.3), we see that (4.2.2) holds whenever $q(y,x) > 0$. Using (4.2.5) for successively larger values of n, and recalling that q is irreducible, it follows that (4.2.2) holds for all $x,y \in S$.

Corollary 4.2.6. Suppose $\mu \in \mathcal{S}$. Then any weak limit of $\mu S(t)$ as $t \to \infty$ is in $\tilde{\mathcal{G}}$. In particular, $\mathcal{J} \cap \mathcal{S} = \tilde{\mathcal{G}}$.

Proof. For any $\rho \in [0,1]$, let ν be the product measure $\mu \times \gamma_\rho$. By theorem 4.2.1, any weak limit of $\nu\bar{S}(t)$ as $t \to \infty$ concentrates on $\{(\eta,\zeta) : \eta \le \zeta$ or $\eta \ge \zeta\}$. Therefore, if μ_∞ is any weak limit of $\mu S(t)$ as $t \to \infty$, there exists $\alpha(\rho),\beta(\rho) \ge 0$ and $\mu_1^\rho,\mu_2^\rho \in \mathcal{S}$ so that $\alpha(\rho) + \beta(\rho) = 1$, $\mu_\infty = \alpha(\rho)\mu_1^\rho + \beta(\rho)\mu_2^\rho$, and $\mu_1^\rho \le \gamma_\rho \le \mu_2^\rho$. Since $\mu_\infty \in \mathcal{S}$, $L = \lim_{n\to\infty} (1/n)\sum_{x=1}^n \eta(x)$ exists a.s. with respect to μ_∞ by the ergodic theorem. Since $L = \rho$ a.s. with respect to γ_ρ, it follows that $\mu_1^\rho(\cdot) = \mu_\infty(\cdot\,|L \le \rho)$, $\mu_2(\cdot) = \mu_\infty(\cdot\,|L \ge \rho)$, and $\alpha(\rho) = \mu_\infty\{L \le \rho\}$ for all ρ for which $\mu_\infty\{L = \rho\} = 0$. Therefore, since γ_ρ is ergodic,

$$\gamma_{\rho_1} \le \mu_\infty(\cdot \mid \rho_1 < L < \rho_2) \le \gamma_{\rho_2}$$

whenever $\rho_1 < \rho_2$ and $\mu_\infty\{\rho_1 < L < \rho_2\} > 0$. Since γ_ρ is continuous in ρ, it then follows that $\mu_\infty(\cdot\,|L) = \gamma_L$ a.s. with respect to μ_∞. Therefore, $\mu_\infty = \int \gamma_\rho \mu_\infty(L \in d\rho)$, so that $\mu_\infty \in \tilde{\mathcal{G}}$.

Lemma 4.2.7. Suppose ν is a probability measure on $X \times X$ such that both marginals of ν are in $\tilde{\mathcal{G}}$ and ν concentrates on $\{\eta \le \zeta\}$ or on $\{\eta \ge \zeta\}$. Then for each $x,y \in S$,

$$\int [\eta(x) + \eta(y) - 1]\{c(x,\zeta)\zeta(x)[1 - \zeta(y)] - c(y,\zeta)\zeta(y)[1 - \zeta(x)]\}d\nu$$
$$+ \int [\zeta(x) + \zeta(y) - 1]\{c(x,\eta)\eta(x)[1 - \eta(y)] - c(y,\eta)\eta(y)[1 - \eta(x)]\}d\nu$$
$$= \int \eta(x)\zeta(x)[1 - \eta(y)][1 - \zeta(y)]\,|c(x,\eta) - c(x,\zeta)|\,d\nu$$
$$- \int \eta(y)\zeta(y)[1 - \eta(x)][1 - \zeta(x)]\,|c(y,\eta) - c(y,\zeta)|\,d\nu$$

Proof. We may assume that ν concentrates on $\{\eta \le \zeta\}$. Let

$$F(\zeta) = c(x,\zeta)\zeta(x)[1 - \zeta(y)] - c(y,\zeta)\zeta(y)[1 - \zeta(x)] .$$

Since the marginals of ν are in $\tilde{\mathcal{G}}$, $\int F(\eta)\, d\nu = \int F(\zeta)\, d\nu = 0$. Therefore, since $\eta(x) = \eta(y) = 1$ implies $\zeta(x) = \zeta(y) = 1$ and hence $F(\zeta) = 0$, we have

$$\int [\eta(x) + \eta(y) - 1]F(\zeta)d\nu = -\int [1 - \eta(x)][1 - \eta(y)]F(\zeta)d\nu$$

$$= \int [1 - \eta(x)]\eta(y)F(\zeta)d\nu + \int [1 - \eta(y)]\eta(x)F(\zeta)d\nu$$

$$= -\int c(y,\zeta)\eta(y)\zeta(y)[1 - \eta(x)][1 - \zeta(x)]d\nu$$

$$+ \int c(x,\zeta)\eta(x)\zeta(x)[1 - \eta(y)][1 - \zeta(y)]d\nu \; .$$

Similarly,

$$\int [\zeta(x) + \zeta(y) - 1]F(\eta)d\nu = \int c(y,\eta)\eta(y)\zeta(y)[1 - \eta(x)][1 - \zeta(x)]d\nu$$

$$- \int c(x,\eta)\eta(x)\zeta(x)[1 - \eta(y)][1 - \zeta(y)]d\nu \; .$$

By assumption (4.1.1), $c(x,\eta) \leq c(x,\zeta)$ a.s. with respect to ν, so the result follows.

Theorem 4.2.8. If $\nu \in \bar{\mathcal{J}}$, then $\nu\{(\eta,\zeta) : \eta \leq \zeta \text{ or } \eta \geq \zeta\} = 1$.

Proof. Let $\alpha(x,y) = \int[c(x,\zeta) + c(y,\eta)]\zeta(x)\eta(y)[1 - \zeta(y)][1 - \eta(x)]d\nu$ and

$$\beta(x,y) = -\int [\eta(x) + \eta(y) - 1]c(x,\zeta)\zeta(x)[1 - \zeta(y)]d\nu$$

$$- \int [\zeta(x) + \zeta(y) - 1]c(x,\eta)\eta(x)[1 - \eta(y)]d\nu$$

$$+ \int |c(x,\eta) - c(x,\zeta)|\eta(x)\zeta(x)[1 - \eta(y)][1 - \zeta(y)]d\nu \; .$$

By (4.2.3) and $\bar{\Omega} f_y d\nu = 0$, we have for $n \geq 1$

$$(4.2.9) \qquad \sum_{|y| \leq n} \sum_x q(y,x)[\alpha(x,y) + \alpha(y,x)] = \sum_{|y| \leq n} \sum_x q(y,x)[\beta(x,y) - \beta(y,x)]$$

$$= \sum_{|y| \leq n} \sum_x q(y,x)\beta(x,y) - \sum_{|x| \leq n} \sum_y q(y,x)\beta(x,y)$$

$$= \sum_{|y| \leq n} \sum_{|x| > n} q(y,x)\beta(x,y) - \sum_{|x| \leq n} \sum_{|y| > n} q(y,x)\beta(x,y)$$

$$= \sum_{|y| \leq n} \sum_{|x| > n} q(y,x)[\beta(x,y) - \beta(y,x)] \; .$$

Any weak limit $\bar{\nu}$ of Cesaro averages of successive translates of ν is in $\bar{\mathcal{J}} \cap \bar{\mathcal{S}}$. Therefore $\bar{\nu}$ concentrates on $\{\eta \leq \zeta$ or $\eta \geq \zeta\}$ by theorem 4.2.1, and so is a convex combination of two measures, each of which is in $\bar{\mathcal{J}} \cap \bar{\mathcal{S}}$, one of which concentrates on $\{\eta \leq \zeta\}$ and the other of which concentrates on $\{\eta \geq \zeta\}$. Therefore by corollary 4.2.6 and lemma 4.2.7,

$$\lim_{|n| \to \infty} \frac{1}{n} \sum_{k=0}^{n} [\beta(x+n,y+n) - \beta(y+n,x+n)] = 0 .$$

By (4.2.9), it then follows that $\Sigma_{x,y}\, q(x,y)[\alpha(x,y) + \alpha(y,x)] = 0$. Since $c(x,\eta)$ is bounded away from zero, this gives

$$\nu\{(\eta,\zeta) : \eta(x) = \zeta(y) \neq \eta(y) = \zeta(x)\} = 0 \tag{4.2.10}$$

for all x,y such that $q(x,y) > 0$. Using the irreducibility of q and the fact that $\nu \in \bar{\mathcal{J}}$ again, it then follows that (4.2.10) holds for all $x,y \in S$.

Corollary 4.2.11. $\mathcal{J} = \tilde{\mathcal{G}}$.

Proof. Suppose $\mu \in \mathcal{J}_e$ and $\rho \in [0,1]$. Then by theorem 4.2.8 and lemma 3.1.2 applied to this context, there exists $\alpha,\beta \geq 0$ with $\alpha + \beta = 1$ and $\mu_1,\mu_2 \in \mathcal{J}$ so that $\mu_1 \leq \mu \leq \mu_2$ and $\gamma_\rho = \alpha\mu_1 + \beta\mu_2$. Suppose $\alpha,\beta > 0$. Then μ_1 and μ_2 are absolutely continuous with respect to γ_ρ, and therefore

$$\mu_i\left\{\eta : \lim_{n\to\infty} \frac{1}{n} \sum_{x=1}^{n} \eta(x) = \rho\right\} = 1$$

for each i by the ergodic theorem applied to γ_ρ. Since $\mu_1 \leq \mu \leq \mu_2$, it follows that

$$\mu\left\{\eta : \lim_{n\to\infty} \frac{1}{n} \sum_{x=1}^{n} \eta(x) = \rho\right\} = 1 .$$

For a given $\mu \in \mathcal{J}_e$, this can be the case for at most one $\rho \in [0,1]$, so α and β are both positive for at most one $\rho \in [0,1]$. We conclude then that there is a $\rho_0 \in [0,1]$ so that $\mu \leq \nu_\rho$ for $\rho > \rho_0$ and $\mu \geq \nu_\rho$ for $\rho < \rho_0$. Since γ_ρ is weakly continuous in ρ, it follows that $\mu = \gamma_{\rho_0}$. Therefore $\mathcal{J}_e \subset \tilde{\mathcal{G}}$, and the desired result follows from $\tilde{\mathcal{G}} \subset \mathcal{J}$

BIBLIOGRAPHIE

1. A. B. Bortz, M. H. Kalos, J. L. Lebowitz, and J. Marro (1975). Time evolution of a quenched binary alloy II. Computer simulation of a three dimensional model system, Physical Review B, 12, 2000-2011.

2. A. B. Bortz, M. H. Kalos, J. L. Lebowitz, and M. A. Zendejas (1974). Time evolution of a quenched binary alloy. Computer simulation of a two dimensional model system, Physical Review B, 10, 535-541.

3. J. Chover (1975). Convergence of a local lattice process, Stochastic Processes and their Applications, 3, 115-135.

4. D. Dawson (1974). Information flow in discrete Markov systems, Journal of Applied Probability, 11, 594-600.

5. D. Dawson (1975). Synchronous and asynchronous reversible Markov systems, Canadian Mathematical Bulletin, 17, 633-649.

6. R. L. Dobrushin (1971). Markov processes with a large number of locally interacting components, Problems of Information Transmission, 7, 149-164, and 235-241.

7. R. Glauber (1963). The statistics of the stochastic Ising model, Journal of Mathematical Physics, 4, 294-307.

8. H. O. Georgii (1975). Canonical Gibbs states, their relation to Gibbs states, and applications to two-valued Markov chains, Z. Wahrscheinlichkeitstheorie verw. Geb., 32, 277-300.

9. H. O. Georgii (1976). On canonical Gibbs states and symmetric and tail events, Z. Wahrscheinlichkeitstheorie verw. Geb., 33, 331-341.

10. L. Gray and D. Griffeath (1976). On the uniqueness of certain interacting particle systems, Z. Wahrscheinlichkeitstheorie verw. Geb., 35, 75-86.

11. D. Griffeath (1975). Ergodic theorems for graph interactions, Advances in Applied Probability, 7, 179-194.

12. T. E. Harris (1972). Nearest-neighbor Markov interaction processes on multidimensional lattices, Advances in Mathematics, 9, 66-89.

13. T. E. Harris (1974). Contact interactions on a lattice, Annals of Probability, 2, 969-988.

14. T. E. Harris (1976). On a class of set-valued Markov Processes, Annals of Probability, 4, 175-194.

15. L. L. Helms (1974). Ergodic properties of several interacting Poisson particles, Advances in Mathematics, 12, 32-57.

16. Y. Higuchi ( ). A study on the canonical Gibbs states.

17. Y. Higuchi and T. Shiga (1975). Some results on Markov processes of infinite lattice spin systems, Journal of Mathematics of Kyoto University, 15, 211-229.

18. R. Holley (1970). A class of interactions in an infinite particle system, Advances in Mathematics, 5, 291-309.

19. R. Holley (1971). Free energy in a Markovian model of a lattice spin system, Communications in Mathematical Physics, 23, 87-99.

20. R. Holley (1971). Pressure and Helmholtz free energy in a dynamic model of a lattice gas, Proceedings of the Sixth Berkeley Symposium on Mathematical Statistics and Probability, III, 565-578.

21. R. Holley (1972). An ergodic theorem for interacting systems with attractive interactions, Z. Wahrscheinlichkeitstheorie verw. Geb., 24, 325-334.

22. R. Holley (1972). Markovian interaction processes with finite range interactions, The Annals of Mathematical Statistics, 43, 1961-1967.

23. R. Holley (1974). Recent results on the stochastic Ising model, Rocky Mountain Journal of Mathematics, 4, 479-496.

24. R. Holley and T. M. Liggett (1975). Ergodic theorems for weakly interacting infinite systems and the voter model, The Annals of Probability, 3, 643-663.

25. R. Holley and D. Stroock (1976). A martingale approach to infinite systems of interacting processes, The Annals of Probability, 4, 195-228.

26. R. Holley and D. Stroock (1976). Applications of the stochastic Ising model to the Gibbs states, Communications in Mathematical Physics, 48, 249-265.

27. R. Holley and D. Stroock ( ). Dual processes and their application to infinite interacting systems, Advances in Mathematics.

28. R. Holley and D. Stroock (1976). L_2 theory for the stochastic Ising model, Z. Wahrscheinlichkeitstheorie verw. Geb., 35, 87-101.

29. J. Kemeny, J. Snell, and A. Knapp (1966). Denumerable Markov Chains, Van Nostrand, Princeton, N. J.

30. R. Kinderman and J. Snell (1976). Markov random fields.

31. T. Kurtz (1969). Extensions of Trotter's operator semigroup approximation theorems, Journal of Functional Analysis, 3, 354-375.

32. T. M. Liggett (1972). Existence theorems for infinite particle systems, Transactions of the A.M.S., 165,471-481.

33. T. M. Liggett (1973). An infinite particle system with zero range interactions, The Annals of Probability, 1, 240-253.

34. T. M. Liggett (1973). A characterization of the invariant measures for an infinite particle system with interactions, Transactions of the A.M.S., 179, 433-453.

35. T. M. Liggett (1974). A characterization of the invariant measures for an infinite particle system with interactions II, Transactions of the A.M.S., 198,201-213.

36. T. M. Liggett (1974). Convergence to total occupancy in an infinite particle system with interactions, The Annals of Probability, 2, 989-998.

37. T. M. Liggett (1975). Ergodic theorems for the asymmetric simple exclusion process, Transactions of the A.M.S., 213, 237-261.

38. T. M. Liggett (1976). Coupling the simple exclusion process, The Annals of Probability, 4, 339-356.

39. K. Logan (1974). Time reversible evolutions in statistical mechanics, Cornell University, Ph.D. dissertation.

40. N. Matloff ( ). Ergodicity conditions for a dissonant voting model, The Annals of Probability.

41. C. J. Preston (1974). Gibbs states on countable sets, Cambridge University Press.

42. D. Ruelle (1969). Statistical mechanics, W. A. Benjamin.

43. D. Ruelle (1968). Statistical mechanics of a one-dimensional lattice gas, Communications in Mathematical Physics, 9, 267-278.

44. D. Schwartz (1976). Ergodic theorems for an infinite particle system with births and deaths, Annals of Probability, 4

45. F. Spitzer (1970). Interaction of Markov processes, Advances in Mathematics, 5, 246-290.

46. F. Spitzer (1971). Random fields and interacting particle systems, M. A. A. Summer Seminar Notes, Williamstown, Mass.

47. F. Spitzer (1974). Introduction aux processus de Markov a parametres dans Z_ν, Lecture Notes in Mathematics 390, Springer-Verlag.

48. F. Spitzer (1974). Recurrent random walk of an infinite particle system, Transactions of the A.M.S., 198, 191-199.

49. F. Spitzer (1975). Random time evolution of infinite particle systems, Advances in Mathematics, 16, 139-143.

50. O. N. Stavskaya and I. I. Pyatetskii-Shapiro (1971). On homogeneous nets of spontaneously active elements, Systems Theory Res., 20, 75-88.

51. W. G. Sullivan (1973). Potentials for almost Markovian random fields, Communications in Mathematical Physics, 33, 61-74.

52. W. G. Sullivan (1974). A unified existence and ergodic theorem for Markov evolution of random fields, Z. Wahrscheinlichkeitstheorie verw. Geb.,31,47-56.

53. W. G. Sullivan (1976). Processes with infinitely many jumping particles, Proceedings of the A.M.S., 54, 326-330.

54. W. G. Sullivan (1975). Mean square relaxation times for evolution of random fields, Communications in Mathematical Physics, 40, 249-258.

55. W. G. Sullivan (1975). Markov processes for random fields, Communications of the Dublin Institute for Advanced Studies, series A, number 23.

56. W. G. Sullivan ( ). Specific information gain for interacting Markov processes.

57. L. N. Vasershtein (1969). Processes over denumerable products of spaces, describing large systems, Problems of Information Transmission, 3, 47-52.

58. D. Schwartz ( ). Applications of duality to a class of Markov processes,

59. D. Griffeath ( ). An ergodic theorem for a class of spin systems.

60. R. Lang ( ). Unendlich-dimensionale Wienerprozesse mit wechselwirkung I: Existenz.

61. L. N. Vasershtein and A. M. Leontovich (1970). Invariant measures of certain Markov operators describing a homogeneous random medium, Problems of Information Transmission, 6, 61-69.

62. C. Cocozza and C. Kipnis ( ). Existence de processus Markoviens pour des systems infinis de particules.

63. L. Gray and D. Griffeath ( ). On the nonuniqueness of proximity processes.

64. R. Holley and D. Stroock ( ). Nearest neighbor birth and death processes on the real line.

PROCESSUS PONCTUELS

PAR J. NEVEU

CHAPITRE I

GENERALITES SUR LES PROCESSUS PONCTUELS

Sommaire | Pages

Sommaire

Un processus ponctuel sur un espace E est une répartition de points au hasard sur E. Comme ces points aléatoires ne sont généralement pas ordonnés, le mieux est de définir un processus ponctuel comme une variable aléatoire, soit N, à valeurs dans l'espace $M_p(E)$ des mesures dites ponctuelles, c'est-à-dire de la forme $\sum_I \varepsilon_{x_i}$. Nous nous sommes restreints à des espaces E localement compacts à bases dénombrables pour imposer ensuite aux mesures ponctuelles que nous considérons la condition de Radon d'être localement finies. Ce cadre semble suffisamment général pour couvrir la plupart des applications et de toutes façons une condition de σ-finitude est indispensable ; il permet de donner à certains théorèmes, notamment ceux du dernier paragraphe, une forme particulièrement simple et donc agréable ; enfin les quelques fois où nous avons trouvé ce cadre trop restreint, il était immédiat de passer à des hypothèses plus générales !

L'étude d'un processus ponctuel N commence par celle des variables aléatoires entières positives N (F) qui comptent ses nombres de points dans les divers boréliens de E ; le premier paragraphe leur est consacré. Les plus simples et les plus importants des processus ponctuels sont les processus de Poisson que nous définissons et construisons sur un espace général au paragraphe 2 ; le paragraphe suivant développe quelques idées rencontrées lors de cette construction.

Au début du quatrième et dernier paragraphe, nous étudions le problème de la construction d'un processus ponctuel à partir des lois de probabilité de ses variables de comptage ; dans le cadre des mesures de Radon positives aléatoires, la solution de ce problème est très simple. L'importance du résultat du point de vue des applications n'est pourtant pas très grande car comme nous le verrons dans les chapitres ultérieurs, les processus ponctuels sont souvent obtenus d'autre manière. L'étude de la convergence en loi fait l'objet de la fin du paragraphe ; elle aboutit à un théorème de limite centrale et notamment à une "loi des petits nombres" qui fait comprendre l'importance des processus de Poisson. Nous avons abondamment utilisé la transformation de Laplace dans tout ce paragraphe et pour cette raison, nous avons cru bon de rappeler dans un appendice les propriétés qui nous sont utiles des transformées de Laplace des mesures positives sur R_+^n. Ajoutons qu'à moins d'un intérêt particulier pour les problèmes (un peu anciens !) qui y sont traités, le lecteur pourra lire ce paragraphe 4 rapidement, les méthodes de transformation de Laplace ne devant pas avoir d'autres applications dans la suite de l'exposé ...

I.1. LOIS, MOMENTS ET FONCTIONNELLES DE LAPLACE DES PROCESSUS PONCTUELS.

Les processus ponctuels servent à décrire des répartitions de points au hasard sur un espace E. Nous supposerons dans toute la suite que cet espace E est un espace localement compact à base dénombrable dont nous noterons $\mathcal{E}$ la tribu borélienne. Nous noterons aussi $\mathcal{E}_c$ la classe des boréliens relativement compacts de E. Dans les applications l'espace E sera souvent l'un des espaces R^d, Z^d ou T^d (où $T = R/Z$) ou un produit de tels espaces.

Nous appellerons <u>mesure ponctuelle</u> sur E toute mesure positive m sur l'espace mesurable $(E, \mathcal{E})$ qui s'écrive comme une somme localement finie de masses unités, soit $m = \sum_I \varepsilon_{x_i}$ (les x_i $(i \in I)$ ne sont pas nécessairement distincts 2 à 2). Par définition cette somme est localement finie si tout compact de E ne contient qu'un nombre fini de x_i $(i \in I)$; comme E est la réunion d'au moins une suite de compacts, la famille $(x_i, i \in I)$ est nécesairement dénombrable. L'ensemble $S_m = \{x : m(\{x\}) \neq 0\}$ des points de E chargés par la mesure ponctuelle m, c'est-à-dire le sous-ensemble de E formé par les différents x_i si $m = \sum_I \varepsilon_{x_i}$, est fermé car son intersection avec tout compact est finie ; cet ensemble est donc le <u>support</u> de la mesure ponctuelle m au sens habituel où S_m est le plus petit fermé de E portant la mesure m. Pour tout $x \in S_m$, le nombre entier $m(\{x\})$ s'appelle la <u>multiplicité</u> de x et nous pouvons évidemment écrire que $m = \sum_{S_m} m(\{x\})\,\varepsilon_x$; lorsque $m(\{x\}) = 1$ pour tout $x \in S_m$, la mesure ponctuelle m est dite <u>simple</u> et sa donnée équivaut à celle de son support puisqu'une telle mesure vaut $m(F) = \mathrm{Card}(F \cap S_m)$ pour tout $F \in \mathcal{E}$. Notons enfin que parmi toutes les mesures positives sur $(E, \mathcal{E})$ les mesures ponctuelles peuvent être caractérisées comme celles qui ne prennent que des valeurs entières et finies sur les boréliens relativement compacts de E [cf l'exercice 1].

Nous désignerons par $M_p(E)$, ou M_p si aucune confusion n'est à craindre, l'espace des mesures ponctuelles définies sur E et nous introduirons sur cet espace la tribu $\mathcal{M}_p(E)$ ou $\mathcal{M}_p$ engendrée par les applications $m \to m(F)$ de M_p dans $\overline{\mathbb{N}}$ obtenues lorsque F parcourt $\mathcal{E}$. Ceci nous permet de poser la définition de base suivante.

Définition I.1

Etant donné un espace localement compact à base dénombrable E, un processus ponctuel sur E est une application mesurable, soit N, d'un espace de probabilité $(\Omega, \mathcal{A}, P)$ dans l'espace mesurable $(M_p(E), \mathcal{M}_p(E))$ des mesures ponctuelles sur E. La loi de probabilité, soit P_N, de ce processus ponctuel est l'image par N de la probabilité P.

Pour chaque $\omega \in \Omega$, $N(\omega)$ est alors une mesure ponctuelle sur E dont nous noterons $N(\omega, F)$ les valeurs sur les divers boréliens F de E. Par définition de la tribu $\mathcal{M}_p$, pour qu'une application $N : \Omega \to M_p(E)$ soit mesurable sur l'espace de probabilité $(\Omega, \mathcal{A}, P)$ pour la tribu $\mathcal{M}_p$ sur $M_p(E)$ il faut et il suffit que les applications $\omega \to N(\omega, F)$ soient mesurables sur Ω pour tous les boréliens F ; nous noterons $N(F)$ et appellerons variables de comptage ces variables aléatoires entières positives qui peuvent prendre la valeur $+\infty$ lorsque F n'est pas relativement compact. Nous verrons d'ailleurs dans la proposition ci-dessous que sur l'espace E = R par exemple il suffit déjà de vérifier que les applications $N(., I) : \Omega \to \mathbb{N}$ sont mesurables pour tout intervalle réel borné I pour qu'il s'en suive que $N : \Omega \to M_p(R)$ soit mesurable.

Notons que toute probabilité Q sur $(M_p(E), \mathcal{M}_p(E))$ est la loi de probabilité d'au moins un processus ponctuel et notamment du processus défini par l'application identique de l'espace de probabilité $(M_p(E), \mathcal{M}_p(E), Q)$ sur $M_p(E)$; ce dernier processus est généralement appelé le processus ponctuel canonique sur E de loi Q.

Pour toute suite finie $F_1, \ldots, F_k$ de boréliens de E, le vecteur aléatoire $(N(F_1), \ldots N(F_k))$ qui prend ses valeurs dans $\overline{\mathbb{N}}^k$ s'obtient en composant le processus ponctuel $N : \Omega \to M_p(E)$ avec l'application mesurable $m \to (m(F_1), \ldots m(F_k))$ de $M_p(E)$ dans $\overline{\mathbb{N}}^k$; la loi de probabilité du vecteur $(N(F_1), \ldots N(F_k))$ est donc l'image de P_N par cette application et se trouve ainsi déterminée par P_N quels que soient $F_1, \ldots F_k \in \mathcal{E}$. La réciproque de ce résultat est exacte et peut même être formulée de manière plus générale de façon à montrer par exemple que sur R la loi P_N d'un processus ponctuel N est entièrement déterminée par les lois des vecteurs $(N(I_1), \ldots, N(I_k))$ correspondant à des suites finies d'intervalles bornés $I_1, \ldots I_k$ de R.

Proposition I.2

Soit $\mathcal{J}$ une classe de boréliens relativement compacts de E stable par intersection finie et engendrant la tribu borélienne $\mathcal{E}$, qui contienne en outre une suite croissant vers E ou une partition dénombrable de E.

Alors pour qu'une application $N : \Omega \to M_p(E)$ d'un espace de probabilité $(\Omega, \mathcal{A}, P)$ dans $M_p(E)$ soit un processus ponctuel, c'est-à-dire soit mesurable, il suffit déjà que les applications $N(., I)$ de Ω dans $\mathbb{N}$ soient mesurables pour tout $I \in \mathcal{J}$. D'autre part la loi P_N de tout processus ponctuel N sur E est entièrement caractérisée par la donnée des lois des vecteurs $(N(I_1), \ldots N(I_k))$ correspondant à toutes les suites finies $I_1, \ldots I_k$ dans $\mathcal{J}$.

Dans R la classe des intervalles bornés, dans R^d la classe des pavés produits d'intervalles réels bornés, dans E en général la classe des ouverts relativement compacts jouissent des propriétés imposées ci-dessus à la classe $\mathcal{J}$.

Démonstration

Nous nous appuierons pour démontrer ces résultats sur le lemme de théorie de la mesure suivant.

Lemme I.3

Si dans un espace mesurable arbitraire $(E, \mathcal{E})$, la classe $\mathcal{J}$ de parties mesurables de E est stable par intersection finie, engendre $\mathcal{E}$ et contient en outre une suite croissant vers E ou une partition dénombrable de E, alors :

a) la tribu $\mathcal{E}$ est la plus petite classe de parties de E contenant $\mathcal{J}$ qui soit stable par différences et sommes dénombrables ;

b) deux mesures positives m_1, m_2 sur $(E, \mathcal{E})$ telles que $m_1(I) = m_2(I) < \infty$ pour tout $I \in \mathcal{J}$ sont nécessairement identiques.

Pour démontrer la proposition, désignons par $\mathcal{J}$ la classe des parties de M_p (E) de la forme

$$\{m : m(I_1) = n_1, \ldots, m(I_k) = n_k\}$$

où $I_1, \ldots I_k$ est une suite finie quelconque dans $\mathcal{I}$ et où $n_1, \ldots, n_k$ sont des entiers positifs quelconques. La tribu $\tau(\mathcal{J})$ de parties de M_p (E) engendrée par cette classe est la plus petite tribu rendant mesurables les applications $m \to m(I)$ de M_p (E) dans $\mathbb{N}$ obtenues lorsque I parcourt $\mathcal{I}$. Pour établir que sous les hypothèses faites sur $\mathcal{I}$, la tribu $\tau(\mathcal{J})$ coïncide avec $\mathcal{M}_p$ (E), donnons-nous une suite $(I_j, j \in \mathbb{N})$ dans $\mathcal{I}$ qui croisse vers E ou qui soit une partition dénombrable de E et formons la classe des boréliens

$$\mathcal{F} = \{F : F \in \mathcal{E} \text{ et } m \to m(F\, I_j) \text{ est mesurable de } M_p(E) \text{ dans } \mathbb{N} \text{ pour la tribu } \tau(\mathcal{J}), \text{ quel que soit } j \in \mathbb{N}\}$$

Cette classe $\mathcal{F}$ contient $\mathcal{I}$ (car $II_j \in \mathcal{I}$ si $I \in \mathcal{I}$) et est stable par différences et sommes dénombrables ; d'après le lemme rappelé ci-dessus $\mathcal{F} = \mathcal{E}$ et pour tout $F \in \mathcal{E}$ l'application $m \to m(F) = \left(\lim_j \uparrow \text{ ou } \sum_j\right) m(F\, I_j)$ est donc mesurable de M_p (E) dans $\overline{\mathbb{N}}$. Il s'en suit que $\mathcal{M}_p \subset \tau(\mathcal{J})$ et donc que ces deux tribus coïncident.

Pour démontrer le deuxième résultat de la proposition, il s'agit d'établir que deux probabilités P et Q définies sur $(M_p(E), \mathcal{M}_p(E))$ sont égales dès qu'elles coïncident sur la classe $\mathcal{J}$ ci-dessus. Or cette classe $\mathcal{J}$ vérifie sur l'espace M_p (E) les hypothèses du lemme I.3 : elle est manifestement stable par intersection finie, elle engendre $\mathcal{M}_p$ (E) d'après ce que nous venons de démontrer et elle contient l'ensemble M_p (E) [correspondant à la suite vide de I !] ; d'après la seconde partie de ce lemme, deux probabilités ne peuvent coïncider sur $\mathcal{J}$ sans être égales sur toute la tribu $\mathcal{M}_p$ (E).□

La proposition précédente est un résultat d'unicité. Elle ne permet en aucune manière de construire un processus ponctuel dont les lois des vecteurs aléatoires $(N(F_1), \ldots N(F_k))$ seraient données à l'avance. Un tel théorème d'existence existe [voir le paragraphe 4] mais il ne semble pas avoir une grande importance, la plupart des processus ponctuels intéressants se construisant par des méthodes particulières.

Une suite finie ou infinie de processus ponctuels N_1 sur E_1, N_2 sur E_2, ... définis sur un même espace de probabilité $(\Omega, \mathcal{A}, P)$ est dite <u>indépendante</u> si la suite des sous-tribus $\tau(N_i) = N_i^{-1}[\mathcal{M}_p(E_i)]$ de $\mathcal{A}$ est indépendante. Dans le cas de deux processus indépendants N_1 et N_2 par exemple, quels que soient les boréliens $F_1, \ldots F_k$ de E_1 et $G_1, \ldots, G_l$ de E_2, les deux vecteurs aléatoires $(N_1(F_1), \ldots N_1(F_k))$ et $(N_2(G_1), \ldots N_2(G_l))$ seront alors indépendants [Pour une réciproque, voir l'exercice 8].

L'<u>intensité</u> d'un processus ponctuel $N : \Omega \to M_p(E)$ est définie comme la mesure positive sur $(E, \mathcal{E})$, soit μ, donnant les espérances des variables de comptage, soit :

$$(1) \quad \mu(F) = E[N(F)] \equiv \int_\Omega N(\omega, F)\, P(d\omega) \equiv \int_{M_p(E)} m(F)\, P_N(dm)$$

pour tout $F \in \mathcal{E}$. La σ-additivité de cette fonction d'ensembles μ résulte immédiatement de celle des mesures $N(\omega, .)$ et de la propriété de σ-additivité de l'espérance des variables aléatoires positives. Par contre la mesure positive μ n'est pas nécessairement une mesure de Radon sur E, c'est-à-dire n'est pas nécessairement finie sur les boréliens relativement compacts bien que cela soit le plus souvent le cas dans les applications. Lorsque l'intensité est une mesure de Radon, nous dirons que le <u>processus ponctuel est intégrable</u>.

Si $f : E \to [0, \infty]$ est une fonction borélienne, nous poserons

$$N(\omega, f) = \int_E f(x)\, N(\omega, dx) \leq +\infty \quad \text{pour tout } \omega \in \Omega .$$

L'application $N(f) = N(., f)$ de Ω dans $[0, \infty]$ ainsi définie est encore une variable aléatoire comme cela se voit en écrivant f comme limite croissante de fonctions boréliennes positives étagées et la même approximation montre en outre que

$$(2) \quad E[N(f)] = \int_E f(x)\, \mu(dx) \qquad (f \text{ borélienne } \geq 0).$$

Cette formule montre que si $f \in L^1_+(\mu)$, la variable aléatoire $N(f)$ est intégrable et donc p.s. finie sur $(\Omega, \mathcal{A}, P)$; cette propriété permet de définir l'intégrale

$N(\omega, f) = \int_E N(\omega, dx)\, f(x)$ pour presque tout ω lorsque f est une fonction borélienne μ-intégrable de signe quelconque sur E. Les variables aléatoires $N(f)$ ainsi définies à une équivalence presque sûre près pour tout $f \in L^1(\mu)$ jouissent évidemment des propriétés suivantes.

Proposition I.4

L'application $f \to N(f)$ est une application linéaire de $L^1(E, \mu)$ dans $L^1(\Omega, P)$ qui préserve l'intégrale et contracte la norme, soit

(3) $E[N(f)] = \mu(f)$, $\quad E[|N(f)|] \leqslant \mu[|f|]$

Cette proposition n'a réellement d'intérêt que si μ est de Radon.

Après avoir défini le "premier moment" d'un processus ponctuel et l'avoir appelé intensité, voici la bonne manière de définir le second moment d'un tel processus.

La seconde puissance d'un processus ponctuel N sur E est définie comme le processus ponctuel sur E^2 donné par les produits de mesures ponctuelles

$$N^{2\otimes}(\omega, .) = N(\omega, .) \otimes N(\omega, .) \qquad \text{pour tout } \omega \in \Omega .$$

Il est clair que nous définissons ainsi une application $N^{2\otimes}$ de Ω dans $M_p(E^2)$; cette application est mesurable car pout tout couple F_1, F_2 de boréliens relativement compacts $N^{2\otimes}(F_1 \times F_2) = N(F_1) \,.\, N(F_2)$ est mesurable et car la première partie de la proposition I.2 peut être appliquée sur l'espace E^2 à la classe $\mathfrak{J}$ de ces $F_1 \times F_2$ et à $N^{2\otimes}$.

Le moment d'ordre 2 d'un processus ponctuel N sur E est alors défini comme étant l'intensité du processus $N^{2\otimes}$, c'est-à-dire comme la mesure positive μ_2 sur $(E^2, \mathscr{E}^{2\otimes})$ donnée par la formule $\mu_2(G) = E[N^{2\otimes}(G)]$ où $G \in \mathscr{E}^{2\otimes}$ et vérifiant donc la formule plus générale suivante

$$\int_{E^2} g(x, y)\mu_2(dx\, dy) = \int_\Omega P(d\omega) \int_{E^2} g(x, y)\, N(\omega, dx)\, N(\omega, dy)$$

au moins pour toute fonction borélienne positive g sur E^2. En particulier quelles que soient les fonctions boréliennes positives f, f' sur E et quels que soient les boréliens F, F' de E

(4) $$E\left[N(f)\,N(f')\right] = \int_{E^2} f(x)\,f'(y)\,\mu_2(dx\,dy)$$

et

(4') $$E\left[N(F)\,N(F')\right] = \mu_2(F \times F').$$

Lorsque le moment d'ordre 2 d'un processus ponctuel N est une mesure de Radon, c'est-à-dire lorsque $E\left[N(F)^2\right] < \infty$ pour tout borélien relativement compact F, nous dirons que le <u>processus ponctuel N est de carré intégrable.</u>

<u>La transformée de Laplace d'une probabilité</u> Q définie sur l'espace $M_p(E)$ des mesures ponctuelles sur E est définie par la formule

(5) $$\psi(f) = \int_{M_p(E)} \exp\left(-\int_E f\,dm\right) Q(dm)$$

où f parcourt le cône des fonctions boréliennes positives de E dans R. Nous appellerons alors <u>fonctionnelle de Laplace</u> d'un processus ponctuel N la transformée de Laplace de sa loi de probabilité, soit

(5') $$\psi_N(f) = \int_{M_p(E)} \exp\left[-\int_E f\,dm\right] P_N(dm) = \int_\Omega \exp\left[-N(\omega, f)\right] P(d\omega)$$

($f : E \to R_+$ borélienne). La formule suivante qui découle de cette définition

(6) $$E\left[\prod_1^k u_i^{N(F_i)}\right] = \psi_N\left[\sum_1^k \log\left(\frac{1}{u_i}\right) . 1_{F_i}\right] \qquad (0 < u_i \leq 1,\ i = 1, \ldots$$

où $F_1, \ldots F_k$ est une suite finie arbitraire de boréliens de E, montre que la fonctionnelle ψ_N détermine les fonctions génératrices et donc les lois de tous les vecteurs aléatoires $(N(F_1), \ldots N(F_k))$; la proposition I.2 nous montre alors que la fonctionnelle de Laplace ψ_N détermine entièrement la loi P_N du processus ponctuel N.

La fonctionnelle de Laplace $f \to \psi_N(f)$ d'un processus ponctuel, qui n'est évidemment pas une fonctionnelle linéaire, jouit de la propriété de continuité monotone suivante.

Lemme I.5

Pour toute suite croissante de fonctions boréliennes positives $(f_n, n \in \mathbb{N})$:

$$\psi_N (\lim_n \uparrow f_n) = \lim_n \downarrow \psi_N (f_n).$$

Démonstration

En effet $N(f_n) \uparrow N(f)$ sur Ω lorsque $n \uparrow \infty$ si $f = \lim_n \uparrow f_n$ et alors le théorème de Lebesgue s'applique aux espérances des variables aléatoires $\exp[-N(f_n)]$ qui décroissent vers $\exp[-N(f)]$ lorsque $n \uparrow \infty$ tout en étant majorées par 1.□

Il résulte par exemple de ce lemme que la fonctionnelle de Laplace ψ_N est déjà entièrement déterminée par les valeurs qu'elle prend sur les fonctions boréliennes positives étagées, ce qui sera utile dans l'étude des exemples.

Pour terminer, étudions les relations entre la fonctionnelle de Laplace ψ_N et l'intensité μ d'un processus ponctuel N sur E. D'abord pour toute fonction borélienne positive f sur E :

$$\mu(f) = E[N(f)] = \lim_{t \downarrow 0} \uparrow \frac{1}{t} [1 - \psi_N(tf)] \tag{7}$$

car lorsque $t \downarrow 0$ les variables $\frac{1}{t}(1 - \exp[-t N(f)])$ croissent vers $N(f)$; ainsi ψ_N permet de calculer facilement μ.

D'autre part si f, g sont des fonctions boréliennes positives μ-intégrables sur E, l'inégalité élémentaire $|\exp(-a) - \exp(-b)| \leqslant |a - b|$ valable lorsque $a, b \geqslant 0$ entraîne que

$$\begin{aligned} |\psi_N(f) - \psi_N(g)| &\leqslant E\Big[\big|\exp[-N(f)] - \exp[-N(g)]\big|\Big] \\ &\leqslant E\Big[|N(f) - N(g)|\Big] \\ &\leqslant E(N\Big[|f - g|\Big]) = \int_E |f - g| \, d\mu \end{aligned}$$

Ce calcul démontre que <u>la restriction à $L^1_+(E, \mu)$ de la fonctionnelle de Laplace ψ_N d'un processus ponctuel N d'intensité μ est une application lipschitzienne et donc continue de $L^1_+(\mu)$ dans R_+</u>. (Ce résultat n'a vraiment d'intérêt que si μ est de Radon).

I.2. PROCESSUS DE POISSON ET SUPERPOSITIONS DE PROCESSUS PONCTUELS INDEPENDANTS

Les processus ponctuels de Poisson que nous définissons et construisons dans la proposition suivante comptent parmi les plus importants des processus ponctuels.

Proposition I.6

Etant donné un espace localement compact à base dénombrable E et une mesure de Radon positive μ sur E, il existe toujours un processus ponctuel N sur E jouissant des deux propriétés suivantes :

a) pour tout borélien F de E, la variable aléatoire N (F) suit une loi de Poisson de paramètre μ (F) (si μ (F) = ∞, N (F) = + ∞ p.s.) ce qui entraîne que la mesure μ est l'intensité de N,

b) les variables aléatoires N (F_1), ... N (F_k) associées à toute suite finie de boréliens F_1, ... F_k deux à deux disjoints de E sont indépendantes.

Un tel processus ponctuel s'appelle un processus de Poisson d'intensité μ sur E. Sa loi est univoquement déterminée par les propriétés (a - b) ci-dessus. Sa fonctionnelle de Laplace est donnée par la formule

$$(8) \qquad \psi_N (f) = \exp \left[- \int_E \left[1 - e^{-f(x)}\right] \mu (dx)\right]$$

et inversement un processus ponctuel N dont la fonctionnelle de Laplace est de la forme précédente est un processus de Poisson d'intensité μ.

Démonstration

Nous démontrerons d'abord cette proposition lorsque la mesure μ donnée est finie et peut donc s'écrire sous la forme $\mu = c\nu$ pour une constante finie c et une probabilité ν sur E.

Soit $E^* = \sum_N E^n$ l'espace de toutes les suites finies (ordonnées) de points de E, y compris la suite vide auquel l'espace E^0 se réduit par convention. Munissons cet espace de la tribu naturelle $\mathcal{E}^*$ formée par les parties G de E^* telles que

$G \cap E^n \in \mathcal{E}^{n\otimes}$ pour tout $n \in \mathbb{N}^*$. L'application $N^* : E^* \to M_p(E)$ définie par la formule

$$N^*(\omega, .) = \sum_1^n \varepsilon_{x_i} \qquad \text{si } \omega = (x_1, \dots x_n) \in E^*$$

est mesurable pour les tribus $\mathcal{E}^*$ et $\mathcal{M}_p$ car pour tout $F \in \mathcal{E}$, tout $k \in \mathbb{N}$ et tout $n \in \mathbb{N}$

$$\{N^*(., F) = k\} \cap E^n = \{(x_1, \dots x_n) : \sum_1^n 1_F(x_i) = k\} \in \mathcal{E}^{n\otimes}.$$

Remarquons aussi que ce processus ponctuel N^* est tel que $N^*(., E) < \infty$ sur E^* et que $\{N^*(., E) = n\} = E^n$ dans E^* pour tout $n \in \mathbb{N}$.

Définissons ensuite une probabilité P sur $(E^*, \mathcal{E}^*)$ en posant pour tout $G \in \mathcal{E}^*$:

$$P(G) = \sum_{\mathbb{N}} e^{-c} \frac{c^n}{n!} \nu^{n\otimes}(G \cap E^n)$$

Cette probabilité est caractérisée par les deux propriétés intéressantes suivantes :

a) $P[N^*(E) = n] \equiv P(E^n)$ vaut $e^{-c} \frac{c^n}{n!}$ pour tout $n \in \mathbb{N}$, autrement dit la variable aléatoire $N^*(E)$ suit une loi de Poisson de paramètre c ;

b) comme pour tout n fixé, $P[G / N^*(E) = n] = \nu^{n\otimes}(G)$ pour tout $G \in \mathcal{E}^*$ contenu dans E^n, la probabilité $P[N^* E . / N^*(E) = n]$ coïncide sur $M_p(E)$ avec la loi de probabilité de $\sum_1^n \varepsilon_{X_m}$ où $(X_1, \dots X_n)$ sont n points choisis au hasard indépendamment suivant la probabilité ν sur E.

Il s'en suit que si $F_o, \dots F_k$ est une partition finie de E et si $n_o, \dots n_k$ sont k entiers de somme $n = n_o + \dots + n_k$

$$P[N^*(F_o) = n_o, \dots N^*(F_k) = n_k]$$

$$= e^{-c} \frac{c^n}{n!} P[N^*(F_o) = n_o, \dots N^*(F_k) = n_k / N^*(E) = n]$$

$$= e^{-c} \frac{c^n}{n!} . n! \prod_0^k \frac{[\nu(F_i)]^{n_i}}{n_i!}$$

puisque comme il est bien connu les nombres de $X_.$ tombant respectivement dans F_o, ... F_k suivent une loi multinomiale de paramètres $\nu(F_o), \dots \nu(F_k)$. Le résultat précédent peut encore s'écrire

$$P\left[N^*(F_o) = n_o, \ldots, N^*(F_k) = n_k\right]$$

$$= \prod_{0}^{k} e^{-\mu(F_i)} \frac{[\mu(F_i)]^{n_i}}{n_i!}$$

puisque $\mu(F_i) = c\,\nu(F_i)$ et $\sum_0^k \mu(F_i) = c$. Ceci montre que les variables aléatoires $N(F_o), \ldots N(F_k)$ sont indépendantes et suivent des lois de Poisson de paramètres respectifs $\mu(F_o), \ldots \mu(F_k)$. Les propriétés (a-b) de la proposition en découlent, la première en considérant la partition (F, F^c), la seconde en formant la partition $(F_o = (\sum_1^k F_i)^c, F_1, \ldots F_k)$.

[Une manière plus analytique de démontrer que le processus ponctuel N^* construit ci-dessus est un processus de Poisson d'intensité μ consiste à écrire que pour toute suite finie $F_1, \ldots F_k$ de boréliens de E

$$E\left[u_1^{N^*(F_1)} \ldots u_k^{N^*(F_k)}\right] \qquad (0 < u_i \leqslant 1\,;\ i = 1, \ldots k)$$

$$= \int_{E^*} \exp\left[-N^*(f)\right] dP \qquad \text{si } f = \sum_1^k \log\left(\frac{1}{u_i}\right) 1_{F_i}$$

$$= \sum_{\mathbb{N}} e^{-c} \frac{c^n}{n!} \int_{E^n} \exp\left[-(f(x_1) + \ldots + f(x_n))\right] \nu^{n\otimes}(dx_1 \ldots dx_n)$$

par définition de P

$$= \sum_{\mathbb{N}} e^{-c} \frac{c^n}{n!} \left[\int_E \exp\left[-f(x)\right] \nu(dx)\right]^n$$

$$= \exp\left(-\int_E \left[1 - e^{-f(x)}\right] \mu(dx)\right);$$

lorsque les F_i sont disjoints 2 à 2 le dernier membre vaut encore

$$= \exp\left[-\sum_1^k (1 - u_i)\,\mu(F_i)\right]$$

ce qui est égal à la fonction génératrice du produit de k lois de Poisson de paramètres respectifs $\mu(F_1), \ldots \mu(F_k)$.]

Nous construirons plus loin un processus de Poisson d'intensité μ lorsque cette intensité n'est pas finie, par superposition de processus de Poisson indépendants d'intensités finies. Auparavant démontrons la fin de la proposition I.6.

La fonctionnelle de Laplace ψ_N d'un processus de Poisson N d'intensité μ est donnée sur les fonctions boréliennes positives étagées par la formule suivante : si $f = \sum_1^n a_i \, 1_{\{f=a_i\}}$ où $a_1, \dots a_n$ désignent les valeurs possibles de f ,

$$\psi_N (f) = E \left[\exp \left(- \sum_1^n a_i \, N (\{f = a_i\})\right)\right]$$

$$= \exp \left[- \sum_1^n (1 - e^{-a_i}) \, \mu (\{f = a_i\})\right]$$

puisque d'après les propriétés (a-b) du processus N, les variables aléatoires $N (\{f = a_i\})$ sont indépendantes et suivent des lois de Poisson de paramètres $\mu (\{f = a_i\})$. Ceci démontre que

$$\psi_N (f) = \exp \left[- \int_E \left[1 - e^{-f(x)}\right] d\mu (x)\right]$$

lorsque f est borélienne positive étagée et comme les deux membres de cette formule sont continus le long de suites croissantes de fonctions boréliennes positives, la formule précédente reste exacte pour toute fonction borélienne positive. La fonctionnelle de Laplace ψ_N et donc d'après le paragraphe I.1 la loi de probabilité P_N d'un processus de Poisson d'intensité μ sont ainsi univoquement déterminées par les propriétés (a-b) de la proposition.

D'autre part, si la fonctionnelle de Laplace d'un processus ponctuel N est de la forme précédente, la fonction génératrice de tout vecteur aléatoire $N (F_1), \dots N (F_k)$ correspondant à k boréliens 2 à 2 disjoints est celle d'un produit de lois de Poisson de paramètres respectifs $\mu(F_1), \dots \mu (F_k)$ et le processus ponctuel N jouit alors des propriétés (a-b) caractérisant un processus de Poisson d'intensité μ.

Afin notamment de terminer la démonstration de la proposition I.6, étudions la superposition des processus ponctuels. Comme la somme $\sum_1^n m_k$ d'une suite finie de mesures ponctuelles $m_1, \dots m_n$ définies sur le même espace E est encore une mesure ponctuelle sur E, il est évident que la superposition $N_1 + \dots + N_n$ de n processus ponctuels $N_k : (\Omega, \mathcal{A}, P) \to M_p(E)$ $(k = 1, \dots n)$ est encore un processus ponctuel sur E. Mais la superposition d'une infinité de processus ponctuels est aussi intéressante. Or, pour que la série $\sum_{\mathbb{N}} m_k$ de mesures ponctuelles sur un même espace E soit encore une mesure ponctuelle sur E, il faut que pour tout borélien relativement compact F de E la série d'entiers $\sum_N m_k(F)$ converge et il suffit d'ailleurs que pour un recouvrement donné de E par des ouverts relativement compacts G_j les séries $\sum_{k \in \mathbb{N}} m_k(G_j)$ convergent pour tout j. Le lemme de Borel-Cantelli permet de transformer ce résultat sur les mesures ponctuelles en un résultat sur les processus ponctuels.

Proposition I.7

Pour que la superposition $\sum_{\mathbb{N}} N_k$ d'une suite infinie de processus ponctuels $N_k : (\Omega, \mathcal{A}, P) \to M_p(E)$ définisse presque partout sur Ω un processus ponctuel sur E, il suffit que pour un recouvrement de E par des ouverts relativement compacts G_j $(j \in J)$ les séries suivantes soient convergentes :

$$(9) \qquad \sum_{k \in \mathbb{N}} P[N_k(G_j) \neq 0] < \infty \qquad (j \in J)$$

En particulier cette condition est satisfaite si la série $\sum_{\mathbb{N}} \mu_k$ des intensités μ_k des processus ponctuels N_k est une mesure de Radon sur E et cette mesure est alors l'intensité de $\sum_{\mathbb{N}} N_k$.

Réciproquement lorsque les processus ponctuels N_k sont indépendants, la série $\sum_{\mathbb{N}} N_k$ ne définit p.p. un processus ponctuel sur E que si pour tout borélien relativement compact F de E

$$\sum_{\mathbb{N}} P(N_k(F) \neq 0) < \infty .$$

Démonstration

L'hypothèse du début de la proposition et le lemme de Borel-Cantelli entraînent que les événements $A_j = \{\omega : \sum_k N_k (\omega, G_j) < \infty\}$ sont de probabilité 1. Pour tout $\omega \in \bigcap_j A_j$, c'est-à-dire pour presque tout $\omega \in \Omega$, la série $\sum_k N_k (\omega, .)$ définit alors une mesure ponctuelle d'après ce qui fut rappelé avant la proposition et il existe donc un processus ponctuel sur E p.p. égal à la superposition $\sum_k N_k$ (remplacer $\sum_k N_k (\omega, .)$ par 0 lorsque $\omega \notin \cap A_j$).

Comme $P[N_k (G) \neq 0] \leqslant E[N_k (G)] = \mu_k (G)$ puisque $N_k (G)$ est une variable aléatoire entière positive, la deuxième condition suffisante de la proposition est claire.

Inversement soit F un borélien relativement compact de E. Si la superposition $\sum_k N_k$ est p.p. un processus ponctuel, la série de variables $\sum_k N_k (F)$ converge p.p. et comme ces variables sont entières l'événement $\limsup_k \{N_k (F) \neq 0\}$ est négligeable. Donc si les processus N_k sont indépendants, les événements $\{N_k (F) \neq 0\}$ sont indépendants et le lemme de Borel-Cantelli entraîne que $\sum_k P[N_k (F) \neq 0] < \infty$. □

En ce qui concerne les processus de Poisson, nous avons le résultat suivant.

Corollaire I.8

La superposition $\sum_k N_k$ d'une suite finie ou infinie de processus de Poisson indépendants $N_k : (\Omega, \mathcal{A}, P) \to M_p (E)$ d'intensités respectives μ_k est un processus de Poisson d'intensité $\mu = \sum_k \mu_k$ si et seulement si cette série définit encore une mesure de Radon μ sur E.

Démonstration

Si μ est une mesure de Radon sur E, la proposition précédente montre que $N = \sum_k N_k$ est un processus ponctuel sur E dont μ est évidemment l'intensité. La fonctionnelle génératrice de ce processus est donnée par

$$\Psi_N (f) = E \left(\exp [- \sum_k N_k (f)]\right) \qquad (f : E \to R_+ \text{ borélienne})$$

Or, les variables aléatoires positives N_k (f) obtenues en faisant varier k sont indépendantes car elles sont mesurables par rapport aux sous-tribus indépendantes $\tau(N_k)$ de $\mathcal{A}$. Par conséquent

$$\psi_N(f) = \prod_k \psi_{N_k}(f)$$
$$= \prod_k \exp\left[-\int (1 - e^{-f})\, d\mu_k\right]$$
$$= \exp\left[-\int (1 - e^{-f})\, d\mu\right]$$

et d'après la fin de la proposition I.6 que nous avons déjà démontrée, N est un processus de Poisson d'intensité μ.□

Fin de la démonstration de la proposition I.6

Pour construire un processus de Poisson sur E d'intensité égale à une mesure de Radon infinie donnée μ, décomposons la mesure μ en une série $\mu = \sum_{\mathbb{N}} \mu_k$ de mesures de Radon finies μ_k sur E par exemple en posant $\mu_k = 1_{F_k} . \mu$ après nous être donné une partition dénombrable de E par des boréliens relativement compacts F_k. Nous avons vu comment construire des probabilités P_k sur $(E^*, \mathcal{E}^*)$ pour que le processus ponctuel $N^* : E^* \to M_p(E)$ soit un processus de Poisson d'intensité μ_k. Formons maintenant l'espace de probabilité

$$(\Omega, \mathcal{A}, P) = \prod_{\mathbb{N}} (E^*, \mathcal{E}^*, P_k)$$

Les processus ponctuels $N_k : \Omega \to M_p(E)$ définis par

$$N_k(\omega, .) = N^*(\omega_k, .) \quad \text{où } \omega = (\omega_k, k \in \mathbb{N}) \in \Omega$$

sont alors des processus de Poisson indépendants d'intensités respectives μ_k. D'après le corollaire précédent, la superposition

$$N(\omega, .) = \sum_k N_k(\omega, .) = \sum_k N^*(\omega_k, .)$$

définit p.p. sur $(\Omega, \mathcal{A}, P)$ un processus ponctuel de Poisson d'intensité μ sur E.□

I.3. VARIATIONS SUR LES PROCESSUS DE POISSON

Ce paragraphe contient d'autres exemples de processus ponctuels obtenus en amplifiant certaines des idées du paragraphe précédent.

A. Les processus ponctuels finis

Un processus ponctuel $N : \Omega \to M_p(E)$ est dit (p.s.) fini si $N(E) < \infty$ (p.p.) sur Ω, c'est-à-dire s'il prend (p.s.) ses valeurs dans le sous-espace mesurable

$$M_{pf}(E) = \{m \; ; \; m \in M_p(E), \; m(E) < \infty\}$$

de $M_p(E)$ formé par les mesures ponctuelles finies. En particulier un processus ponctuel d'intensité finie, c'est-à-dire tel que $E[N(E)] = \mu(E) < \infty$ est p.s. fini. Dans ce paragraphe nous nous proposons d'étudier les relations entre l'espace E^* des suites finies ordonnées de points de E et l'espace $M_{pf}(E)$ des mesures ponctuelles finies sur E ainsi que celles entre les probabilités définies sur ces deux espaces.

Au début de la démonstration de la proposition I.6, nous avons déjà introduit l'espace $E^* = \sum_N E^n$ des suites finies de points de E, la tribu naturelle $\mathcal{E}^* = \{F : F \cap E^n \in \mathcal{E}^{n\otimes} \; \forall n \in \mathbb{N}^*\}$ ainsi que le processus ponctuel $N^* : E^* \to M_p(E)$ donné par $N^*(\omega, .) = \sum_1^n \varepsilon_{x_i}$ si $\omega = (x_1, \dots x_n)$ dans E^* ; ce processus ponctuel envoie en fait E^* sur $M_{pf}(E)$. Nous avons déjà démontré que $N^* : (E^*, \mathcal{E}^*) \to (M_p(E), \mathcal{M}_p(E))$ est mesurable ; le résultat plus précis suivant est la clef de la suite.

Lemme I.9

Si pour tout $n \in \mathbb{N}$, $\mathcal{E}^{n\otimes}_{sym}$ désigne la sous-tribu de $\mathcal{E}^{n\otimes}$ formée des boréliens invariants par permutation des coordonnées et si $\mathcal{E}^*_{sym}$ désigne la sous-tribu de $\mathcal{E}^*$ définie par $\mathcal{E}^*_{sym} = \{F : F \cap E^n \in \mathcal{E}^{n\otimes}_{sym} \; \forall n \in \mathbb{N}^*\}$, alors

$$(10) \quad (N^*)^{-1} \left[\mathcal{M}_p(E)\right] = \mathcal{E}^*_{sym} \text{ dans } E^* .$$

Démonstration

Pour tout borélien F et tout entier k , $\{N^*(F) = k\}$ appartient à $\mathcal{E}^*_{sym}$ car pour tout $n \in \mathbb{N}^*$

$$\{N^*(., F) = k\} \cap E^n = \{(x_1 \dots x_n) : \sum_1^n 1_F(x_i) = k\} \in \mathcal{E}^{n\otimes}_{sym}.$$

La définition de la tribu $\mathcal{M}_p$ entraîne alors que $(N^*)^{-1}[\mathcal{M}_p(E)] \subset \mathcal{E}^*_{sym}$.

L'inclusion inverse est plus difficile à démontrer. Pour le faire, il suffit de prouver que pour tout n et tout $F \in \mathcal{E}^{n\otimes}$, la fonction symétrisée $\frac{1}{n!} \sum_{S_n} 1_F \circ \theta_\sigma$ est mesurable par rapport à la tribu trace $E^n \cap (N^*)^{-1}(\mathcal{M}_p)$ car cela entraînera que $\mathcal{E}^{n\otimes}_{sym}$ est contenue dans cette tribu trace et donc que $\mathcal{E}^*_{sym} \subset (N^*)^{-1}(\mathcal{M}_p)$. [Dans la formule précédente S_n désigne le groupe des permutations de $\{1, 2, \dots n\}$ et θ_σ désigne la permutation de coordonnées $(x_1, \dots x_n) \to (x_{\sigma 1}, \dots x_{\sigma n})$ de E^n].

Or la classe des $F \in \mathcal{E}^{n\otimes}$ qui ont la propriété de mesurabilité précédente est manifestement stable par différences et sommes dénombrables. Elle contient aussi tout pavé mesurable $F_1 \times \dots \times F_n$ $(F_i \in \mathcal{E})$ de E^n car d'une part un tel pavé est la somme finie de pavés mesurables dont les côtés sont deux à deux disjoints ou égaux, car d'autre part un pavé de ce type particulier s'écrit $G_{i_1} \times \dots \times G_{i_n}$ pour des $G_1, \dots G_k \in \mathcal{E}$ deux à deux disjoints et des entiers $i_1, \dots i_n$ dans $[1, k]$ et vérifie

$$\frac{1}{n!} \sum_{S_n} 1_{G_{i_1} \times \dots \times G_{i_n}} \circ \theta_\sigma = \frac{\prod_j n_j!}{n!} 1_{E^n \cap \{N^*(G_j) = n_j \ \forall j\}}$$

si n_j désigne le nombre de $i_1, \dots i_n$ égaux à j $(1 \leq j \leq k)$. Le lemme I.3 permet donc de conclure que la classe précédente coïncide avec la tribu produit $\mathcal{E}^{n\otimes}$.□

Une probabilité P sur $(E^*, \mathcal{E}^*)$ sera dite symétrique si sa trace sur chacun des sous-espaces $(E^n, \mathcal{E}^{n\otimes})$ de E^* est invariante par les permutations de coordonnées θ_σ $(\sigma \in S_n)$ de E^n. La proposition suivante se déduit alors facilement du lemme précédent.

Proposition I.10

Toute probabilité sur M_p (E) portée par M_{pf} (E) est l'image par le processus ponctuel $N^* : E^* \to M_p$ (E) d'une et d'une seule probabilité symétrique sur $(E^*, \mathcal{E}^*)$.

Démonstration

L'application $(N^*)^{-1}$ établit une bijection entre la tribu $M_{pf}(E) \cap \mathcal{M}_p(E)$ de parties de M_{pf} (E) et la tribu $\mathcal{E}^*_{sym}$ de E^*. En effet puisque N^* envoie E^* sur M_{pf} (E), l'application $(N^*)^{-1}$ est injective si nous la restreignons aux parties de M_{pf} (E) ; d'autre part d'après le lemme précédent elle applique la tribu $\mathcal{M}_p$ (E) et donc aussi sa trace sur $M_{pf}(E) = N^*(E^*)$ sur $\mathcal{E}^*_{sym}$. Cela entraîne que les probabilités sur M_p (E) portées par M_{pf} (E) et les probabilités sur $(E^*, \mathcal{E}^*_{sym})$ se correspondent bijectivement par l'isomorphisme de tribus $(N^*)^{-1}$.

D'autre part à toute probabilité Q sur $(E^*, \mathcal{E}^*_{sym})$ est associée une et une seule probabilité symétrique P définie sur $(E^*, \mathcal{E}^*)$, dont Q soit la trace sur $\mathcal{E}^*_{sym}$. En effet pour construire P, il suffit de poser pour tout n et tout $A \in \mathcal{E}^{n\otimes}$

$$P(A) = \int_{E^n} \left(\frac{1}{n!} \sum_{S_n} 1_A \circ \theta_\sigma \right) dQ$$

puisque la fonction du second membre est mesurable par rapport à la tribu $\mathcal{E}^*_{sym}$. Inversement si P' est une probabilité symétrique sur $\mathcal{E}^*$ de trace Q sur $\mathcal{E}^*_{sym}$, nécessairement

$$P'(A) = \int_{E^n} \left(\frac{1}{n!} \sum_{S_n} 1_A \circ \theta_\sigma \right) dP' = \int_{E^n} \left(\frac{1}{n!} \sum_{S_n} 1_A \circ \theta_\sigma \right) dQ$$

pour tout $A \in \mathcal{E}^{n\otimes}$ et tout n.□

B. Mélanges de processus de Poisson

Soit μ une mesure de Radon positive sur E, par exemple la mesure de Lebesgue sur $E = R^d$, et pour tout réel $\theta \geqslant 0$ soit P_θ la loi de probabilité sur M_p (E) du processus de Poisson d'intensité $\theta . \mu$. Etant donné une probabilité λ sur R_+, la formule

(11) $\qquad P_\lambda (A) = \int_{R_+} \lambda (d\theta)\, P_\theta (A) \qquad\qquad (A \in \mathcal{M}_p (E))$

définit une nouvelle probabilité sur M_p (E) qui est la loi d'un mélange de processus de Poisson d'intensités proportionnelles à μ. La mesurabilité de la fonction $\theta \to P_\theta (A)$ est facile à vérifier sur les formules explicitant P_θ lorsque A est de la forme $A = \{N (F_1) = n_1, \ldots N (F_k) = n_k\}$; il ne reste plus alors qu'à appliquer le lemme I.3 à la classe des $A \in \mathcal{M}_p (E)$ rendant la fonction $P_. (A)$ mesurable. [Pour une notion plus générale de mélanges de processus de Poisson, voir la 2ème application du paragraphe I.4] .

Les probabilités P_λ ainsi construites ont comme les lois de processus de Poisson qu'elles généralisent la propriété intéressante suivante que nous avons déjà utilisée dans la construction des processus de Poisson [propriété (b) page 12]

Proposition I.11

Soit μ une mesure de Radon positive non nulle sur E. Les probabilités P_λ $(\lambda \neq \varepsilon_o)$ construites ci-dessus ont alors la propriété suivante : pour tout borélien F tel que $0 < \mu (F) < \infty$ et tout entier $n \geqslant 0$, la loi de probabilité conditionnelle par rapport à l'évènement (non négligeable) $\{N (F) = n\}$ de la trace sur F du processus ponctuel N, soit de $1_F . N$ est égale à la loi de la mesure ponctuelle $\sum_1^n \varepsilon_{X_i}$ où $X_1, \ldots X_n$ sont n points aléatoires de F choisis au hasard indépendamment et suivant la probabilité $1_F . \mu / \mu(F)$.

Réciproquement dans le cas où $\mu (E) = + \infty$, une probabilité P sur M_p (E) ne peut jouir de la propriété précédente relativement à la mesure de Radon μ que si elle est de la forme P_λ $(\lambda \neq \varepsilon_o)$.

Démonstration

Soit $Q_{n,F}$ la loi de probabilité sur M_p (F) de $\sum_1^n \varepsilon_{X_i}$ où $X_1, \dots X_n$ sont n points choisis au hasard indépendamment suivant la loi $1_F \mu / \mu(F)$ sur F. Pour montrer que pour tout $A \in \mathcal{M}_p(F)$

$$P_\lambda \left[1_F . N \in A / N(F) = n \right] = Q_{n,F}(A),$$

il suffira, grâce au lemme I.3, de considérer le cas de A de la forme suivante $A = \{m : m(F_i) = n_i \ (i = 1, \dots k)\}$ où $(F_1, \dots F_k)$ est une partition borélienne finie arbitraire de F et où $n_1, \dots n_k$ sont des entiers positifs quelconques de somme $\sum_1^k n_i = n$. Or

$$P_\lambda \left[N(F_i) = n_i \ (i = 1, \dots k) \right] = \int_{R_+} \lambda(d\theta) \, P_\theta \left[N(F_i) = n_i \ (i=1, \dots k) \right]$$

$$= \int_{R_+} \lambda(d\theta) \prod_1^k \exp\left[-\theta \mu(F_i)\right] \frac{\left[\theta \mu(F_i)\right]^{n_i}}{n_i !}$$

$$= \int_{R_+} \lambda(d\theta) \exp\left[-\theta \mu(F)\right] \frac{\left[\theta \mu(F)\right]^n}{n!} . n! \prod_i \frac{\left[\mu(F_i)/\mu(F)\right]^{n_i}}{n_i !}$$

$$= P_\lambda \left[N(F) = n\right] . Q_{n,F} \left[\{ m : M(F_i) = n_i \ (i = 1 \dots k)\}\right]$$

La première partie de la proposition est ainsi établie.

Réciproquement si $\mu(E) = \infty$, soit F un borélien tel que $0 < \mu(F) < \infty$ et soit G un second borélien contenu dans F. L'hypothèse entraîne que pour tous m et n ($m \leqslant n$) dans $\mathbb{N}$

$$P\left[N(G) = m / N(F) = n\right] = \binom{n}{m} \frac{\mu(G)^m \, \mu(F-G)^{n-m}}{\mu(F)^n}$$

ce qui implique par un calcul facile que pour tout $u \in [0, 1]$

$$E\left[u^{N(G)}\right] = E\left(\left[1 - \frac{(1-u)\,\mu(G)}{\mu(F)}\right]^{N(F)} \right).$$

Ceci suggère d'introduire les fonctions continues φ_F appliquant l'invervalle $[0, \mu(F)]$ dans $[0, 1]$ par la formule

$$\varphi_F(\theta) = E\left(\left[1 - \frac{\theta}{\mu(F)}\right]^{N(F)} \right)$$

car alors la relation que nous venons de démontrer peut se réécrire

$$\varphi_G\ \left[(1-u)\ \mu\ (G)\right] = E\ \left[u^{N(G)}\right] = \varphi_F\ \left[(1-u)\ \mu\ (G)\right] .$$

$(0 \leq u \leq 1)$. Il s'en suit que les fonctions φ_F que nous venons d'introduire ne dépendent pas de F si ce n'est pas leur domaine de définition ; comme il existe des boréliens relativement compacts F de mesure μ arbitrairement grande, il existe donc une fonction continue $\varphi : R_+ \to [0, 1]$ telle que $\varphi_F = \varphi$ sur $[0, \mu\ (F)]$ pour tout F.

Montrons que cette fonction φ est la transformée de Laplace d'une probabilité sur R_+. A cet effet choisissons une suite F_j de boréliens relativement compacts croissant vers E et étudions la transformée de Laplace des lois des variables aléatoires $N\ (F_j)\ /\ \mu\ (F_j)$; d'après ce qui précède, il vient pour tout $\theta \in R_+$

$$E\left[\exp\left(-\ \theta\ \frac{N\ (F_j)}{\mu\ (F_j)}\right)\right] = \varphi\left(\left[1 - e^{\frac{-\theta}{\mu(F_j)}}\right] \mu\ (F_j)\right)$$

et le deuxième membre tend vers $\varphi(\theta)$ lorsque $j \uparrow \infty$ puisque $\lim_j \uparrow \mu\ (F_j) = \mu\ (E) = +\infty$. D'après un critère classique (voir l'appendice) les lois de probabilité des variables aléatoires $N\ (F_j)\ /\ \mu\ (F_j)$ tendent donc étroitement sur R_+ vers une probabilité limite, soit λ, dont φ est la transformée de Laplace.

Il est maintenant facile de montrer que $P = P_\lambda$. En effet si $F_1, \ldots F_k$ sont des boréliens relativement compacts deux à deux disjoints et si $F = \sum_1^k F_i$, l'hypothèse faite sur P entraîne que pour tous $u_i \in [0, 1]$ $(i=1,\ldots k)$ et tout entier positif n

$$E\left[\prod_1^k u_i^{N(F_i)}\ /\ N\ (F) = n\right] = v^n \quad \text{si } v = \sum_1^k u_i\ \frac{\mu\ (F_i)}{\mu\ (F)} \ ;$$

par conséquent, d'après ce qui précède

$$E\left[\prod_1^k u_i^{N(F_i)}\right] = E\left[v^{N(F)}\right] = \varphi\left[(1-v)\ \mu\ (F)\right]$$

$$= \int_{R_+} e^{-\theta \sum_1^k (1-u_i)\ \mu(F_i)}\ \lambda\ (d\theta)$$

$$= E_\lambda \left[\prod_1^k u_i^{N(F_i)} \right]$$

Les vecteurs aléatoires $(N(F_1), \dots N(F_k))$ ayant donc les mêmes lois pour P et P_λ, il s'en suit bien que $P = P_\lambda$. □

C. Processus poissonniens de nuages

Un processus ponctuel de Poisson d'intensité finie sur E est un modèle de répartition d'un nombre aléatoire N (E) de points de E que l'on choisit chacun au hasard suivant la loi $\mu / \mu(E)$; une interprétation analogue est valable pour un processus ponctuel de Poisson N d'intensité μ infinie, ses traces $1_F \cdot N$ sur les boréliens μ-intégrables F étant des processus de Poisson d'intensités finies. Au lieu de répartir des points au hasard de cette manière, nous allons plus généralement dans ce paragraphe répartir des processus ponctuels (ou nuages) au hasard.

Commençons par remarquer que dans la définition et la construction des processus de Poisson, la structure topologique de l'espace de base ne sert pas. Si $(F, \mathfrak{F})$ est un espace mesurable, si $(F_j, j \in \mathbb{N})$ est une suite (ou un ensemble filtrant croissant à base dénombrable) dans $\mathfrak{F}$ croissant vers F, si $M_p(F)$ est défini comme l'ensemble des mesures sur F de la forme $\sum_I \varepsilon_{x_i}$ qui sont finies sur les F_j, alors les démonstrations du paragraphe I.2 se modifient aisément pour établir l'existence pour toute mesure positive μ sur $(F, \mathfrak{F})$ finie sur les F_j d'un processus ponctuel de Poisson d'intensité μ à valeurs dans $M_p(F)$.

Nous appliquerons cette remarque au cas où l'espace $(F, \mathfrak{F})$ est obtenu en enlevant 0 à $(M_p(E), \mathcal{M}_p(E))$ associé à un E localement compact à base dénombrable et où

$$F_j = \{ m : m \in M_p(E), m(E_j) \neq 0 \}$$

pour une suite donnée d'ouverts relativement compacts E_j de E croissant vers E. [La définition de $M_p(F)$ ne dépend pas alors de la suite (E_j) choisie].

Si Q est une mesure positive sur $F = M_p(E) - \{0\}$, finie sur les F_j, il existe donc un espace de probabilité $(\Omega, \mathcal{A}, P)$ et un processus de Poisson $N : \Omega \to M_p(F)$ d'intensité Q.

Soit maintenant $\sum_I \varepsilon_{m_i}$ $\left[m_i \in F = M_p(E)\right]$ un élément de $M_p(F)$. D'après la définition de ce dernier espace, l'ensemble $\{i : i \in I,\ m_i(E_j) \neq 0\}$ est fini pour tout j, ce qui permet de considérer la mesure ponctuelle $\sum_I m_i$ sur E. Nous définissons ainsi une application $\phi : M_p(F) \to M_p(E)$ par

$$\phi\left(\sum_I \varepsilon_{m_i}\right) = \sum_I m_i$$

dont il n'est pas difficile de vérifier qu'elle est mesurable ; en effet si $A \in \mathcal{E}$, soit $\chi_A : F \to \overline{\mathbb{N}}$ la fonction $\mathcal{F}$-mesurable définie par $\chi_A(m) = m(A)$, alors pour tout $n \in M_p(F)$ nous avons $\int_F \chi_A\, dn = \phi n(A)$.

En composant le processus de Poisson $N : \Omega \to M_p(F)$ et l'application mesurable $\phi : M_p(F) \to M_p(E)$, nous obtenons un processus ponctuel $N^\bullet : \Omega \to M_p(E)$ dont nous allons calculer la fonctionnelle de Laplace. A cet effet si $f : E \to R_+$ est une fonction borélienne, définissons la fonction $\chi_f : F \to R_+$ par $\chi_f(m) = \int f\, dm$ si $m \in F \subset M_p(E)$; cette fonction étant mesurable par rapport à $\mathcal{F} = \mathcal{M}_p(E)$, nous pouvons écrire que $\int_E N^\bullet(\omega, dx)\, f(x) = \int_F N(\omega, dm)\ \chi_f(m)$ et donc que

$$\psi_{N^\bullet}(f) = E\left(\exp\left[-N^\bullet(f)\right]\right) = E\left(\exp\left[-N(\chi_f)\right]\right)$$

$$= \exp\left[-\int_F \left[1 - e^{-\chi_f(m)}\right] Q(dm)\right]$$

D'où la formule

$$(12) \qquad \psi_{N^\bullet}(f) = \exp\left[-\int_{M_p(E) - \{0\}} \left[1 - e^{-\int f\, dm}\right] Q(dm)\right].$$

Comme il est naturel, cette formule se réduit à celle donnant la fonctionnelle de Laplace d'un processus de Poisson ordinaire lorsque la mesure Q ne charge dans $M_p(E)$ que les masses unités, c'est-à-dire lorsque Q est l'image par l'application $x \to \varepsilon_x$ de E dans $M_p(E) - \{0\}$ d'une mesure de Radon μ sur E.

I.4. PROBLEMES D'EXISTENCE ET DE CONVERGENCE DE MESURES ALEATOIRES

Comme nous allons nous en rendre compte, les problèmes d'existence et de convergence de lois de processus ponctuels se posent plus naturellement dans le cadre un peu plus général des mesures aléatoires. En nous plaçant dans le cadre des mesures de Radon, nous aboutirons à des résultats d'une très grande simplicité.

Commençons par généraliser aux mesures aléatoires les notions générales du paragraphe 1 ; cela se fait sans difficultés. Etant donné toujours un espace E localement compact à base dénombrable, le théorème de Riesz nous apprend que les mesures positives définies sur la tribu borélienne $\mathcal{E}$ de E et finies sur les boréliens relativement compacts correspondent bijectivement par la théorie de l'intégration avec les formes linéaires positives définies sur l'espace C_k (E) des fonctions continues de E dans R à supports compacts. Nous noterons M_+ (E) le cône convexe de ces mesures et nous le munirons de la plus petite tribu, soit $\mathcal{M}_+$ (E), rendant les applications $m \to m(f) \equiv \int f\,dm$ $(f \in C_k(E))$ mesurables de M_+ (E) dans R.

La tribu $\mathcal{M}_+$ (E) est aussi celle engendrée par les applications $m \to m(G)$ de M_+ (E) dans R_+ obtenues lorsque G parcourt les ouverts relativement compacts de E, ou lorsque G parcourt tous les boréliens de E [Pour la première équivalence noter que pour tout ouvert G, la fonction 1_G est la limite d'une suite croissante de fonctions de C_k^+ (E) et que toute fonction $f \in C_k^+$ (E) peut s'écrire sous la forme $\sum_N a_n 1_{G_n}$ pour des $a_n \in R_+$ et des G_n ouverts relativement compacts. Pour la deuxième équivalence, voir le § I.1 et le lemme I.3]. L'espace M_p (E) des mesures ponctuelles est un sous-ensemble borélien de M_+ (E) ; en effet si $(G_j, j \in J)$ est une base dénombrable de la topologie de E formée d'ouverts relativement compacts, il est aisé de vérifier que [cf exercice 1]

$$(13) \qquad M_p(E) = \{m : m \in M_+(E)\ ,\ m(G_j) \in \mathbb{N} \qquad \forall j \in J\}$$

Il est alors clair d'après ce qui précède que

$$\mathcal{M}_p (E) = M_p (E) \cap \mathcal{M}_+ (E)$$

<u>Une mesure aléatoire positive sur</u> E, soit N, est définie comme une application mesurable d'un espace de probabilité $(\Omega, \mathcal{A}, P)$ dans $M_+ (E)$; une mesure aléatoire positive prenant ses valeurs dans $M_p (E)$ est un processus ponctuel d'après l'alinéa précédent. La loi et la fonctionnelle de Laplace d'une mesure aléatoire positive se définissent comme dans le cas des processus ponctuels. Nous ne considérerons souvent cette fonctionnelle de Laplace qui est définie sur les fonctions boréliennes positives $f : E \to R_+$ par

$$\psi_N (f) = \int_\Omega \exp \left[- \int_E N(\omega, dx) f(x)\right] P(d\omega)$$

$$= \int_{M_+(E)} \exp \left[- \int_E m(dx) f(x)\right] P_N (dm)$$

que par sa restriction au cône $C_k^+ (E)$. Il résulte clairement de cette définition que pour toute suite finie $f_1, \ldots f_n$ dans $C_k^+ (E)$, si $\mu_{f_1,\ldots f_n}$ désigne la loi du vecteur aléatoire $(N(f_1), \ldots N(f_n))$ sur R_n^+, la transformée de Laplace de cette loi est donnée par

$$\int_{R_n^+} \exp \left(- \sum_1^n t_m y_m \right) \mu_{f_1 \ldots f_n} (dy_1 \ldots dy_n) = \psi_N \left(\sum_1^n t_m f_m \right)$$

Il est remarquable que cette propriété caractérise les fonctionnelles de Laplace comme le montre le <u>théorème d'existence et d'unicité suivant</u> :

<u>Proposition I.12</u>

<u>Soit</u> ψ <u>une application de</u> $C_k^+ (E)$ <u>dans</u> R <u>telle que pour toute suite finie</u> $(f_1, \ldots f_n)$ <u>dans</u> $C_k^+ (E)$ <u>la fonction</u>

$$(t_1, \ldots t_n) \to \psi \left(\sum_1^n t_j f_j \right)$$

<u>soit sur</u> R_+^n <u>la transformée de Laplace d'une probabilité</u>. <u>Il existe alors une probabilité unique</u> P <u>sur</u> $M_+ (E)$ <u>dont</u> ψ <u>soit la transformée de Laplace</u>.

Une manière équivalente de formuler cette proposition est la suivante. Etant donné, pour toute suite finie $f_1, \dots f_n$ dans $C_k^+(E)$ une probabilité $\mu_{f_1,\dots,f_n}$ sur R_+^n, si ces probabilités vérifient la condition de compatibilité

(C) : quelles que soient $f_1, \dots f_n \in C_k^+(E)$ et quels que soient $a_1, \dots a_n \in R_+$ l'image de $\mu_{f_1,\dots,f_n}$ par l'application $(y_1 \dots y_n) \to \sum_1^n a_j y_j$ est égale à μ_f si $f = \sum_1^n a_j f_j$,

il existe une probabilité unique P sur $M_+(E)$ telle que $\mu_{f_1,\dots,f_n}$ soit pour toute suite finie $f_1, \dots f_n$ dans $C_k^+(E)$, l'image de P par l'application $m \to (m(f_1), \dots m(f_n))$ de $M_+(E)$ dans R_n^+.

Cette reformulation de la proposition précédente s'obtient en remarquant que la condition (C) est exactement celle qui permet de définir une application $\psi : C_k^+(E) \to [0, 1]$ telle que $\psi(t_1 f_1 + \dots + t_n f_n)$ soit la transformée de Laplace de $\mu_{f_1,\dots,f_n}$. [Pour définir ψ, poser $\psi(f) = \int_{R_+} e^{-y} \, d\mu_f(y)$ et noter alors que (C) entraîne que $\psi(t_1 f_1 + \dots + t_n f_n) =$

$$= \int \exp\left(-\sum_1^n t_m y_m\right) \mu_{f_1\dots f_n}(dy_1 \dots dy_n)\Big] .$$

Démonstration classique.□

Dans le cas des processus ponctuels, c'est sans doute la proposition suivante qui est la plus utile.

Proposition I.13

Soit B_{ke}^+ le cône des fonctions boréliennes étagées et positives nulles hors d'un compact. Toute application ψ de B_{ke}^+ dans $[0, 1]$ telle que

a) pour toute suite finie $F_1, \dots F_n$ de boréliens relativement compacts deux à deux disjoints la fonction

$$(u_1, \dots u_n) \to \psi\left[\sum_1^n \log \frac{1}{u_i} \cdot 1_{F_i}\right] \qquad (0 < u_1, \dots u_n \leqslant 1)$$

soit la fonction génératrice d'une probabilité sur $\mathbb{N}^n$,

b) $\lim_{n\to\infty} \psi(1_{F_n}) = 1$ pour toute suite $(F_n, n \in \mathbb{N})$ de boréliens relativement compacts décroissant vers $\emptyset$,

il existe une probabilité unique P sur $M_p(E)$ telle que

$$\psi(f) = \int_{M_p(E)} \exp\left[-\int_E f\, dm\right] P(dm) \qquad (f \in B^+_{ke}).$$

Démonstration

Elle consiste à se ramener à la proposition précédente. A cet effet, remarquons d'abord que les hypothèses entraînent que la fonction ψ jouit des trois propriétés suivantes :

1) $\psi(f) \geq \psi(g)$ si $f \leq g$ dans B^+_{ke},

2) $1 - \psi(f+g) \leq [1 - \psi(f)] + [1 - \psi(g)]$ quels que soient $f, g \in B^+_{ke}$

et $1 - \psi(cf) \leq c\,[1 - \psi(f)]$ si $c \geq 1$ et $f \in B^+_{ke}$,

3) $\lim_n \uparrow \psi(f_n) = 1$ pour toute suite (f_n) dans B^+_{ke} qui décroît vers 0.

Pour démontrer (1-2), écrivons f et g sous la forme $f = \sum_1^n a_i 1_{F_i}$, $g = \sum_1^n b_i 1_{F_i}$ pour des boréliens relativement compacts deux à deux disjoints $F_1, \dots F_n$ et des réels $a_1, \dots b_n \geq 0$. Comme par hypothèse il existe une probabilité μ sur R^n_+ (et même sur $\mathbb{N}^n$) telle que

$$\psi\left(\sum_1^n t_i 1_{F_i}\right) = \int_{R^+_n} \exp(-t \,.\, y)\, \mu(dy),$$

les inégalités (1-2) s'obtiennent en intégrant par rapport à μ les inégalités élémentaires suivantes valables pour tout $y \in R^n_+$

$$\exp(-a.y) \geq \exp(-b.y) \qquad \text{si } a \leq b,$$

$$1 - \exp[-(a+b).y] \leq [1 - \exp(-a.y)] + [1 - \exp(-b\,.\,y)],$$

$$1 - \exp[-c\,(a.y)] \leq c\,[1 - \exp(-a.y)] \text{ si } c \geq 1.$$

La propriété (3) s'obtient alors en majorant f_0 par $c\,1_F$ pour une constante $c \geqslant 1$ et un borélien relativement compact F et en écrivant que $f_n \leqslant \varepsilon\, 1_F + c\, 1_{\{f_n > \varepsilon\}}$ de façon à en déduire, en utilisant les inégalités (1-2) que

$$1 - \psi(f_n) \leqslant [1 - \psi(\varepsilon\, 1_F)] + c\,[1 - \psi(1_{\{f_n > \varepsilon\}})] \quad ;$$

ainsi

$$1 - \psi(f_n) \to 0 \text{ lorsque } n \uparrow \infty \text{ et puis } \varepsilon \downarrow 0,$$

car d'une part l'hypothèse (b) de la proposition s'applique aux ensembles $\{f_n > \varepsilon\}$ qui décroissent vers Ø et d'autre part $\psi(t\,1_F)$ est une transformée de Laplace de probabilité.

Ces propriétés préliminaires permettent de prolonger la fonctionnelle ψ au cône B_k^+ des fonctions boréliennes positives bornées nulles hors d'un compact de manière à ce que pour toute suite $(f_n,\ n \in N)$ de B_k^+ croissant ou décroissant vers une fonction f de B_k^+, on ait $\lim_n \downarrow$ ou $\uparrow \psi(f_n) = \psi(f)$.
[La démonstration est laissée au lecteur] .

Notons que pour toute suite finie $f_1, \ldots f_n$ dans B_k^+, la fonction $\psi(\sum_1^n t_i f_i)$ est encore la transformée de Laplace d'une probabilité sur R_+^n. En effet si les f_i sont étagées donc de la forme $f_i = \sum_1^m a_j^i\, 1_{F_j}$ pour des boréliens relativement compacts deux à deux disjoints $F_1, \ldots F_m$, alors la fonction $\psi(\sum_1^n t_i f_i)$ est la transformée de Laplace de l'image par l'application $(y_1, \ldots y_m) \to (\sum_{j=1}^m a_j^i y_j\ ;\ i = 1 \ldots n)$ de la probabilité sur R_m^+ de transformée de Laplace $\psi(\sum_1^m s_j\, 1_{F_j})$. Dans le cas général écrivons les fonctions f_i comme limites croissantes de fonctions étagées f_i^p $(i = 1 \ldots n\ ;\ p \uparrow \infty)$; alors la fonction $\psi(\sum_1^n t_i f_i)$ de $(t_1, \ldots t_n)$ est la limite des fonctions $\psi(\sum_1^n t_i f_i^p)$ qui sont des transformées de Laplace de probabilités et en outre cette fonction est continue en 0 car si $f_i \leqslant c\, 1_F$ $(i = 1, \ldots n)$ pour une constante c et un borélien relativement compact F

$$1 - \psi(\sum_1^n t_i f_i) \leqslant 1 - \psi\left[(\sum_1^n t_i)\, c\, 1_F\right] \to 0$$

lorsque $\sum_1^n t_i \to 0$; d'après un critère classique (voir l'appendice) cela entraîne que $\psi(\sum_1^n t_i f_i)$ est la transformée de Laplace d'une probabilité sur R_+^n.

La restriction à $C_k^+(E)$ de la fonctionnelle ψ vérifie les hypothèses de la proposition I.12. Il existe donc une probabilité sur $M_+(E)$ telle que

$$\psi(f) = \int_{M_+(E)} \exp[-m(f)]\; P(dm)$$

pour tout $f \in C_k^+(E)$. Mais les deux membres sont définis plus généralement sur B_k^+ et ont sur ce cône les propriétés de continuité monotone croissante et décroissante ; comme d'autre part B_k^+ est la plus petite classe de fonctions stables par limites monotones contenant $C_k^+(E)$, l'égalité ci-dessus reste valable pour toute fonction f de $B_k^+(E)$.

Enfin si (G_j) est une base de la topologie formée d'ouverts relativement compacts, nous avons

$$P[M_p(E)] = P(\{m : m(G_j) \in \mathbb{N} \quad \forall j\}) = 1$$

car pour tout j, $\psi(t\, 1_{G_j})$ est par hypothèse la transformée de Laplace d'une probabilité sur $\mathbb{N}$ qui puisque ψ est la fonctionnelle de Laplace de P, ne peut être que l'image de P par $m \to m(G_j)$. La proposition est ainsi complètement démontrée.□

Pour étudier les problèmes de convergence de lois de probabilité de processus ponctuels ou plus généralement de mesures positives aléatoires, munissons l'espace $M_+(E)$ des mesures de Radon positives sur E de la topologie de la convergence vague. Notons que pour cette topologie le sous-espace $M_p(E)$ des mesures ponctuelles est un fermé de $M_+(E)$; nous le munirons bien entendu de la topologie induite. La convergence vague dans $M_p(E)$ peut d'ailleurs se décrire explicitement de la manière suivante.

Lemme I.14

Si la suite $(m_n, n \in N)$ converge vaguement vers m dans $M_p(E)$, tout borélien relativement compact F de frontière négligeable pour m est tel que $m(F) = \lim_n m_n(F)$ et de plus il existe un entier n_F tel que si $n \geqslant n_F$ les mesures ponctuelles m_n et m restreintes à F puissent s'écrire sous la forme

$$m_n = \sum_1^p \varepsilon_{x_q^n} \quad , \quad m = \sum_1^p \varepsilon_{x_q} \quad \text{sur } F \qquad (n \geqslant n_F)$$

pour p suites de points $(x_q^n, n \geqslant n_F)$ dans F $(q = 1, \dots p)$ convergeant respectivement vers x_q.

Démonstration

La première affirmation du lemme est un résultat général de convergence vague de mesures de Radon positives. Pour démontrer la seconde partie, écrivons la restriction de m à F sous la forme $1_F.m = \sum_1^s c_r \varepsilon_{y_r}$ où $y_1, \dots y_s$ sont des points de F, par hypothèse dans $\mathring{F}$ car $m(\partial F) = 0$, et où $c_1, \dots c_s \in \mathbb{N}^*$. Choisissons un voisinage ouvert G_r dans $\mathring{F}$ de y_r pour tout $r = 1 \dots s$ de sorte que les G_r soient deux à deux disjoints ; les frontières des G_r sont alors m-négligeables et par conséquent $\lim_n m_n(G_r) = m(G_r)$; pour un entier n_F suffisamment grand nous aurons donc $m_n(G_r) = m(G_r) = c_r$ $(r = 1, \dots s)$ et $m_n(F) = m(F)$ si $n \geqslant n_F$. Par suite la restriction de m_n à F est exactement formée de c_r points dans chacun des G_r $(r = 1, \dots s)$. Il reste à numéroter ces points et à faire décroître ensuite les G_r vers les $\{y_r\}$ pour obtenir le lemme.□

Posons maintenant la définition fondamentale suivante.

Définition I.15

Une suite $(P_n, n \in \mathbb{N})$ de probabilités définies sur l'espace $(M_+(E), \mathcal{M}_+(E))$ est dite converger étroitement vers une probabilité P_∞ définie sur ce même espace si pour toute fonction continue bornée $F : M_+(E) \to R$

(14) $$\lim_{n\to\infty} \int_{M_+(E)} F(m)\, P_n\,(dm) = \int_{M_+(E)} F\,(m)\, P\,(dm).$$

La proposition suivante donne un critère fort simple de convergence étroite. Notons parce que cela nous sera utile au paragraphe suivant que la définition précédente ainsi que la proposition ci-dessous s'étendent immédiatement à des suites de mesures positives finies sur M_+ (E).

Proposition I.16

Pour qu'une suite (P_n, $n \in N$) de probabilités sur M_+ (E) converge étroitement vers une probabilité P_∞, il faut et il suffit que leurs transformées de Laplace convergent sur C_k^+ (E), c'est-à-dire que

(15) $$\lim_{n\to\infty} \psi_{P_n}\,(f) = \psi_P\,(f) \quad \text{pour tout} \quad f \in C_k^+\,(E)$$

Démonstration

Puisque l'application $m \to \exp\,[-\,m\,(f)]$ est continue et bornée sur M_+ (E) pour tout $f \in C_k^+$ (E), la condition est évidemment nécessaire. Montrons maintenant qu'elle est suffisante.

Etant donné une suite finie $f_1, \dots f_d$ dans C_k^+ (E) et une fonction continue bornée $\phi : R^d \to R$, établissons pour commencer que

$$\lim_{n\to\infty} \int_{M_+(E)} \phi\,[m(f_1), \dots m\,(f_d)]\; P_n\,(dm) = \int_{M_+(E)} \phi\,[m(f_1), \dots m(f_d)]\; P_\infty(dm$$

Ceci revient à montrer que

$$\lim_{n\to\infty} \int_{R_+^d} \phi\,(x)\, P_n^{f_1 \dots f_d}\,(dx) = \int_{R_+^d} \phi\,(x)\, P_\infty^{f_1,\dots f_d}\,(dx) \quad .$$

si $P_n^{f_1 \dots f_d}$ ($n \leqslant \infty$) désigne l'image de P_n par l'application $m \to (m\,(f_1), \dots m\,(f_d))$ de M_+ (E) dans R_+^d. Or pour établir cette convergence quelle que soit la fonction continue bornée ϕ il suffit d'après un critère classique (voir l'appendice) de prouver la convergence des transformées de Laplace de ces mesures images ; mais cette convergence est assurée par l'hypothèse puisque

$$\int_{R_+^d} \exp\left(-\sum_1^d t_j x_j\right) P_n^{f_1,\dots,f_d}(dx) = \psi_{P_n}\left[\sum_1^d t_j f_j\right] \qquad (n \leqslant \infty)$$

et que $\sum_1^d t_j f_j \in C_k^+(E)$.

La famille A de toutes les fonctions réelles sur $M_+(E)$ de la forme $m \to \phi\ (m(f_1), \dots m(f_d))$ obtenues lorsque $(f_1, \dots f_d)$ parcourt les suites finies de $C_k^+(E)$ et lorsque ϕ parcourt les fonctions continues bornées sur R^d, constitue une sous-algèbre unitaire de l'algèbre $C_b[M_+(E)]$ des fonctions continues bornées sur $M_+(E)$. De plus cette sous-algèbre sépare les points de $M_+(E)$ car si m_1, m_2 sont deux éléments distincts de $M_+(E)$ il existe au moins une fonction $f \in C_k^+(E)$ telle que $m_1(f) \neq m_2(f)$ et une fonction continue bornée $\phi : R \to R$ telle que $\phi[m_1(f)] \neq \phi[m_2(f)]$. Le théorème de Stone-Weierstrass montre alors que pour tout compact K de $M_+(E)$, la sous-algèbre de C (K) formée par les restrictions à K des fonctions de A est dense pour la norme sup dans C (K). Par conséquent, K étant fixé, pour toute fonction $F \in C_b[M_+(E)]$, il existe une suite de fonctions F_p $(p \in N)$ dans A telle que $\sup_K |F(m) - F_p(m)| \to 0$ lorsque $p \uparrow \infty$; en outre en tronquant au seuil $\sup_{M_+(E)} |F(m)|$ les fonctions ϕ_p intervenant dans la définition des F_p si c'est nécessaire, nous pouvons supposer que la suite F_p $(p \in N)$ approchant F sur K est telle que

$$\sup_{M_+(E)} |F_p(m)| \leqslant \sup_{M_+(E)} |F(m)| .$$

Nous montrerons ci-dessous que pour tout $\varepsilon > 0$ donné, il existe au moins un compact K_ε de $M_+(E)$ tel que $P_n(K_\varepsilon^c) \leqslant \varepsilon$ pour tout $n \leqslant \infty$. Cela permettra d'achever la démonstration en écrivant que

$$\left|\int F\,dP_n - \int F\,dP_\infty\right|$$

$$\leqslant \left|\int F_p\,dP_n - \int F_p\,dP_\infty\right| + \int |F - F_p|\,dP_n + \int |F - F_p|\,dP_\infty$$

$$\leqslant \left|\int F_p\,dP_n - \int F_p\,dP_\infty\right| + 2\sup_{K_\varepsilon} |F(m) - F_p(m)| + 4\,\varepsilon \sup_{M_+(E)} |F(m)|$$

$\to 0$ lorsque successivement $n \uparrow \infty$, $p \uparrow \infty$ et $\varepsilon \downarrow 0$.

Pour construire les compacts K_ε précédents, donnons-nous une suite $(g_k, k \in \mathbb{N}^*)$ dans $C_k^+(E)$ croissant vers 1. Quelle que soit alors la suite de réels positifs a_k $(k \in \mathbb{N}^*)$, l'ensemble $K = \{m : m(g_k) \leq a_k \ (k \in \mathbb{N}^*)\}$ est compact pour la topologie de la convergence vague. Comme

$$\sup_n P_n(\{m : m(g_k) > a\}) \leq e^{-a} \sup_n \psi_{P_n}(g_k)$$

nous pouvons choisir $a = a_k$ pour que le premier membre soit majoré par $\varepsilon/2^k$ $(k \in \mathbb{N}^*)$ et alors nous avons bien

$$P_n(K^c) = P_n\left[\bigcup_{k \in \mathbb{N}^*} \{m(g_k) > a_k\}\right] \leq \sum_{\mathbb{N}^*} \varepsilon/2^k = \varepsilon$$

pour tout $n \in \mathbb{N}$. □

Remarquons que la proposition précédente montre que quels que soient les processus ponctuels N_n $(n \in \mathbb{N})$ et N_∞ sur E, les conditions suivantes sont équivalentes :

a) la loi P_n de N_n converge étroitement vers celle de P_∞ au sens de la définition I.15 ,

b) pour toute suite finie $(f_1, \ldots f_k)$ dans $C_k^+(E)$, la loi de $(N_n(f_1), \ldots N_n(f_k))$ converge étroitement sur R_+^k vers celle de $(N_\infty(f_1), \ldots, N_\infty(f_k))$,

c) pour tout $f \in C_k^+(E)$, la loi de $N_n(f)$ converge étroitement sur R_+ vers celle de $N_\infty(f)$.

[En effet (a) $\Longrightarrow$ (b) $\Longrightarrow$ (c) $\Longrightarrow \lim_n \psi_n(f) = \psi(f)$ pour tout $f \in C_k^+(E)$ et la proposition précédente achève d'établir la boucle.]

<u>Corollaire I.17</u>

<u>Si la suite</u> $(\psi_n, n \in \mathbb{N})$ <u>des transformées de Laplace de probabilités</u> P_n <u>sur</u> $M_+(E)$ <u>converge simplement sur</u> $C_k^+(E)$ <u>vers une fonction</u> $\psi : C_k^+(E) \to [0,1]$, <u>soit si</u> $\lim_{n\to\infty} \psi_n(f) = \psi(f)$ <u>pour tout</u> $f \in C_k^+(E)$ <u>et si de plus</u> $\lim_{t\downarrow 0} \psi(tf) = 1$

pour tout $f \in C_k^+$ (E) alors la suite des probabilités P_n converge étroitement sur M_+ (E) vers une probabilité P dont ψ est la transformée de Laplace.

Démonstration

Pour toute suite finie $f_1, \ldots f_k$ dans C_k^+ (E), la fonction $\psi\,(t_1 f_1 + \ldots + t_k f_k)$ est la transformée de Laplace d'une probabilité sur R_+^k car d'une part c'est la limite des transformées de Laplace $\psi_n\,(t_1 f_1 + \ldots + t_k\, f_k)$ lorsque $n \uparrow \infty$ et car d'autre part $\psi\,(t_1\, f_1 + \ldots + t_k f_k) \to 1$ lorsque $\sup\limits_i t_i \to 0$ puisque

$$\psi\,(t_1 f_1 + \ldots + t_k f_k) \geq \psi\left[(\sup_i t_i)\,(f_1 + \ldots + f_k)\right] \to 1 \text{ par hypothèse.}$$

Grâce à la proposition I.12, la fonction ψ est donc la transformée de Laplace d'une probabilité P sur M_+ (E). Il ne reste plus alors qu'à appliquer la proposition I.16 précédente.□

Terminons ce paragraphe par une remarque utile dans les applications des processus ponctuels.

Proposition I.18

Si les lois des processus ponctuels N_n $(n \in \mathbb{N})$ sur E convergent étroitement vers celle du processus ponctuel N et si μ désigne l'intensité de ce dernier processus, alors quelle que soit la suite finie de boréliens relativement compacts $F_1, \ldots F_k$ de frontières μ-négligeables dans E, la loi du vecteur aléatoire $(N_n\,(F_1), \ldots, N_n\,(F_k))$ tend étroitement vers celle du vecteur $(N\,(F_1), \ldots N\,(F_k))$.

Démonstration

D'après le critère de convergence étroite donné par la transformation de Laplace, il s'agit de démontrer que

$$\psi_n\,(t_1\, 1_{F_1} + \ldots + t_k\, 1_{F_k}) \to \psi\,(t_1\, 1_{F_1} + \ldots + t_k\, 1_{F_k})$$

lorsque $n \uparrow \infty$ et quels que soient $t_1, \ldots t_k \in R_+$. Or si $f_1, \ldots f_k$ sont des fonctions de C_k^+ (E) majorées respectivement par $1_{F_1}, \ldots 1_{F_k}$, la convergence de ψ_n vers ψ sur C_k^+ (E) et la décroissance de ces fonctions entraînent que

$$\limsup_n \psi_n (t_1 1_{F_1} + \dots + t_k 1_{F_k}) \leqslant \psi (t_1 f_1 + \dots t_k f_k) ;$$

en faisant tendre f_j vers les fonctions indicatrices des ouverts $\mathring{F}_1, \dots \mathring{F}_k$ nous trouvons ainsi que

$$\limsup_n \psi_n (t_1 1_{F_1} + \dots + t_k 1_{F_k}) \leqslant \psi (t_1 1_{\mathring{F}_1} + \dots t_k 1_{\mathring{F}_k})$$

De manière analogue, nous avons

$$\liminf_n \psi_n (t_1 1_{F_1} + \dots + t_k 1_{F_k}) \geqslant \psi (t_1 1_{\overline{F}_1} + \dots t_k 1_{\overline{F}_k})$$

Or, d'après les hypothèses, les boréliens relativement compacts F_j ont des frontières μ-négligeables, soit $E [N (\overline{F}_j - \mathring{F}_j)] = 0$ $(j = 1, \dots k)$; par suite $N (\mathring{F}_j) = N (F_j) = N (\overline{F}_j)$ p.s. et ceci entraîne que

$$\psi (t_1 1_{\mathring{F}_1} + \dots t_k 1_{F_k}) = \psi (t_1 1_{F_1} + \dots t_k 1_{F_k}) = \psi (t_1 1_{\overline{F}_1} + \dots + t_k 1_{\overline{F}_k}),$$

ce qui permet de conclure.□

Application 1. Lois indéfiniment divisibles

Le problème de la limite centrale et des lois indéfiniment divisibles tel qu'on le pose classiquement pour des variables aléatoires se généralise comme nous allons le voir aux processus ponctuels. De la même manière que dans le cas classique la théorie se développe très facilement dans le cadre de la transformation de Laplace lorsqu'on se limite à ne considérer que des variables aléatoires positives, dans le cas des processus ponctuels l'utilisation des fonctionnelles de Laplace va rendre l'étude simple.

Pour tout $n \in \mathbb{N}$, soit $(N_n^1, \dots N_n^{k_n})$ une suite finie de processus ponctuels indépendants sur le même espace E dont nous noterons les lois et les fonctionnelles de Laplace P_n^j et ψ_n^j $(j = 1, \dots k_n)$ respectivement. Nous supposons que chacun de ces processus est rarement différent de 0 sur tout compact donné et ceci uniformément en j ; de manière plus précise nous ferons l'hypothèse

$$(*) \qquad \sup_j P\left[N_n^j (F) \neq 0\right] \to 0 \qquad \text{lorsque } n \to \infty \qquad (F \in \mathcal{C}_c)$$

Le problème posé est alors celui de trouver les lois limites possibles des processus ponctuels $N_n = \sum_{j=1}^{k_n} N_n^j$ sur E et les conditions de convergence.

Or, pour que la loi de N_n converge étroitement sur M_p (E) vers une probabilité Q de fonctionnelle de Laplace Ψ, il faut et il suffit d'après la proposition I.16 que

$$\lim_{n\to\infty} \prod_{j=1}^{k_n} \psi_n^j (f) = \psi (f) \qquad \text{pour tout } f \in C_k^+ (E)$$

puisque le premier membre est bien d'après l'indépendance des N_n^j $(j = 1 \dots k_n)$ la fonctionnelle de Laplace de N_n. Prenons les logarithmes des deux membres. Comme l'hypothèse (*) entraîne que pour tout $f \in C_k^+ (E)$,

$$\sup_j \left[1 - \psi_n^j (f)\right] \leqslant \sup_j P\left[N_n^j (S) \neq 0\right] \to 0 \quad \text{si } S = \{f > 0\} \text{ lorsque } n \to \infty$$

les inégalités élémentaires $1 - a \leqslant \log (1/a) \leqslant (1-a)/a$ où $a \in \left]0, 1\right]$ montrent que la condition nécessaire et suffisante précédente peut encore s'écrire

$$\lim_{n\to\infty} \sum_{j=1}^{k_n} \left[1 - \psi_n^j (f)\right] = \log \left[1 / \psi (f)\right] \qquad (f \in C_k^+ (E)).$$

Remarquons que les sommes du premier membre s'écrivent encore

$$\sum_{j=1}^{k_n} \left[1 - \psi_n^j (f)\right] = \int_{M_p(E)} (1 - \exp\left[- m (f)\right]) \, Q_n (dm)$$

à condition de définir la mesure positive et finie Q_n par $Q_n = 1_{\{o\}^c} \sum_{j=1}^{k_n} P_n^j$

sur M_p (E). Or la convergence ci-dessus ne peut avoir lieu que si ces mesures Q_n convergent vers une mesure Q au sens précis suivant.

<u>Proposition I.19</u>

<u>Sous les hypothèses faites, pour que les lois de probabilité des processus</u> $N_n = \sum_{j=1}^{k_n} N_n^j$ <u>convergent étroitement lorsque</u> $n \to \infty$ <u>, il est nécessaire et suffisant qu'il existe une mesure positive</u> Q <u>sur</u> M_p (E)-{O} <u>telle que</u> $Q(\{m:m(F)\neq O\}) < \infty$ <u>pour tout</u> $F \in \mathcal{C}_c$ <u>pour laquelle</u>

$$(16) \qquad \int_{M_p(E)} \left[1 - e^{-m(g)}\right] . \sum_{j=1}^{k_n} P_n^j (dm) \to \int_{M_p(E)} \left[1 - e^{-m(g)}\right] Q (dm)$$

<u>lorsque</u> $n \to \infty$ <u>et pour tout</u> $g \in C_k^+$ (E). <u>Et alors les lois des</u> N_n <u>convergent vers la loi du processus poissonien de nuages associé à</u> Q.

<u>Démonstration</u>

Désignons par Q_n^g les mesures positives finies $Q_n^g (dm) = \left[1-e^{-m(g)}\right]. Q_n (dm)$; leurs transformées de Laplace ψ_n^g sont telles que

$$\psi_n^g (f) = \int \left[e^{-m(f)} - e^{-m(f+g)}\right] Q_n (dm)$$
$$= \sum_1^{k_n} \left[\psi_n^j (f) - \psi_n^j (f + g)\right] \to \log \left[\frac{\psi (f)}{\psi (f+g)}\right]$$

lorsque $n \uparrow \infty$, pour tout $f \in C_k^+$ (E). D'après le corollaire I.18 dûment étendu aux mesures positives finies, cette convergence entraîne la convergence étroite des mesures Q_n^g vers des mesures positives et finies Q^g de fonctionnelle de Laplace égale à

$$\int e^{-m(f)} \quad Q^g \;(dm) = \log \quad \frac{\psi\,(f)}{\psi\,(f+g)}$$

Montrons que comme pour les mesures Q^g_n , les mesures positives et finies Q^g $(g \in C^+_k\;(E))$ peuvent s'écrire sous la forme

$$Q^g\;(dm) = \left[1 - e^{-m(g)}\right] \;.\; Q\;(dm)$$

pour une mesure positive Q sur M_p (E). Or quelles que soient $g,\ g' \in C^+_k\;(E)$, nous avons l'égalité

$$\left[1 - e^{-m(g)}\right] \;\; Q^{g'}\;(dm) = \left[1 - e^{-m(g')}\right] Q^g\;(dm)$$

car les mesures des deux membres ont la même fonctionnelle de Laplace à savoir $\log \;\left[\psi\,(f)\;\psi\,(f+g+g')\;/\;\psi\,(f+g)\;\psi\,(f+g')\right]$. Si $(g_j,\ j \in \mathbb{N})$ est alors une suite dans C^+_k (E) croissant vers 1, il reste à définir la mesure Q par morceaux sur $M_p(E) - \{0\}$ en posant

$$Q\;(dm) = \left[1 - e^{-m(g_j)}\right]^{-1} \;.\; Q^{g_j}\;(dm) \quad \text{sur} \quad \{m : m\;(g_j) \neq 0\}\;,$$

ce que nous venons de démontrer assurant d'une part la compatibilité de ces définitions sur les intersections deux à deux des ensembles $\{m\;(g_j) \neq 0\}$ et d'autre part l'exactitude de la formule posée au début. Bien entendu la mesure positive Q n'est pas finie en général mais

$$Q\;(\{m : m\;(F) \neq 0\}\;) \leqslant c \quad \int \left[1 - e^{-m(f)}\right] \;\; Q\;(dm) \;\; < \infty$$

si $f \in C^+_k$ (E) est choisie pour que $f \geqslant 1_F$ et si $c\;(1 - e^{-1}) = 1$.

Soit alors P la loi de probabilité sur M_p (E) du processus poissonien de nuages associé à la mesure positive Q. Sa fonctionnelle de Laplace ψ_P est donnée par la formule

$$\log\;\left[1\;/\;\psi_P(f)\right] = \int_{M_p(E)} \left[1 - e^{-m(f)}\right] Q\;(dm) \;\; = \int_{M_p(E)} Q^f(dm) = \log\;\left[1\;/\;\psi(f)\right],$$

c'est-à-dire par ψ . La condition nécessaire de la proposition est ainsi établie ainsi que la forme de la limite.

Réciproquement la condition de la proposition est suffisante car elle exprime que $\sum_{j=1}^{k_n} \left[1 - \psi_n^j (g)\right] \to \log \left[1 / \psi_p (g)\right]$ pour tout $g \in C_k^+ (E)$ et le début de la démonstration permet alors de conclure.□

Application 2. Les processus de Cox et la Raréfaction

Les processus de Cox s'obtiennent en mélangeant des lois de processus de Poisson. Pour toute mesure de Radon positive μ sur E, désignons par P^μ la loi de probabilité sur M_p (E) d'un processus de Poisson d'intensité μ sur E. Comme nous allons le voir, pour tout $A \in \mathcal{M}_p (E)$, l'application $\mu \to P^\mu (A)$ de $M_+ (E)$ dans R_+ est mesurable ; par conséquent la formule

(17) $$P_\pi (A) = \int_{M_+(E)} \pi (d\mu)\, P^\mu (A) \qquad (A \in \mathcal{M}_p (E))$$

associe à toute probabilité π définie sur $M_+ (E)$ une probabilité P_π sur $M_p (E)$, qui est un mélange de lois de Poisson. Dans le cas particulier où la probabilité π est concentrée sur la famille $(\theta\mu_o, \theta \in R_+)$ des multiples d'une mesure $\mu_o \in M_+ (E)$ fixée, c'est-à-dire lorsque π est l'image d'une probabilité λ sur R_+ par l'application $\theta \to \theta \mu_o$ de R_+ dans $M_+ (E)$, la probabilité P_π coïncide avec celle de la formule (11) du paragraphe I.3 (B). Un processus ponctuel est appelé processus de Cox lorsque sa loi est de la forme (17) précédente.

Pour établir la mesurabilité des applications $\mu \to P^\mu (A)$ $(A \in \mathcal{M}_p (E))$, il suffit de considérer le cas où A est de forme $A = \{m : m (F_1) = n_1, \ldots m(F_q) = n_q\}$ pour des $F_1, \ldots F_q$ boréliens relativement compacts et des entiers $n_1, \ldots n_q$ $(q \in N)$, car d'après la démonstration de la proposition I.2, les sous-ensembles de $M_p (E)$ de cette forme constitue une classe $\mathcal{J}$ stable par intersection, engendrant $\mathcal{M}_p (E)$ et contenant $M_p (E)$. Il suffit même de considérer les A de la forme précédente pour lesquels $F_1, \ldots F_q$ sont disjoints deux à deux puisque tout A de la forme ci-dessus est la somme d'un nombre fini de A de cette forme plus particulière. Or, pour de tels A :

$$P^{\mu}\left(\{m : m(F_1) = n_1, \dots, m(F_q) = n_q\}\right) = \prod_1^q \exp\left[-\mu (F_j)\right] \frac{\left[\mu (F_j)\right]^{n_j}}{n_j\,!}$$

($F_i \in \mathcal{J}_c$ (i=1, ... q) ; $F_i F_j = \emptyset$ ($i \neq j$) ; $n_i \in N^*$ (i=1,...q)) et le second membre de cette formule est manifestement une fonction $\mathcal{M}_+$ (E)-mesurable de μ.

La transformée de Laplace $\widetilde{P}_\pi$ de P_π s'exprime à partir de la transformée de Laplace $\widetilde{\pi}$ de π par la formule simple

(18) $$\widetilde{P}_\pi (f) = \widetilde{\pi}\left[1 - \exp(-f)\right]$$

puisque pour les lois de Poisson : $\widetilde{P^{\mu}} (f) = \exp\left\{-\mu\left[1 - \exp(-f)\right]\right\}$. Notons que comme $1 - \exp(-f) \in C_k^+(E)$ lorsque $f \in C_k^+(E)$, la proposition I.16 montre que $\lim_{n\to\infty} P_{\pi_n} = P_{\pi_\infty}$ dès que $\lim_{n\to\infty} \pi_n = \pi_\infty$, les limites étant prises au sens de la convergence étroite des probabilités sur M_+ (E).

Pour toute fonction borélienne $g : E \to [0, 1[$, la fonction $f = \log\left[1/(1-g)\right]$ est une fonction borélienne de E dans R_+ telle que $g = 1 - \exp(-f)$; la formule (18) s'inverse donc en la formule

(18') $$\widetilde{\pi}(g) = \widetilde{P}_\pi\left(\log\left[1 / (1-g)\right]\right)$$

valable pour toute fonction borélienne $g : E \to [0, 1[$. Ceci permet de montrer que <u>la correspondance</u> $\pi \to P_\pi$ <u>est injective</u>. En effet pour toute fonction borélienne $g : E \to R_+$ bornée et à support compact, la fonction $\widetilde{\pi}(tg)$ de $t \in R_+$ est la transformée de Laplace de la probabilité π_g sur R_+, image de π par l'application $m \to m(g)$ de M_+ (E) dans R_+ ; cette fonction est donc analytique sur $]0, \infty[$ et continue sur $[0, \infty[$ de sorte qu'elle peut être obtenue sur R_+ par prolongement analytique de sa restriction à $[0, a[$ pour n'importe quel réel $a > 0$ fixé. D'après la formule (18'), la fonction $\widetilde{\pi}(tg)$ de t ($t \in R_+$) est donc pour toute fonction borélienne $g : E \to R_+$ bornée et à support compact, le prolongement analytique à R_+ de la fonction $\widetilde{P}_\pi\left\{\log\left[1 / (1-tg)\right]\right\}$ de t qui ne peut être définie que sur $[0, (\sup_E g)^{-1}[$. L'injectivité de la correspondance $\pi \to P_\pi$ est donc claire.

Les processus de Cox sont intimement liés au procédé de raréfaction suivant. Soit $N = \sum_n \varepsilon_{T_n}$: $\Omega \to M_p$ (E) un processus ponctuel dont les points ont été numérotés T_o, T_1, ... et soit

$$N^* = \sum_n Y_n \varepsilon_{T_n}$$

le processus ponctuel obtenu en effaçant au hasard les points de N avec la probabilité 1-p ($0 < p < 1$). De manière plus précise, considérons l'espace de probabilité produit $\Omega^* = \Omega \times \{0, 1\}^N$ dont nous appellerons Y_n ($n \in N$) les dernières coordonnées et munissons le de la probabilité produit $P^* = P \otimes \{1-p, p\}^{N \otimes}$; la formule ci-dessus définit alors un processus ponctuel N^* : $\Omega^* \to M_p$ (E). Calculons sa transformée de Laplace ; pour toute fonction borélienne $f : E \to R_+$ nous avons

$$\int_{\Omega^*} \exp\left[-N^*(f)\right] dP^* = \int_{\Omega^*} \prod_n \left\{(1-Y_n) + Y_n \exp\left[-f(T_n)\right]\right\} dP^*$$

$$= \int_\Omega \prod_n (1-p) + p \exp\left[-f(T_n)\right]\Big\} dP$$

$$= \int_\Omega \exp\left[-N(g)\right] dP$$

si g est la fonction borélienne $\log \left\{1 / 1-p \left[1 - \exp(-f)\right]\right\}$ qui est telle que $\exp(-g) = (1-p) + p \exp(-f)$.

Le calcul de transformées de Laplace que nous venons de faire montre d'abord que la loi $P^*_{N^*}$ de N^* ne dépend que de p et de la loi P_N de N. Pour tout $p \in]0, 1[$, nous pouvons donc définir une transformation R_p de l'espace des probabilités sur M_p (E) qui fait passer de P_N à $P^*_{N^*}$ et qui est telle, d'après le calcul précédent, que

$$(R_p Q)^{\sim} (f) = \widetilde{Q} \left[\log \left\{1 / 1 - p \left[1 - \exp(-f)\right]\right\}\right].$$

pour toute fonction borélienne $f : E \to R_+$. En posant $g = 1 - \exp(-f)$, cette formule peut encore s'écrire

$$(19) \qquad (R_p Q)^{\sim}(\log \left[1 / (1-g)\right] = \widetilde{Q}(\log \left[1 / (1-pg)\right])$$

et est valable pour toute fonction borélienne $g : E \to [0, 1[$.

Le p-raréfié d'un processus de Poisson d'intensité μ est un processus de Poisson d'intensité $p\mu$; autrement dit $R_p P^{\mu} = P^{p\mu}$ quels que soient $p \in]0, 1[$ et $\mu \in M_+(E)$. Ceci résulte en effet immédiatement de la formule (19) compte tenu de ce que $\widetilde{P}^{\mu}(\log [1/(1-g)]) = \exp [-\mu(g)]$, pour toute fonction borélienne $g : E \to [0, 1[$. Cette propriété des processus de Poisson entraîne pour les processus de Cox que

$$(20) \qquad R_p P_{\pi} = P_{\rho_p \pi} \qquad\qquad (0 < p < 1)$$

si π est une probabilité sur $M_+(E)$ et si $\rho_p\,\pi$ désigne l'image de π par l'application $\mu \to p\mu$ de $M_+(E)$ dans lui-même. Notons, car cela va nous être utile, que l'application ρ_p précédente est bien définie pour tout $p \in R_+$ et non seulement si $p \in]0, 1[$.

Les processus de Cox et le procédé de raréfaction sont liés par les propriétés suivantes.

Proposition I.21

a) Si la suite $(P_{\pi_n}, n \in N)$ de lois de processus de Cox converge étroitement vers une probabilité P sur $M_p(E)$, nécessairement la suite $(\pi_n, n \in N)$ de probabilités sur $M_+(E)$ converge étroitement vers une probabilité π et $P = P_{\pi}$.

b) Soit $(p_n, n \in N)$ une suite de réels dans $]0, 1[$ tendant vers 0. Si la suite $(R_{p_n} P_n, n \in N)$ des p_n-raréfiées de probabilités P_n sur $M_p(E)$ converge étroitement vers une probabilité P lorsque $n \uparrow \infty$, nécessairement la suite des probabilités $\rho_{p_n} P_n$ $(n \in N)$ sur $M_+(E)$ converge étroitement vers une probabilité π et $P = P_{\pi}$.

Il résulte de la première partie de cette proposition et de l'injectivité de la correspondance $\pi \to P_\pi$ que la convergence étroite $\lim_n P_{\pi_n} = P_\pi$ ne peut avoir lieu que si $\lim_n \pi_n = \pi$ étroitement. D'autre part, comme la formule (20) ci-dessus appliquée à $\rho_{1/p}\ \pi$ montre que P_π est la p-raréfiée de $P_{\rho_{1/p}\pi}$, la seconde partie de la proposition établit que les lois de processus de Cox sont exactement les lois de processus ponctuels qui, pour tout $p \in]0, 1[$ (ou pour une suite de $p_n \in]0, 1[$ tendant vers 0) s'obtiennent comme p-raréfiées d'autres lois.

Démonstration

a) Si $\lim_n P_{\pi_n} = P$ étroitement, il résulte de la formule (18') que pour toute fonction $g \in C_k(E)$ à valeurs dans $[0, 1[$ nous avons

$$\widetilde{\pi}_n(g) = \widetilde{P}_{\pi_n}(\log [1 / (1-g)]) \to \widetilde{P}(\log [1 / (1-g)])$$

lorsque $n \nearrow \infty$. Il s'ensuit que pour toute fonction $h \in C_k^+(E)$, la limite $\psi(h) = \lim_n \widetilde{\pi}_n(h)$ existe et $\lim_{t \downarrow 0} \psi(th) = 1$. En effet pour une telle fonction h, les transformées de Laplace des limites faibles par rapport à $C_0(R_+)$ des probabilités $\pi_{n,h}$ images des π_n par $m \to m(h)$, sont des fonctions analytiques t sur $]0, \infty[$ toutes égales sur l'intervalle $]0, (\sup h)^{-1}[$ à $\lim_n \widetilde{\pi}_n(th) = \widetilde{P}(\log [1 / (1-th)])$; il ne peut donc y avoir qu'une seule limite faible des $\pi_{n,h}$ lorsque $n \nearrow \infty$, ce qui établit l'existence de ψ avec $\psi(th) = \widetilde{P}(\log [1 / (1-th)])$ lorsque $0 < t < (\sup h)^{-1}$ et il est alors clair que $\lim_{t \downarrow 0} \psi(th) = 1$.

D'après le corollaire I.17, il existe alors une probabilité π sur $M_+(E)$ vers laquelle les π_n convergent étroitement lorsque $n \nearrow \infty$; alors P_{π_n} converge vers P_π et par conséquent $P = P_\pi$.

b) Soit g une fonction de $C_k(E)$ à valeurs dans $[0, 1[$; les fonctions $g_n = p_n^{-1}[1 - \exp(-p_n g)]$ ont alors les mêmes propriétés et de plus g_n

croit vers g lorsque n $\nearrow \infty$. D'après la formule 19 :

$$(\rho_{p_n} P_n)^{\sim} (g) = \widetilde{P}_n (p_n g) = \widetilde{P}_n (\log [1/(1-p_n g_n)]) = (R_{p_n} P_n)^{\sim} (\log [1/(1-g_n)])$$

Comme la suite de probabilités $(R_{p_n} P_n , n \in N)$ converge étroitement vers une probabilité P lorsque n $\uparrow \infty$, nous obtenons que

$$\lim_{n\to\infty} (R_{p_n} P_n)^{\sim} (\log [1 / (1-g_n)]) = \widetilde{P} (\log [1 / (1-g)])$$

en encadrant g_n par g_m et g $(n \geqslant m)$ puis en faisant tendre m $\nearrow \infty$.

La suite des probabilités $\pi_n = \rho_{p_n} P_n$ $(n \in N)$ sur M_+ (E) est donc telle que $\lim_n \widetilde{\pi}_n (g) = \widetilde{P} (\log [1 / (1-g)])$ pour toute fonction $g \in C_k$ (E) à valeurs dans $[0, 1[$. Comme nous l'avons vu dans la première partie de la démonstration, cela n'est possible que si cette suite $(\pi_n , n \in N)$ converge étroitement vers une probabilité π sur M_+ (E). De plus $\widetilde{\pi} (g) = \widetilde{P} (\log [1 / (1-g)])$ pour toute fonction $g \in C_k$ (E) à valeurs dans $[0, 1[$ ce qui suffit déjà à entraîner que $P = P_\pi$.□

APPENDICE. TRANSFORMATION DE LAPLACE SUR R_+^d

La transformation de Laplace d'une mesure positive et finie m sur R_+^d est la fonction continue et bornée de R_+^d dans R_+ définie par l'intégrale

$$\tilde{m}(t) = \int_{R_+^d} \exp\left(-\sum_1^d t_j x_j\right) m(dx) \qquad (t \in R_+^d).$$

D'après le théorème de Stone-Weierstrass cette transformation $m \to \tilde{m}$ est injective. (En effet d'après ce théorème les exponentielles e_t $\left[e_t(x) = \exp\left(-\sum_1^d t_j x_j\right)\right]$ de paramètre $t_1, \ldots t_d > 0$ strictement forment un ensemble générateur de l'espace de Banach $C_o(R_+^d)$ des fonctions réelles continues nulles à l'infini définies sur R_+^d, de sorte que $\tilde{m}$ détermine la forme linéaire $f \to m(f)$ définie sur $C_o(R_+^d)$; d'après le théorème de Riesz la mesure m est alors complètement déterminée).

Proposition I.22

<u>Pour qu'une suite</u> $(m_n, n \in \mathbb{N})$ <u>de mesures positives et finies sur</u> R_+^d <u>converge étroitement lorsque</u> $n \to \infty$, <u>c'est-à-dire pour qu'il existe une mesure positive et finie</u> m <u>sur</u> R_+^d <u>telle que</u> $\lim_{n\to\infty} m_n(f) = m(f)$ <u>pour toute fonction continue et bornée</u> $f : R_+^d \to R$, (il faut et) <u>il suffit que</u> $\lim_{n\to\infty} \tilde{m}_n(t)$ <u>existe pour tout</u> $t \in R_+^d$ <u>et que la limite</u> φ <u>ainsi obtenue soit continue au point</u> 0. <u>Cette fonction est alors la transformée de Laplace de la mesure limite</u> m.

Il résulte en particulier de ce théorème que la suite de mesures positives et finies $(m_n, n \in \mathbb{N})$ converge étroitement vers la mesure positive et finie m si et seulement si $\lim_{n\to\infty} \tilde{m}_n(t) = \tilde{m}(t)$ pour tout $t \in R_+^d$.

Démonstration succinte

Dès que $\varphi(t) = \lim_{n\to\infty} \tilde{m}_n(t)$ existe pour tout t strictement positif $(t_1, \ldots, t_d > 0)$ et que $\sup_n m_n(R_+^d) = \sup_n \tilde{m}_n(0)$ est fini, il est possible de définir une forme linéaire positive m sur $C_o(R_+^d)$ par $m(f) = \lim_n m_n(f)$ en commençant par les fonctions exponentielles e_t et en passant ensuite à toutes les fonctions de C_o grâce au théorème de Stone Weierstrass ; cette mesure de Radon positive m est telle donc que $\tilde{m}(t) = \lim_n \tilde{m}_n(t) = \varphi(t)$ lorsque t est strictement positif. Dans la proposition nous supposons de plus que $\lim_n m_n(R^d_+) = \lim_n \tilde{m}_n(o)$ $= \varphi(o)$ existe et que $\varphi(o) = \lim_{t>0,\, t\downarrow 0} \uparrow \varphi(t) = \lim_{t>0;\, t\downarrow 0} \uparrow \tilde{m}(t) = \tilde{m}(o) = m(R^d_+)$ de sorte que $m_n(R^d_+) \to m(R^d_+)$ lorsque $n \to \infty$. Ceci suffit à entraîner la convergence étroite de m_n vers m ; en effet si $(g_k,\ k \in \mathbb{N})$ est une suite croissant vers 1 dans $C_o^+(R^d_+)$, pour toute fonction continue bornée positive $f : R^d \to R$ nous pouvons écrire que

$$|m_n(f) - m(f)| \leqslant |m_n(f\,g_k) - m(f\,g_k)| + \|f\| \quad [m_n(1-g_k) + m(1-g_k)]$$

$$\to \quad 2\,\|f\|\ m(1-g_k) \text{ lorsque } n \to \infty$$

puisque $f\,g_k \in C_o$ et $m_n(1-g_k) = m_n(R^d_+) - m_n(g_k) \to m(1-g_k)$ lorsque $n \to \infty$; ceci nous montre en faisant tendre $k \to \infty$ que

$$|m_n(f) - m(f)| \quad \to \quad 0 \qquad \text{lorsque } n \to \infty\ .$$

La proposition est ainsi démontrée.□

COMPLEMENTS ET EXERCICES

1. Etablir que les seules mesures de Radon positives sur E qui ne prennent que des valeurs entières sur une base d'ouverts relativement compacts sont les mesures ponctuelles. [Montrer que cette condition entraîne pour tout $x \in E$ l'existence d'un voisinage ouvert V tel que $m(\{x\}) = m(V)$ et en déduire que pour tout compact K :
$1_K \cdot m = \sum_K m(\{x\})\, \varepsilon_x$].

[Si les mesures ponctuelles m_α convergent vaguement vers la mesure de Radon positive m, les ouverts relativement compacts G de frontières m-négligeables forment une base de la topologie et sont tels que $m(G) = \lim_\alpha m_\alpha(G)$].

2. Munissons l'espace P (E) des parties S localement finies de E de la tribu $\mathcal{P}(E)$ engendrée par les ensembles $\{S : S \cap F = \emptyset\}$ obtenus lorsque F parcourt $\mathcal{E}$. Si σ est une partie localement finie aléatoire de E, c'est-à-dire si σ est une application mesurable d'un espace de probabilité $(\Omega, \mathcal{A}, P)$ dans $(P(E), \mathcal{P}(E))$, montrer qu'il existe un processus ponctuel simple unique $N : \Omega \to M_p(E)$ tel que $\sigma(\omega)$ soit le support de $N(\omega, .)$ pour tout $\omega \in \Omega$.

Inversement montrer que pour tout processus ponctuel $N : \Omega \to M_p(E)$, l'application "support" : $\omega \to \operatorname{supp}[N(\omega, .)]$ est mesurable de Ω dans $\mathcal{P}(E)$.

3. Soit δ l'application $\delta(x) = (x, x)$ de E sur la diagonale $\Delta = \delta(E)$ de E^2. Remarquer que toute mesure ponctuelle m sur E vérifie l'inégalité
$1_\Delta \cdot m^{2\otimes} \geq \delta \circ m$ sur E^2 et que cette mesure est simple si et seulement si cette inégalité est une égalité.

Montrer qu'un processus ponctuel de carré intégrable est p.s. simple si et seulement si son second moment μ_2 et son intensité μ sont tels que $1_\Delta \cdot \mu_2 = \delta \circ \mu$.

4. Soit N un processus ponctuel de Poisson d'intensité μ sur E. Calculer son second moment. Montrer que N est simple p.s. si et seulement si μ est diffuse sur E. Pour tout borélien F, établir que $1_F \cdot N$ est encore un processus de Poisson. Si l'image $\mu' = \varphi \circ \mu$ de μ par une application borélienne $\varphi : E \to E'$ (où E' est un second espace localement compact à base dénombrable) est encore de Radon, prouver que $\varphi \circ N$ est un processus de Poisson d'intensité μ' sur E'.

5. Moments et moments factoriels des processus ponctuels

a) Montrer que le second moment μ_2 d'un processus ponctuel N est une mesure inva-invariante par $(x, y) \to (y, x)$ sur E^2 et en déduire lorsque cette mesure est de Radon qu'elle est entièrement déterminée par ses valeurs $\mu_2(F \times F) = E[N(F)^2]$ $(F \in \mathcal{E})$ [Exprimer $\mu_2(F_1 \times F_2)$ en fonction des $\mu_2(F \times F)$]. Le $n^{\text{ème}}$ moment μ_n du processus ponctuel N est défini comme l'intensité de $N^{n\otimes}$ sur E^n. Etendre à ces mesures les propriétés démontrées pour μ_2.

b) Soit $N^{(2)}$ le processus ponctuel sur E^2 obtenu en effaçant de $N^{2\otimes}$ les points sur la diagonale ; montrer que son intensité $\mu_{(2)}$ vaut $1_{\Delta^c} \cdot \mu_2$ et que lorsque le processus ponctuel N est simple

$\mu_{(2)}(F \times F') = E[N(F)\, N(F') - N(F \cap F')]$, $\mu_{(2)}(F \times F) = E(N(F)[N(F)-1])$

quels que soient $F, F' \in \mathcal{E}$. La dernière formule fait appeler $\mu_{(2)}$ le second moment factoriel de N.

Le $n^{\text{ème}}$ moment factoriel de N est défini comme l'intensité de la restriction de $N^{n\otimes}$ au sous-ensemble $\bigcap_{1 \leq l < m \leq n} \{x_l \neq x_m\}$ de E^n. Généraliser à ces mesures les formules démontrées pour $\mu_{(2)}$. Calculer les moments factoriels d'un processus de Poisson.

c) si l est un entier ⩾ 2 et si $(J_1, \ldots, J_k)$ est une partition de $\{1, 2, \ldots l\}$ en sous-ensembles non vides, définissons une application $\phi : E^k \to E^l$ en posant $(\phi x)_j = x_i$ si $j \in J_i$ $(j = 1, 2, \ldots l)$. Montrer alors que si m est une mesure ponctuelle simple, $\phi \circ m^{k\otimes} = 1_{\phi(E^k)} \cdot m^{l\otimes}$. En déduire que les moments d'un processus ponctuel simple sont tels que $\phi \circ \mu_k = 1_{\phi(E^k)} \cdot \mu_l$. Etablir une réciproque.

6. L'intensité μ d'un processus ponctuel N est telle que : $\mu(F) = 0 \Longleftrightarrow N(F) = 0$ p.s. pour tout borélien F , mais ce n'est pas nécessairement une mesure de Radon. Construire une mesure de Radon positive et finie sur E satisfaisant à l'équivalence précédente.

7. <u>La fonctionnelle caractéristique</u> d'un processus ponctuel N sur E d'intensité de Radon μ est définie sur l'espace $L^1_R(E, \mu)$ par

$$\phi_N(f) = E(\exp[-2\pi i N(f)]).$$

Etablir que l'application $f \to \phi_N(f)$ est lipschitzienne, qu'elle détermine la loi P_N de N et qu'elle permet par un développement de Taylor de trouver l'intensité et éventuellement les moments de N. Calculer ϕ_N pour un processus de Poisson.

8. Soit $\mathfrak{J}$ une classe de boréliens de E vérifiant les hypothèses du lemme I.3. Montrer alors que les deux processus ponctuels $N_i : \Omega \to M_p(E)$ $(i = 1, 2)$ sont indépendants dès que pour tous $I^1_1, \ldots I^1_k, I^2_1, \ldots I^2_l$ dans $\mathfrak{J}$ les vecteurs aléatoires $(N_1(I^1_1), \ldots N_1(I^1_k))$ et $(N_2(I^2_1), \ldots N_2(I^2_l))$ sont indépendants. En déduire que N_1 et N_2 sont indépendants si et seulement si en termes de fonctionnelles de Laplace

$$E(\exp[-N_1(f_1) - N_2(f_2)]) = \psi_{N_1}(f_1)\, \psi_{N_2}(f_2)$$

pour tout couple f_1, f_2 de fonctions boréliennes positives.

Généraliser à plus de deux processus ponctuels et au cas où ces processus prennent leurs valeurs dans des espaces $M_p(E)$ différents.

Montrer que la somme de deux processus poissoniens de nuages indépendants est encore un processus de ce type. Comment ce résultat s'étend-t-il à une série $\sum_{\mathbb{N}} N_k$ de processus poissoniens de nuages indépendants ?

9. Pour le mélange $P = \int_0^\infty \varphi(\theta)\, P_\theta\, d\theta$ de lois P_θ de processus de Poisson d'intensité $\theta\mu$ $[\mu \in M_+(E)]$ effectué à l'aide de la densité gamma $\varphi(\theta) = \exp(-\lambda\theta)\, \lambda^a\, \theta^{a-1} / \Gamma(a)$ (où $\lambda, a \in]0, \infty[$ et Γ est la fonction d'Euler), montrer que pour tout borélien μ-intégrable F :

$$P[N(F) = n] = \frac{\Gamma(n+a)}{n!\,\Gamma(a)} \; \frac{\lambda^a\ \mu(F)^n}{[\lambda + \mu(F)]^{a+n}} \quad (n \in \mathbb{N}).$$

et $E(u^{N(F)}) = [\lambda / \lambda + (1-u)\,\mu(F)]^a$ $(0 \leqslant u \leqslant 1)$.

[Pour a = 1, on trouve une loi géométrique ; pour a entier $\geqslant 1$, on trouve les lois binomiales négatives. Pour a > 0 quelconque, ces lois sont dites de Polya].

10.a) Etablir que les raréfactions : $P \to R_p P$ $(0 < p < 1)$ définissent des applications continues et injectives sur l'espace des probabilités sur $M_p(E)$. Si P_n $(n \in N)$ et Q sont des probabilités sur $M_p(E)$ telles que $\lim_n R_p P_n = Q$ étroitement pour un $p \in]0, 1[$ fixé, établir que la suite $(P_n, n \in N)$ converge étroitement vers une probabilité P et que $Q = R_p P$.

b) Pour toute loi P_π de processus de Cox, montrer que $\pi = \lim_{p \to 0} \rho_p (R_p)^{-1} P_\pi$.

11. Soit $\pi = (\pi(x, F) ; x \in E, F \in \mathcal{E})$ une probabilité de transition sur E admettant une mesure de Radon positive et infinie, soit μ, comme mesure invariante : $\mu\pi = \mu$; par exemple $E = R^d$, $\pi(x, F) = \lambda(F - x)$ pour une probabilité λ sur R^d auquel cas la mesure de Lebesgue convient pour mesure μ.

Sur un espace de probabilité convenable $(\Omega, \mathcal{A}, P)$ $[\Omega = (E^{\mathbb{N}})^{\mathbb{N}}$ par exemple$]$, donnons nous une suite infinie $\{(X_n^j, n \in \mathbb{N}) ; j \in \mathbb{N}\}$ de chaînes de Markov indépendantes de probabilité de transition π et formons les mesures aléatoires

$$N_n(\omega, .) = \sum_{j \in \mathbb{N}} \varepsilon_{X_n^j(\omega)} \qquad (n \in \mathbb{N})$$

qui ne sont pas nécessairement de Radon.

Montrer que si $N_o(\omega, .)$ est p.s. de Radon et suit une loi de processus de Poisson d'intensité μ sur E, il en est encore de même de chacun des processus N_n. [Utiliser les fonctionnelles de Laplace]. Comment faudrait-il modifier ce qui précède pour tenir compte du cas où la mesure μ serait finie ?

12. a) Soit $(\mu_n, n \in \mathbb{N}^*)$ une suite de mesures positives diffuses sur E telle que la mesure $\sum_{\mathbb{N}^*} \mu_n$ soit de Radon. Soit $N : \Omega \to M_p(E \times \mathbb{N}^*)$ un processus ponctuel de Poisson d'intensité μ sur $E \times \mathbb{N}^*$ donnée par $\mu(F \times \{n\}) = \mu_n(F)$ $(F \in \mathcal{E})$. Montrer que la formule

$$\tilde{N}(\omega, F) = \sum_{\mathbb{N}^*} n \; N[\omega, F \times \{n\}] \qquad (F \in \mathcal{E})$$

définit un processus ponctuel sur E tel que

1°) $N(\{x\}) = 0$ p.s. pour tout point $x \in E$

2°) pour toute suite $F_1, \dots F_k$ de boréliens relativement compacts 2 à 2 disjoints les variables $\tilde{N}(F_1), \dots \tilde{N}(F_k)$ soient indépendantes.

Calculer l'intensité et le moment d'ordre deux de ce processus $\tilde{N}$ en fonction des mesures μ_n ; établir que pour tout $u \in [0, 1]$ il existe une mesure de Radon positive et diffuse λ_u sur E que l'on exprimera en fonction des μ_n qui soit telle que

$$E\left[u^{N(F)}\right] = \exp\left[-\lambda_u(F)\right] \qquad (F \in \mathcal{E}).$$

A quelle condition sur les μ_n, le processus ponctuel $\tilde{N}$ est-il de Poisson ?

b) Réciproquement soit $\tilde{N} : \Omega \to M_p(E)$ un processus ponctuel sur E jouissant des propriétés 1-2 ci-dessus. Alors la formule

$$N(\omega, F \times \{n\}) = \int_F \tilde{N}(\omega, dx) \frac{1}{n} 1_{(\tilde{N}(\omega,\{x\})=n)}$$

définit un processus ponctuel simple sur $E \times \mathbb{N}^*$ tel que

$$\tilde{N}(\omega, F) = \sum_{\mathbb{N}^*} n \; N(\omega, F \times \{n\}).$$

Montrer que N est un processus de Poisson sur $E \times \mathbb{N}^*$ d'intensité diffuse et finie sur tout $F \times \mathbb{N}^*$ ($F \in \mathcal{E}_c$).

[A cet effet, on établira d'abord que pour toute suite $u = (u(n), n \in \mathbb{N}^*)$ de réels positifs, la formule

$$E\left[\exp - \int_F \tilde{N}(\omega, dx)\, u\left[\tilde{N}(\omega, \{x\})\right]\right] = \exp\left[-\lambda_u(F)\right] \qquad (F \in \mathcal{E})$$

définit une mesure de Radon positive et diffuse sur E. En écrivant alors que $\lambda_u(F) = \lim \sum_j 1 - \exp\left[-\lambda_u(F_j)\right]$ le long de suites de partitions $F_1, \ldots F_k$ de F de plus en plus fines, montrer que les limites $\mu_n(F) = \lim \sum_j P\left[\tilde{N}(F_j) = n\right]$ existent et définissent des mesures de Radon positives telles que $\lambda_u = \sum_n \left[1 - e^{-n\,u(n)}\right] \mu_n$. Conclure.]

c) Généraliser les résultats précédents à un processus de Poisson N sur $E \times]0, \infty[$ et à la mesure aléatoire associée $\tilde{N}(\omega, dx) = \int_0^\infty t\, N(\omega, dx\, dt)$.

13. Soit $N : \Omega \to M_p(E)$ un processus poissonien de nuages associé à une mesure positive Q sur $M_p(E) - \{0\}$ finie sur les ensembles $\{m : m(F) \neq 0\}$ ($F \in \mathcal{E}_c$) [cf § 3.C].

a) Si $Q(\{m : m(E) \geq 2\}) = 0$, montrer que N est un processus de Poisson ; quelle est son intensité ? Inversement si pour une suite $(F_j, j \in \mathbb{N})$ dans $\mathcal{E}_c$

croissant vers E les variables aléatoires N (F_j) suivent des lois de Poisson ou si pour tout couple de boréliens F_1, F_2 disjoints N (F_1) et N (F_2) sont indépendantes, montrer que Q vérifie la condition précédente et donc que N est un processus de Poisson. [Utiliser la fonctionnelle de Laplace].

b) En supposant que les mesures $\nu = \int m \, Q \, (dm)$ et $\nu_2 = \int m^{2 \otimes} Q \, (dm)$ sont des mesures de Radon sur E et E^2 respectivement, trouver l'intensité et le second moment de N en fonction de ν et ν_2. Montrer que N est simple p.s. si et seulement si ν est diffuse sur E et si de plus $1_\Delta \cdot \nu_2 = \delta \circ \nu$ sur E^2 [notations de l'exercice 3].

c) Etablir que N (E) = ∞ p.s., si la mesure Q est infinie. Au contraire si la mesure Q est finie, montrer que

$$P \, [N \, (E) = 0] = \exp \, (- \, Q \, [M_p \, (E)]) > 0$$

$$P \, [N \, (E) < \infty] = \exp \, [- \, Q \, (\{m : m \, (E) = \infty\})] \leqslant 1.$$

A quelles conditions nécessaires et suffisantes sur Q, la variable aléatoire N (E) est-elle p.s. finie, resp. p.s. infinie ?

d) Si Q_F désigne l'image de Q sur M_p (F) par l'application $m \to 1_F \cdot m$ pour un F borélien donné, montrer que le processus $1_F \cdot N$ est un processus ponctuel de nuages associé à la mesure Q_F. Si $\varphi : E \to E'$ est borélienne, que peut-on dire de $\varphi \circ N$?

e) Montrer que $P \, [N \, (F) = 0] = \exp \, [- \, Q \, (\{m : m \, (F) \neq 0\})] > 0$ pour tout $F \in \mathcal{E}_c$. Pour un tel F, établir que la loi de probabilité de N conditionnelle en $\{N \, (F) = 0\}$ est encore la loi d'un processus poissonien de nuages dont on déterminera la mesure associée Q^F.

Etablir que pour tout couple de F, G $\in \mathcal{E}_c$ disjoints,

$$P \left[N \, (G) = 0 \, / \, N \, (F) = 0 \right] = \exp \left[- \, Q \, (\{m : m \, (F) = 0, \, m \, (G) \neq 0\}) \right]$$

En déduire que lorsque la mesure Q est infinie, il est nécessaire et suffisant que $Q \, [m \, (E) < \infty] = 0$ pour que pour tout $G \in \mathcal{E}_c$

$$\lim_{F \in \mathcal{E}_c,\ F \uparrow G^c} \uparrow P\left[N(G) = 0 / N(F) = 0\right] = 1.$$

En général montrer que la loi conditionnelle de $1_G . N$ lorsque $N(F) = 0$ tend vers une limite lorsque $F \uparrow G^c$.

14. a) Soit μ une mesure de Radon positive et finie sur E. Si $N : \Omega \to M_p(E)$ est un processus ponctuel de Poisson d'intensité μ, montrer que pour toute fonction $f \in L^1_+(\mu)$ la formule

$$M_f(\omega) = \exp\left\{\int_E \log f(x)\, N(\omega, dx) - \int_E \left[f(x) - 1\right] d\mu(x)\right\}$$

définit une variable positive sur Ω [que l'on posera égale à 0 si $N\left[\omega; \{f = 0\}\right] > 0$] telle que $M_f . P$ soit une probabilité sur $(\Omega, \mathcal{A})$ pour laquelle N est un processus de Poisson d'intensité $f . \mu$.

b) Soient μ et μ' deux mesures de Radon positives sur E et supposons μ finie. Si P_μ et $P_{\mu'}$ désignent les lois sur $M_p(E)$ de processus de Poisson d'intensités respectives μ et μ', montrer qu'il est nécessaire et suffisant que μ' soit de la forme $\mu' = f.\mu$ avec $f \in L^1_+(\mu)$ pour que $P_{\mu'}$ soit absolument continue par rapport à P_μ (Utiliser la décomposition de Lebesgue de μ' par rapport à μ pour démontrer la nécessité).

c) Etendre les résultats précédents au cas où μ est infini, la condition nécessaire et suffisante pour que $P_{\mu'} \ll P_\mu$ devenant maintenant que $\mu' = f.\mu$ pour une fonction mesurable positive telle que $\sqrt{f} - 1 \in L^2(\mu)$. Utiliser le théorème de Kakutani [proposition III.1.2 de Neveu, Martingales à temps discret].

CHAPITRE II

PROCESSUS PONCTUELS STATIONNAIRES

Sommaire Pages

Sommaire

Après un paragraphe préliminaire où la notion de flot est introduite pour définir les processus ponctuels stationnaires sur un sous groupe fermé E de R^d, la mesure de Palm de ces processus est construite au paragraphe 2 et son étude prend tout le reste du chapitre. En fait trois définitions de cette mesure de Palm sont possibles : nous prenons pour point de départ la plus ingénieuse due à Mecke, qui est aussi la plus simple puisqu'elle ne s'appuie que sur le théorème de Fubini. La seconde définition possible qui dans cet exposé devient un théorème, est la plus intuitive : elle affirme que à une normalisation près, la mesure de Palm est la probabilité conditionnée par l'évènement (négligeable en général !) " le processus ponctuel charge l'origine "(du moins si le processus est simple) ; mais cette définition ne va pas sans quelques problèmes techniques qui imposent des restrictions à la généralité. La troisième définition n'est donnée ici que dans le cas de R ; très importante dans les applications, elle définit la mesure de Palm à partir de la translation de la probabilité au premier point à gauche de O du processus ponctuel. Dans ces trois premiers paragraphes, nous montrons aussi comment passer de la mesure de Palm d'un processus ponctuel à celle d'un autre processus ponctuel et résultat non moins important, comment reconnaître qu'une mesure est une mesure de Palm pour reconstruire ensuite la probabilité de départ.

La notion de processus marqué est introduite et étudiée au paragraphe 4 dans le cadre de la théorie des processus stationnaires ; peu de processus stationnaires s'introduisent en pratique qui ne soient marqués. Le cinquième paragraphe consacré à l'ergodisme et à l'analyse spectrale des processus stationnaires est présenté ici comme une application de la notion de mesure de Palm : il est moins important. Enfin le sixième paragraphe est une ébauche d'étude de la partie de la géométrie aléatoire qui fait intervenir des processus ponctuels : elle demande que la notion de mesure de Palm soit étendue aux espaces d'états qui sont des espaces homogènes de groupes.

II.1. FLOTS ET PROCESSUS PONCTUELS STATIONNAIRES.

Dans tout ce chapitre à l'exception du paragraphe 6 où des espaces plus généraux seront considérés, E désignera un sous-groupe fermé de R^d, par exemple Z^d ou R^d $(d \geq 1)$. Nous noterons $(\tau_t\ ,\ t \in E)$ le groupe des translations définies sur E par $\tau_t\ (s) = s - t$; nous pourrons écrire alors en faisant attention aux signes que pour tout borélien $F \in \mathcal{E}$

(1) $(\tau_t)^{-1}\ (F) = \tau_{-t}\ (F) = F + t$ si $F + t = \{s + t\ ;\ s \in F\}$.

Pour toute mesure positive m sur E, la mesure image τ_t o m de m par τ_t est donc donnée par τ_t o m $(F) = m\ (F + t)$. Il existe sur E une mesure de Radon positive unique à une constante multiplicative près qui est invariante par les translations ; nous la noterons λ.

Observons que les translations τ_t appliquent l'espace M_p (E) des mesures ponctuelles sur E sur lui-même puisque en particulier $\tau_t\ \varepsilon_s = \varepsilon_{s-t}$; ceci nous permet donc de considérer aussi $(\tau_t,\ t \in E)$ comme un groupe de transformations sur M_p (E). Ce groupe est un flot mesurable au sens de la définition suivante.

Définition II.1

Un flot mesurable indexé par E sur un espace mesurable $(\Omega, \mathcal{A})$ est une famille $(\theta_t,\ t \in E)$ de transformations $\theta_t : \Omega \to \Omega$ telle que

a) $\theta_s \circ \theta_t = \theta_{s+t}$ sur Ω $(s,\ t \in E)$ et θ_o = identité,

b) l'application $t,\ \omega \to \theta_t\ (\omega)$ de $E \times \Omega$ dans Ω est mesurable pour les tribus $\mathcal{E} \times \mathcal{A}$ et $\mathcal{A}$ respectivement.

Un tel flot est dit préserver la probabilité P définie sur $(\Omega, \mathcal{A})$ si $\theta_t \circ P = P$ pour tout $t \in E$.

Dans la suite, sauf mention du contraire, nous conviendrons d'appeler en abrégé flot tout "flot mesurable préservant la mesure".

Pour démontrer que le groupe $(\tau_t, t \in E)$ sur $M_p(E)$ est un flot mesurable, il suffit par exemple de démontrer que pour toute fonction $f \in C_k^+(E)$ l'application $m, t \to \tau_t m(f) = m(f \circ \tau_t)$ de $M_p(E) \times E$ dans R_+ est mesurable ; or cette application est mesurable sur $M_p(E)$ en son premier argument et continue sur E en son deuxième argument.

La définition précédente conduit à celle des processus ponctuels stationnaires.

Définition II.2

Un processus ponctuel $N : \Omega \to M_p(E)$ sur E défini sur un espace de probabilité $(\Omega, \mathcal{A}, P)$ est dit stationnaire s'il existe un flot $(\theta_t, t \in E)$ sur cet espace tel que

(2) $N(\theta_t \omega, .) = \tau_t [N(\omega, .)] \qquad (\omega \in \Omega, t \in E).$

Un processus ponctuel $N : \Omega \to M_p(E)$ sur E est dit stationnaire en loi si pour toute suite finie $F_1, \dots F_n$ de boréliens relativement compacts de E, la loi de probabilité du vecteur aléatoire $(N(F_1+t), \dots N(F_n+t))$ ne dépend pas de t $(t \in E)$.

Un processus ponctuel stationnaire est stationnaire en loi. En effet pour un tel processus $N(\theta_t \omega, F) = N(\omega, F+t)$ puisque $N(\theta_t \omega, .) = \tau_t N(\omega, .)$ et alors

$$P[N(F_j+t) = n_j ; j = 1, \dots k] = P[N(F_j) \circ \theta_t = n_j ; j = 1 \dots k]$$

$$= \theta_t \circ P[N(F_j) = n_j ; j = 1 \dots k]$$

ne dépend pas de t puisque $\theta_t \circ P = P$ $(t \in E)$.

D'autre part un processus ponctuel $N : \Omega \to M_p(E)$ est stationnaire en loi au sens de la définition précédente si et seulement si

$$\tau_t \circ P_N = P_N \qquad (t \in E)$$

sur M_p (E). Cela résulte en effet du lemme I.3 (b) appliqué à la classe des parties $\{m : m(F_j) = n_j \ (j = 1 \dots k)\}$ de M_p (E) obtenues lorsque $(F_1, \dots F_k)$ parcourt les suites finies de boréliens relativement compacts et $n_1, \dots n_k$ parcourent N^* et de ce que en vertu des définitions

$$\tau_t \circ P_N \left[\{m : m(F_j) = n_j \ (j = 1 \dots k)\}\right]$$

$$= \quad P_N \left[\{m : m(F_j + t) = n_j \ (j = 1 \dots k)\}\right]$$

$$= \quad P \left[N(F_j + t) = n_j \qquad (j = 1 \dots k)\right].$$

Observons enfin que si Q est une probabilité sur M_p (E) invariante par les translations τ_t $(t \in E)$, le processus ponctuel identique $I : (M_p(E), \mathcal{M}_p(E), Q) \to (M_p(E), \mathcal{M}_p(E))$ admet la loi Q et est stationnaire par le flot $(\tau_t, t \in E)$ sur M_p (E).

Il est facile de vérifier qu'un processus ponctuel N est stationnaire en loi si et seulement si sa fonctionnelle de Laplace est invariante par les τ_t, soit $\psi(f \circ \tau_t) = \psi(f)$ pour tout $t \in E$ et toute fonction f borélienne positive. En particulier <u>un processus de Poisson sur</u> E <u>est stationnaire en loi</u> si et seulement si son intensité est proportionnelle à la mesure invariante λ du début.

Bien que cela ne soit pas postulé dans sa définition, tout flot possède des propriétés de continuité, comme le montre le lemme suivant qui nous servira dans la suite.

<u>Lemme II.3</u>

<u>Pour tout flot</u> $(\theta_t, t \in E)$ <u>sur l'espace de probabilité</u> $(\Omega, \mathcal{A}, P)$, <u>les fonctions réelles mesurables bornées</u> $h : \Omega \to R$ <u>telles que pour tout</u> $\omega \in \Omega$ <u>l'application</u> $t \to h(\theta_t \omega)$ <u>de</u> E <u>dans</u> R <u>soit continue</u>, <u>forment une algèbre de fonctions sur</u> Ω, <u>soit</u> H, <u>qui est dense dans</u> $L^2(\Omega, \mathcal{A}, P)$. <u>Il s'en suit que</u>

<u>pour tout réel</u> $p \in [1, \infty[$ <u>et toute fonction</u> $f \in L^p$, <u>l'application</u> $t \to f \circ \theta_t$ <u>de</u> E <u>dans</u> L^p <u>est continue.</u>

<u>Démonstration</u>

La famille H que nous venons de définir est manifestement une algèbre. D'autre part pour toute fonction mesurable bornée $g : \Omega \to R$ et toute fonction λ-intégrable $u : E \to R$, la formule

$$(3) \qquad h_{g,u}(\omega) = \int_E g(\theta_s \omega)\, u(s)\, d\lambda(s)$$

définit une fonction de H. En effet cette fonction $h_{g,u}$ est bien définie et est mesurable bornée sur Ω car l'application $s, \omega \to g(\theta_s \omega)$ est mesurable bornée sur $E \times \Omega$ et car u est λ-intégrable. De plus $h_{g,u} \circ \theta_t = h_{g,u \circ \tau_t}$ pour tout $t \in E$ et par conséquent pour tous $t, t' \in E$

$$|h_{g,u}(\theta_{t'}\omega) - h_{g,u}(\theta_t \omega)| = \left| \int_E g(\theta_s \omega) [u \circ \tau_{t'}(s) - u \circ \tau_t(s)]\, d\lambda(s) \right|$$

$$\leq \sup_\Omega |g| \, . \int_E |u \circ \tau_{t'} - u \circ \tau_t| \, d\lambda \quad ;$$

or le dernier membre tend vers 0 lorsque $|t'-t| \to 0$ d'après la continuité uniforme de l'action du groupe des translations sur l'espace $L^1(E, \lambda)$. Nous avons ainsi montré que les applications $t \to h_{g,u}(\theta_t \omega)$ sont uniformément continues sur E et trouvé un module de continuité uniforme indépendant de ω ; cela suffit évidemment à entraîner que $h_{g,u} \in H$.

Pour établir que l'algèbre H est dense dans $L^2(\Omega)$, montrons que toute fonction $f \in L^2$ orthogonale aux $h_{g,u}$ est nulle. Une telle fonction f vérifie les égalités :

$$\int_{E\times\Omega} f(\theta_{-s}\omega)\, u(s)\, g(\omega)\, d\lambda(s)\, dP(\omega) = \int_E u(s) \left[\int_\Omega f(\theta_{-s}\omega)\, g(\omega)\, dP(\omega) \right] d\lambda(s)$$

$$= \int_E u(s) \left[\int_\Omega f(\omega)\, g(\theta_s \omega)\, dP(\omega) \right] d\lambda(s)$$

$$= \int_\Omega f(\omega)\, h_{g,u}(\omega)\, dP(\omega) = 0$$

quelles que soient la fonction λ-intégrable $u : E \to R$ et la fonction mesurable bornée $g : \Omega \to R$. Cela n'est possible que si la fonction $f(\theta_{-s}\omega)$ de (s, ω) est négligeable sur $E \times \Omega$ pour la mesure produit $\lambda \otimes P$; mais alors il existe au moins un $s_o \in E$ tel que $f(\theta_{-s_o}\omega)$ soit P-négligeable sur Ω et cela entraîne que $||f||_2 = ||f \circ \theta_{-s_o}||_2 = 0$ comme nous voulions le montrer.

Une algèbre de fonctions mesurables bornées, H en particulier, ne peut être dense dans L^2 sans l'être aussi dans tout L^p $(p \in [1, \infty[)$ [Voir Neveu [4] p. 5] D'autre part pour toute fonction $h \in H$, l'application $t \to h \circ \theta_t$ de E dans L^p est continue grâce à la continuité des applications $t \to h(\theta_t \omega)$ $(\omega \in \Omega)$ et au théorème de convergence dominée qui est bien applicable puisque la fonction h est bornée. Comme l'ensemble des fonctions $f \in L^p$ telles que $t \to f \circ \theta_t$ soit continue de E dans L^p est fermé dans cet espace, la dernière partie de la proposition est démontrée.□

II.2. MESURES DE PALM DES PROCESSUS PONCTUELS STATIONNAIRES

La mesure de Palm que la proposition fondamentale suivante attache à un processus ponctuel stationnaire est intimement liée à la probabilité conditionnelle $P(\cdot / N\{0\} \neq 0)$, du moins pour un processus simple d'intensité de Radon. Malgré cette interprétation probabiliste simple, ce sont pourtant plutôt les formules "à la Fubini" ci-dessous qui nous serviront dans les applications que nous tirerons de l'existence de ces mesures.

Théorème II.4

A tout processus ponctuel $N : \Omega \to M_p(E)$ stationnaire pour le flot $(\theta_t,\ t \in E)$ défini sur l'espace de probabilité $(\Omega, \mathcal{A}, P)$ est associée une mesure positive σ-finie unique $\widehat{P}$ sur $(\Omega, \mathcal{A})$ telle que

$$(4) \quad \theta\,[P(d\omega)\,N(\omega, ds)] = \widehat{P}(d\omega)\,\lambda(ds)$$

sur $\Omega \times E$ si θ désigne la transformation de $\Omega \times E$ définie par

$$(5) \quad \theta\,[(\omega, s)] = (\theta_s\,\omega, s).$$

Cette mesure $\widehat{P}$ est appelée la mesure de Palm du processus ponctuel stationnaire N.

La formule du théorème signifie que pour toute fonction mesurable positive $f : \Omega \times E \to R_+$

$$(4') \quad \int_\Omega P(d\omega) \int_E N(\omega, ds)\, f(\theta_s\,\omega, s) = \int_\Omega \widehat{P}(d\omega) \int_E \lambda(ds)\, f(\omega, s)$$

Comme la transformation $\theta : \Omega \times E \to \Omega \times E$ admet l'inverse $\theta^{-1}(\omega,s) = (\theta_{-s}\,\omega, s)$, cette dernière formule peut aussi s'écrire :

$$(4'') \quad \int_\Omega P(d\omega) \int_E N(\omega, ds)\, g(\omega,s) = \int_\Omega \widehat{P}(d\omega) \int_E \lambda(ds)\, g(\theta_{-s}\,\omega, s)$$

pour toute fonction mesurable positive $g : \Omega \times E \to R_+$.

Démonstration

La formule

$$Q(B) = \int_\Omega P(d\omega) \int_E N(\omega, ds)\, 1_B(\omega, s) \qquad (B \in \mathcal{A} \otimes \mathcal{E})$$

définit une mesure positive Q sur $\Omega \times E$ que nous avons d'ailleurs désignée par $P(d\omega)\, N(\omega, ds)$ dans l'énoncé du théorème. Cette mesure est invariante par les transformations produits $\theta_t \otimes \tau_t$ de $\Omega \times E$ $(t \in E)$. En effet, pour toute fonction mesurable positive $f : \Omega \times E \to R_+$ et pour tout $t \in E$, nous avons

$$\int_E N(\omega, ds)\, f(\theta_t\, \omega, \tau_t\, s) = \int_E \tau_t\, N(\omega, ds)\, f(\theta_t \omega, s) = \int_E N(\theta_t \omega, ds)\, f(\theta_t \omega, s)$$

ce qui, avec l'invariance de la probabilité P par le flot, entraîne que

$$\begin{aligned}\int_{\Omega\times E} Q(d\omega\, ds)\, f(\theta_t\, \omega, \tau_t\, s) &= \int_\Omega P(d\omega) \int_E N(\theta_t\, \omega, ds)\, f(\theta_t\, \omega, s)\\ &= \int_\Omega P(d\omega) \int_E N(\omega, ds)\, f(\omega, s)\\ &= \int_\Omega Q(d\omega\, ds)\, f(\omega, s).\end{aligned}$$

Observons maintenant que la transformation θ du théorème qui est définie par $\theta\,[(\omega, s)] = (\theta_s\, \omega, s)$ sur $\Omega \times E$ vérifie les identités

(6) $$\theta \circ (\theta_t \times \tau_t) = (i_\Omega \times \tau_t) \circ \theta$$

où i_Ω désigne la transformation identique de Ω. En effet, pour tous $s, t \in E$ et $\omega \in \Omega$, nous avons

$$\theta \circ (\theta_t \times \tau_t)\,[(\omega, s)] = \theta\,[(\theta_t \omega, \tau_t s)] = (\theta_{\tau_t s}\, \theta_t\, \omega, \tau_t\, s) = (i_\Omega \times \tau_t)\,[(\theta_s\, \omega, s)]$$

car $\theta_{\tau_t s}\, \theta_t = \theta_s$ puisque $\tau_t\, s + t = s$. De l'invariance de la mesure positive Q par les transformations $\theta_t \times \tau_t$ $(t \in E)$, découle alors celle de la mesure image $\theta \circ Q$ par les transformations $i_\Omega \times \tau_t$ $(t \in E)$. Pour tout $A \in \mathcal{A}$, la mesure $\theta \circ Q\,(A \times .)$ est donc une mesure positive sur $(E, \mathcal{E})$ invariante par les translations τ_t $(t \in E)$; si cette mesure est de Radon, elle est alors nécessairement égale à un multiple de la mesure λ sur E. Pour achever la démonstration du théorème, il ne nous reste plus qu'à prouver l'existence dans $\mathcal{A}$ d'une suite $(A_n, n \in N)$ croissant vers Ω telle que $\theta \circ Q\,(A_n \times .)$ soit une mesure de Radon sur E car ceci prouvé, en posant $\hat{P}\,(B) = \theta \circ Q\,(B \times G_o) / \lambda\,(G_o)$ pour un ouvert relativement compact G_o fixé, nous obtiendrons une mesure positive $\hat{P}$ sur $(\Omega, \mathcal{A})$, finie sur les A_n $(n \in N)$ et donc σ-finie ; de plus pour tout B contenu dans un des A_n, la mesure de Radon positive $\theta \circ Q\,(B \times .)$ qui est proportionnelle à λ d'après ce qui précède, sera exactement égale à $\hat{P}\,(B)\,\lambda$ et il s'en suivra que $\theta \circ Q = \hat{P} \otimes \lambda$ sur $\Omega \times E$.

Si G est un voisinage relativement compact arbitraire de O dans E, montrons que la suite $A_n = \{N\,(G) \leqslant n\}$, qui croît en effet vers Ω, convient toujours. Pour cela choisissons un ouvert relativement compact non vide F dans E tel que $F - F \subset G$ (F existe en vertu de la continuité à l'origine de $x, y \to x - y$) et écrivons que

$$\theta \circ Q\,(A_n \times F) = \int_\Omega P\,(d\omega) \int_F N\,(\omega, ds)\ 1_{A_n}(\theta_s\,\omega) \qquad (n \in N).$$

Mais $N\,(\omega, F) \leqslant n$ si $s \in F$ et $\theta_s\,\omega \in A_n$ car $N\,(\omega, F) = N\,(\theta_s\,\omega, F - s) \leqslant N\,(\theta_s\,\omega, G) \leqslant n$ puisque $F - s \subset G$ si $s \in F$; de l'égalité $1_{A_n}(\theta_s .) \leqslant 1_{\{N(F) \leqslant n\}}$ valable sur Ω si $s \in F$, il résulte finalement que

$$\theta \circ Q\,(A_n \times F) \leqslant \int_\Omega dP\ N\,(F)\ 1_{\{N(F) \leqslant n\}} \leqslant n\,.$$

Comme la mesure $\theta \circ Q\,(A_n \times .)$ est invariante par translation, elle ne peut être finie sur un ouvert non vide F sans être finie sur tout compact (qui est recouvert par un nombre fini de translatés de F) et donc sans être de Radon.

La proposition est ainsi démontrée et de plus nous avons établi que $\hat{P}\,(N\,(G) \leqslant n) < \infty$ <u>pour tout voisinage relativement compact</u> G <u>de</u> O <u>et tout entier</u> n.□

<u>Remarque</u>. La démonstration qui précède n'utilise ni le caractère abélien du groupe E (si ce n'est dans la notation additive), ni la propriété des mesures $N(\omega, .)$ d'être ponctuelles. Cette remarque servira à la fin du chapitre.

Il résulte de la définition de la mesure de Palm que <u>l'intensité d'un processus ponctuel stationnaire</u> vaut

$$(7) \qquad \mu(F) = \int_\Omega P(d\omega)\, N(\omega, F) = \hat{P}(\Omega)\, \lambda(F) \qquad (F \in \mathcal{E}).$$

Lorsque $\hat{P}(\Omega) < \infty$, cette intensité est proportionnelle à la mesure invariante λ ; lorsque $\hat{P}(\Omega) = \infty$, $\mu(F) = 0$ ou $+\infty$ selon que $\lambda(F) = 0$ ou > 0. Dans tous les cas, $N(F) = 0$ P p.s. si et seulement si $\lambda(F) = 0$.

Si $N : \Omega \to M_p(E)$ est un processus ponctuel stationnaire, sa loi P_N est invariante par le flot $(\tau_t, t \in E)$ des translations sur $M_p(E)$. Le processus ponctuel canonique N' défini sur l'espace de probabilité $(M_p(E), \mathcal{M}_p(E), P_N)$ comme l'application identique de $M_p(E)$ sur lui-même, est donc un processus ponctuel stationnaire par le flot $(\tau_t, t \in E)$. Il est facile de vérifier que sa mesure de Palm $(P_N)^\wedge$ est l'image de la mesure de Palm $\hat{P}$ sur Ω de N par l'application $N : \Omega \to M_p(E)$, soit

$$(8) \qquad (P_N)^\wedge = N \circ \hat{P} \quad \text{sur} \quad M_p(E).$$

Contrairement à la probabilité P, la mesure de Palm $\hat{P}$ d'un processus ponctuel stationnaire n'est pas invariante par le flot $(\theta_t, t \in E)$, mais la famille $(\theta_t \hat{P}, t \in E)$ des images de $\hat{P}$ par le flot jouit de la propriété intéressante suivante qui exprime que pour tout $A \in \mathcal{A}$, la fonction $t \to \theta_{-t} \hat{P}(A)$ est la densité sur E de la mesure $\int_A N(.)\, dP$ par rapport à la mesure λ .

<u>Corollaire II.5</u>

<u>Pour toute fonction borélienne positive</u> $u : E \to R_+$ <u>et pour tout</u> $A \in \mathcal{A}$

$$(9) \qquad \int_E \lambda(ds)\, u(s)\, \theta_{-s} \hat{P}(A) = \int_A N(u)\, dP.$$

Démonstration

Le premier membre de cette formule s'écrit encore

$$\int_{\Omega\times E} \hat{P}\,(d\omega)\ \lambda\,(ds)\ u\,(s)\ 1_A\,(\theta_{-s}\,\omega) = \int_{\Omega\times E} P\,(d\omega)\ N\,(\omega,ds)\ u\,(s)\ 1_A\,(\omega)$$

grâce au théorème précédent [prendre $g\,(\omega,\,s) = 1_A\,(\omega)\ u\,(s)$] .□

La formule précédente exprime encore que la mesure $N\,(u)\,.\,P$ sur $(\Omega,\,\mathcal{A})$ est l'intégrale par rapport à $u\,(s)\ \lambda\,(ds)$ des mesures $\theta_{-s}\hat{P}$. Lorsque $E = Z^d$ ou plus généralement lorsque E est un sous-groupe discret de R^d, cette formule donne explicitement $\hat{P}$ en fonction de P puisque pour $u = 1_{\{o\}}$ il vient

$$\hat{P} = N\,(\{o\})\,.\,P \text{ sur } (\Omega,\,\mathcal{A}) \qquad \text{(E discret)},$$

du moins si dans la définition de $\hat{P}$ nous avons pris pour mesure invariante λ sur E, la mesure de comptage qui est telle que $\lambda\,(\{o\}) = 1$.

En particulier si le processus ponctuel N est simple, la variable aléatoire $N\,(\{o\})$ ne peut prendre que les deux valeurs 0 et 1 et, en dehors du cas trivial où $N\,(\{o\}) = 0$ p.s. et par suite $N = 0$ p.s., $\hat{P} = 0$, nous pouvons écrire que

$$(10)\quad P\,(.\ /\ N\,(\{o\}) \neq 0) = \frac{1}{\hat{P}\,(\Omega)}\,\hat{P} \quad \text{sur } (\Omega,\,\mathcal{A})$$

puisque $P\,[N\,(\{o\}) \neq 0] = E\,[N\,(\{o\})] = \hat{P}\,(\Omega)$.

Lorsque E n'est pas discret, les formules précédentes ne sont plus exactes parce que $\lambda\,(\{o\}) = 0$ et $N\,[\{o\}] = 0$ P p.s. [prendre $u = 1_{\{o\}}$] ; néanmoins si $\hat{P}\,(\Omega) < \infty$, il est possible de redonner un sens à ces formules par un passage à la limite. Nous aurons besoin à cet effet du lemme préliminaire suivant.

Lemme II.6

La mesure de Palm $\hat{P}$ d'un processus ponctuel stationnaire N défini sur un sous-groupe fermé E de R^d est telle que $\hat{P}\,[N\,(\{o\}) = 0] = 0$; si E n'est pas discret dans R^d , cette mesure $\hat{P}$ est donc étrangère à P. De plus le processus N est p.s. simple si et seulement si $\widehat{P}\,[N(\{o\}) \neq 1] = 0$.

Démonstration

Etant donné une fonction borélienne positive $u : E \to R_+$ de λ-intégrale égale à 1, le théorème II.4 entraîne que, pour tout entier positif k,

$$\widehat{P}\,[N(\{o\}) = k] = \int_\Omega \widehat{P}(d\omega) \int_E \lambda(ds)\; 1_{\{N(\omega,\{o\})=k\}}\, u(s)$$

$$= \int_\Omega P(d\omega) \int_E N(\omega, ds)\; 1_{\{N(\omega,\{s\})=k\}}\, u(s)$$

car $N(\theta_s\,\omega, \{o\}) = N(\omega, \{s\})$. Mais l'intégrale $\int_E m(ds)\; 1_{(m(\{s\})=k)}$ est nulle pour k = 0 quelle que soit la mesure ponctuelle m et pour tout $k \geqslant 2$ si et seulement si la mesure ponctuelle m est simple.□

Voici alors lorsque $\widehat{P}(\Omega) < \infty$, le lien existant entre $\widehat{P}$ et certaines probabilités conditionnelles définies à partir de P.

Proposition II.7

Soit N un processus ponctuel stationnaire défini sur un sous-groupe fermé non discret E de R^d, dont la mesure de Palm $\widehat{P}$ soit finie. L'algèbre H de fonctions mesurables bornées $h : \Omega \to R$ définie dans le lemme II.3 est alors dense dans $L^2(\widehat{P})$ et, pour toute fonction h de cette algèbre,

$$\int_\Omega h\, d\widehat{P} = \lim_{n\to\infty} \int_\Omega h\, \frac{N(V_n)}{\lambda(V_n)}\, dP$$

quelle que soit la suite décroissante $(V_n, n \in N)$ de voisinages relativement compacts de 0 dont les fermetures $\overline{V}_n \downarrow \{o\}$. De plus si le processus N est p.s. simple, il est exact aussi dans les mêmes conditions que

$$(11)\quad \frac{1}{\widehat{P}(\Omega)} \int_\Omega h\, d\widehat{P} = \lim_{n\to\infty} \int_\Omega h(\omega)\, P(d\omega \,/\, N(V_n) \neq 0)$$

Démonstration

Pour montrer que l'algèbre H est dense dans $L^2(\hat{P})$, il suffit d'après la démonstration du lemme II.3 d'établir que toute fonction $f \in L^2(\hat{P})$ orthogonale aux fonctions $h_{g,u}$ de ce lemme est p.s. nulle. Or pour une telle fonction f, la fonction $f(\theta_s\, \omega)$ de (ω, s) est p.s. nulle sur $\Omega \times E$ pour la mesure $P(d\omega)\, N(\omega, ds)$, car quelles que soient la fonction mesurable bornée $g : \Omega \to R$ et la fonction borélienne λ-intégrable $u : E \to R$, la définition de la mesure de Palm entraîne que

$$\int_{\Omega\times E} f(\theta_s\, \omega)\, g(\omega)\, u(-s)\, P(d\omega)\, N(\omega, ds) = \int_{\Omega} f(\omega)\, h_{g,u}(\omega)\, \hat{P}(d\omega) = 0$$

Mais pour toute fonction borélienne positive $v : E \to R_+$ de λ-intégrale égale à 1, nous pouvons alors écrire que

$$\int_{\Omega} f^2(\omega)\, \hat{P}(d\omega) = \int_{\Omega\times E} f^2(\omega)\, v(s)\, \hat{P}(d\omega)\, \lambda(ds)$$

$$= \int_{\Omega\times E} P(d\omega)\, N(\omega, ds)\, f^2(\theta_s\, \omega)\, v(s) = 0\ ;$$

autrement dit $f = 0$ dans $L^2(\hat{P})$.

D'après le théorème II.4 définissant la mesure de Palm, nous avons

$$\frac{1}{\lambda(V_n)} \int_{\Omega} P(d\omega)\, h(\omega)\, N(\omega, V_n) = \int_{\hat{\Omega}} \hat{P}(d\omega)\, h_n(\omega)$$

$$\text{si } h_n(\omega) = \frac{1}{\lambda(V_n)} \int_{V_n} h(\theta_{-s}\, \omega)\, \lambda(ds)$$

Or, pour tout $\omega \in \Omega$: $\lim_{n\to\infty} h_n(\omega) = h(\omega)$ parce que $h \in H$; par convergence dominée il s'ensuit lorsque $\hat{P}(\Omega) < \infty$, que $\lim_{n\to\infty} \int h_n\, d\hat{P} = \int h\, d\hat{P}$, ce qui prouve la première partie de la proposition.

Remarquons ensuite que si N est un processus simple et si $\hat{P}(\Omega) < \infty$, nous avons par le même théorème

$$\frac{1}{\lambda(V_n)}\, E\left[N(V_n)\, 1_{\{N(V_n)\geq 2\}}\right] = \frac{1}{\lambda(V_n)} \int_{\Omega} d\hat{P} \int_{V_n} \lambda(ds)\; 1_{\{N(V_n - s)\geq 2\}}$$

$$\leq \hat{P}\left[N(V_n - V_n) \geq 2\right]$$

$$\to \hat{P}\,[N(\{o\}) \geq 2] = 0 \quad \text{lorsque } n \to \infty$$

en tenant compte successivement des relations

$N(\theta_{-s}\,\omega, V_n) = N(\omega, V_n - s) \leq N(\omega, V_n - V_n)$ et $N(\omega, V_n - V_n) \downarrow N(\omega, \{o\})$,

où l'hypothèse $\overline{V}_n \downarrow \{o\}$ sert à assurer que $V_n - V_n \downarrow \{o\}$. Cette remarque permet d'écrire que

$$\frac{1}{\lambda(V_n)} \left| \int_\Omega h\, N(V_n)\, dP - \int_{\{N(V_n) \neq 0\}} h\, dP \right| \to 0 \quad \text{lorsque } n \uparrow \infty$$

pour toute fonction mesurable bornée $h : \Omega \to R$ et en particulier pour toute $h \in H$; pour $h = 1$, cela montre que

$$\lim_{n\to\infty} \frac{1}{\lambda(V_n)}\, P\,[N(V_n) \neq 0] = \hat{P}(\Omega)$$

puisque $\int_\Omega N(V_n)\, dP = \hat{P}(\Omega)\, \lambda(V_n)$ pour tout $n \in N$. Il ne nous reste plus maintenant qu'à écrire en tenant compte de tout ce qui précède que si $h \in H$

$$\lim_{n\to\infty} \int_\Omega h(\omega)\, P\,[d\omega \,/\, N(V_n) \neq 0] = \lim_{n\to\infty} \frac{1}{P\,[N(V_n) \neq 0]} \int_{\{N(V_n)\neq 0\}} h\, dP$$

$$= \frac{1}{\hat{P}(\Omega)} \lim_{n\to\infty} \frac{1}{\lambda(V_n)} \int_\Omega h\, N(V_n)\, dP$$

$$= \frac{1}{\hat{P}(\Omega)} \int_\Omega h\, d\hat{P}. \square$$

<u>Corollaire II.8</u>

<u>Si</u> N <u>est un processus ponctuel stationnaire simple de mesure de Palm</u> $\hat{P}$ <u>finie, sur un sous-groupe fermé non discret</u> E <u>de</u> R^d, <u>pour toute suite finie</u> $F_1, \ldots F_k$ <u>de boréliens relativement compacts dont les frontières sont telles que</u> $\hat{P}\,[N(\partial F_j) \neq 0] = 0$ $(j = 1, \ldots k)$, <u>nous avons</u>

$$\lim_{n\to\infty} P\,[N(F_j) = n_j \ (j = 1 \ldots k) \,/\, N(V_n) \neq 0] = \frac{1}{\hat{P}(\Omega)}\, \hat{P}\,[N(F_j) = n_j \ (j=1\ldots k)]$$

Démonstration

Si $f_1, \dots f_k$ sont k fonctions de $C_k^+(E)$, la proposition précédente s'applique aux fonctions $h = \exp\left[-\sum_1^k t_j N(f_j)\right]$ $(t_1, \dots t_k \in R_+)$ qui appartiennent bien à H. En encadrant la fonction $\exp\left[-\sum_1^k t_j N(F_j)\right]$ par de telles fonctions, nous obtiendrons alors que sous les hypothèses du corollaire

$$\lim_{n\to\infty} E\left[\exp\left[-\sum_1^k t_j N(F_j)\right] / N(V_n) \neq 0\right] = \frac{1}{\hat{P}(\Omega)} \int d\hat{P} \exp\left[-\sum_1^k t_j N(F_j)\right]$$

et ce résultat équivaut à celui énoncé dans le corollaire.□

Dans l'étude d'un processus ponctuel stationnaire, la donnée de sa mesure de Palm $\hat{P}$ équivaut essentiellement à celle de la mesure P. Nous avons défini $\hat{P}$ à partir de P ; le lemme suivant va nous permettre de retrouver P à partir de $\hat{P}$ au moins sur l'évènement $\{N(\omega, E) \neq 0\}$. Nous verrons ultérieurement comment ceci permet de construire aisément les probabilités P de processus ponctuels stationnaires.

Lemme II.9

Si l'application $N : (\Omega, \mathcal{A}) \to (M_p(E), \mathcal{M}_p(E))$ est mesurable et si Q est une mesure positive σ-finie sur $(\Omega, \mathcal{A})$, il existe au moins une fonction mesurable $a : \Omega \times E \to R$ strictement positive telle que

$$\int_E N(\omega, ds)\, a(\omega, s) = 1_{\{N(\omega,E)\neq 0\}} \quad Q \text{ p.p.}$$

Démonstration

Soit Q' une mesure positive finie équivalente à Q et soit $(f_j, j \in \mathbb{N})$ une suite de fonctions continues positives à supports compacts sur E croissant vers 1. Alors comme $N(f_j) < \infty$ sur Ω et comme Q' est finie, il est possible de trouver des entiers $n_j \geq 1$ tels que

$$\sum_j Q'\left[N(f_j) > n_j\right] < \infty$$

et donc tels d'après le lemme de Borel-Cantelli que pour Q'-presque tout ω, nous ayons $N(\omega, f_j) \leq n_j$ sauf pour un nombre fini de j. La série $f = \sum_j \frac{1}{2^j n_j} f_j$ définit alors une fonction continue strictement positive sur E tandis que la variable aléatoire $N(f) = \sum_j \frac{1}{2^j n_j} N(f_j)$ est strictement positive sur $\{N(E) \neq 0\}$ et finie Q'-presque partout. Il ne reste plus alors qu'à poser

$$a(\omega, s) = \begin{cases} f(s) / N(\omega, f) & \text{si } 0 < N(\omega, f) < \infty \\ 1 & \text{sinon} \end{cases}$$

pour obtenir une fonction a satisfaisant aux conditions du lemme.□

Etant donné un processus ponctuel stationnaire N, si $a : \Omega \times E \to R_+$ est une fonction mesurable strictement positive telle que $\int_E N(\omega, dt)\, a(\omega, t) = 1$ P p.p. sur $\{N(E) \neq 0\}$ comme le lemme précédent montre qu'il en existe, la seconde formule du théorème II.4 montre que pour toute fonction mesurable $f : \Omega \to R_+$

$$\int_{\{N(E) \neq 0\}} f(\omega)\, P(d\omega) = \int_{\Omega \times E} P(d\omega)\, N(\omega, ds)\, f(\omega)\, a(\omega, s)$$

$$= \int_{\Omega \times E} \hat{P}(d\omega)\, \lambda(ds)\, f(\theta_{-s}\omega)\, a(\theta_{-s}\omega, s)$$

La probabilité P est donc bien déterminée sur $\{N(E) \neq 0\}$ par $\hat{P}$.

La proposition suivante donne le lien existant entre les mesures de Palm de deux processus ponctuels stationnaires définis sur les mêmes espaces.

<u>Proposition II.10</u>

<u>Soient</u> N_1 et N_2 <u>deux processus ponctuels sur</u> E <u>définis sur le même espace de probabilité</u> $(\Omega, \mathcal{A}, P)$ <u>et stationnaires pour le même flot</u> $(\theta_t, t \in E)$. <u>Les mesures de Palm</u> $\hat{P}_1$ <u>et</u> $\hat{P}_2$ <u>de ces deux processus sont alors liées par la relation</u>

$$(12) \qquad \hat{P}_2(d\omega)\, N_1(\omega, ds) = \rho\left[\hat{P}_1(d\omega)\, N_2(\omega, ds)\right]$$

<u>où</u> ρ <u>désigne l'involution mesurable de</u> $\Omega \times E$ <u>définie par</u> $\rho(\omega, s) = (\theta_s\omega, -s)$.

Démonstration

Soit $\overline{Q}$ la mesure positive sur $\Omega \times E^2$ définie par

$$\overline{Q}\,(B) = \int_{\Omega} P\,(d\omega) \int_{E^2} N_1\,(\omega, ds)\, N_2\,(\omega, dt)\, 1_B\,(\omega, s, t)$$

($B \in \mathcal{A} \otimes \mathcal{E}^{2\otimes}$) que nous désignerons aussi par $P\,(d\omega)\, N_1\,(\omega, ds)\, N_2\,(\omega, dt)$.
Si $\overline{\theta}$ désigne l'application mesurable de $\Omega \times E^2$ dans lui-même définie par $\overline{\theta}\,[(\omega, s, t)] = (\theta_t\,\omega, \tau_t\, s, t)$, un calcul semblable à celui du début de la démonstration du théorème II.4 établit que

$$\overline{\theta} \circ \overline{Q} = \widehat{P}_2\,(d\omega)\, N_1\,(\omega, ds)\, \lambda\,(dt) \qquad \text{sur} \quad \Omega \times E^2.$$

En effet pour toute fonction mesurable positive $f : \Omega \times E^2 \to R_+$, la stationnarité du processus ponctuel N_1 entraîne que

$$\int_E N_1\,(\omega, ds)\, f \circ \overline{\theta}\,(\omega, s, t) = \int_E N_1\,(\theta_t\,\omega, ds)\, f\,(\theta_t\,\omega, s, t)$$

si bien que

$$\int_{\Omega\times E^2} f \circ \overline{\theta}\; d\overline{Q} = \int_{\Omega} P\,(d\omega) \int_{E^2} N_1\,(\omega, ds)\, N_2\,(\omega, dt)\, f \circ \overline{\theta}\,(\omega, s, t)$$

$$= \int_{\Omega\times E} P\,(d\omega)\, N_2\,(\omega, dt) \int_E N_1\,(\theta_t\,\omega, ds)\, f\,(\theta_t\,\omega, s, t)$$

$$= \int_{\Omega\times E} \widehat{P}_2\,(d\omega)\, \lambda\,(dt) \int_E N_1\,(\omega, ds)\, f\,(\omega, s, t)$$

d'après la définition de la mesure de Palm de N_2.

D'autre part considérons la symétrie σ de E^2 définie par $\sigma[(s,t)] = (t, s)$; il est clair que le produit de la transformation identique i_Ω de Ω et de la symétrie σ de E^2 échange le rôle de N_1 et N_2 dans la définition de la mesure $\overline{Q}$ ci-dessus, soit

$$(i_\Omega \times \sigma)\, [P\,(d\omega)\, N_2\,(\omega, ds)\, N_1\,(\omega, dt)] = P\,(d\omega)\, N_1\,(\omega, ds)\, N_2\,(\omega, dt).$$

Comme le calcul du début de cette démonstration montre aussi bien que

$$\overline{\theta}\left[P\ (d\omega)\ N_2\ (\omega,\ ds)\ N_1\ (\omega,\ dt)\right] = \widehat{P}_1\ (d\omega)\ N_2\ (\omega,\ ds)\ \lambda\ (dt),$$

il est donc clair que

$$\widehat{P}_2\ (d\omega)\ N_1\ (\omega,\ ds)\ \lambda\ (dt) = \phi\ \left[\widehat{P}_1\ (d\omega)\ N_2\ (\omega,\ ds)\ \lambda\ (dt)\right]$$

si ϕ désigne la transformation mesurable de $\Omega \times E^2$ définie par

$$\phi = \overline{\theta}\ o\ (i_\Omega \times \sigma)\ o\ \overline{\theta}^{\,-1}\ ;$$

notons que la transformation $\overline{\theta}$ admet effectivement une inverse mesurable donnée par $\overline{\theta}^{\,-1}\ (\omega,\ s,\ t) = (\theta_{-t}\ \omega,\ \tau_{-t}\ s,\ t)$. Or, la transformation ϕ vaut

$$\phi\ \left[(\omega,\ s,\ t)\right] = (\theta_s\ \omega,\ -s,\ s+t)$$

car

$$\phi\ \left[(\omega,\ s,\ t)\right] = \overline{\theta}\ o\ (i_\Omega \times \sigma)\ \left[(\theta_{-t}\ \omega,\ s+t,\ t)\right]$$

$$= \overline{\theta}\ \left[(\theta_{-t}\ \omega,\ t,\ s+t)\right] = (\theta_s\ \omega,\ -s,\ s+t)$$

pour tous $\omega \in \Omega$, $s,\ t \in E$.

L'invariance par translation de la mesure λ entraîne alors la proposition. En effet si $f : \Omega \times E \to R_+$ est une fonction mesurable positive, nous pourrons écrire après avoir choisi une fonction borélienne positive $g : E \to R_+$ de λ-intégrale égale à 1, que

$$\int_{\Omega\times E} \widehat{P}_1\ (d\omega)\ N_2\ (\omega,\ ds)\ f\ o\ \rho\ (\omega,\ s)$$

$$= \int_{\Omega\times E} \widehat{P}_1\ (d\omega)\ N_2\ (\omega,\ ds)\ \ f\ (\theta_{-s}\ \omega,\ -s)\ \int_E \lambda\ (dt)\ \ g\ (s+t)$$

$$= \int_{\Omega\times E^2} \widehat{P}_1\ (d\omega)\ N_2\ (\omega,\ ds)\ \lambda\ (dt)\ (f \otimes g)\ o\ \phi\ (\omega,\ s,\ t)$$

$$= \int_{\Omega\times E^2} \widehat{P}_2\ (d\omega)\ N_1\ (\omega,\ ds)\ \lambda\ (dt)\ f\ (\omega,\ s)\ g\ (t)$$

$$= \int_{\Omega\times E} \widehat{P}_2\ (d\omega)\ N_1\ (\omega,\ ds)\ f\ (\omega,\ s). \square$$

La proposition précédente permet de calculer directement $\widehat{P}_2$ à partir de $\widehat{P}_1$ au moins sur l'évènement $\{\omega : N_1(\omega, .) \neq 0\}$ grâce au lemme II.9. Si $a : \Omega \times E \to R_+$ est une fonction mesurable strictement positive telle que $\int_E N_1(\omega, ds)\, a(\omega, s) = 1_{\{N_1(\omega,E)\neq 0\}}$ $\widehat{P}_2$ p.p., comme il en existe d'après le lemme cité, la formule de la proposition précédente montre que pour toute fonction mesurable positive $f : \Omega \to R_+$

$$(13) \quad \int_{\{N_1(E)\neq 0\}} f(\omega)\, \widehat{P}_2(d\omega) = \int_{\Omega\times E} \widehat{P}_1(d\omega)\, N_2(\omega, ds)\, f(\theta_s\, \omega)\, a(\theta_s\, \omega, -s)$$

ce qui détermine $\widehat{P}_2$ sur $\{N_1(E) \neq 0\}$ en fonction de $\widehat{P}_1$.

En prenant dans la proposition précédente les processus N_1 et N_2 identiques à un même processus N, nous obtenons une propriété d'invariance de la mesure de Palm $\widehat{P}$ de N qui comme la proposition suivante le montre, est caractéristique des mesures de Palm.

Proposition II.11

Soit $N : (\Omega, \mathcal{A}) \to (M_p(E), \mathcal{M}_p(E))$ une application mesurable stationnaire par le flot mesurable $(\theta_t, t \in E)$ défini sur Ω, au sens où $N(\theta_t\, \omega, .) = \tau_t\, [N(\omega, .)]$.

Pour toute probabilité P sur Ω invariante par le flot $(\theta_t, t \in E)$, la mesure de Palm $\widehat{P}$ de N relativement à P est portée par l'évènement $\{N(E) \neq 0\}$ et la mesure $\widehat{P}(d\omega)\, N(\omega, ds)$ sur $\Omega \times E$ qui lui est associée est invariante par l'involution $\rho(\omega, s) = (\theta_s\, \omega, -s)$ de $\Omega \times E$. Réciproquement pour toute mesure positive σ-finie $\widehat{P}$ sur Ω portée par $\{N(E) \neq 0\}$ telle que la mesure $\widehat{P}(d\omega)\, N(\omega, ds)$ sur $\Omega \times E$ soit invariante par ρ, il existe une probabilité P sur Ω portée par $\{N(E) \neq 0\}$ invariante par $(\theta_t, t \in E)$ telle que $\widehat{P}$ soit la mesure de Palm de N relativement à P.

Démonstration

La première partie de la proposition découle immédiatement de la proposition II.10 précédente ; démontrons donc la réciproque. Soit $a : \Omega \times E \to R_+$ une fonction mesurable strictement positive telle que $\int_E N(\omega, dt)\, a(\omega, t) = 1$ sur $\{N(., E) \neq 0\}$ presque partout pour la mesure $\int_E \lambda(dt)\, \theta_{-t} \hat{P}$, c'est-à-dire presque partout pour $\theta_{-t} \hat{P}$ pour λ presque tout t. Définissons alors une mesure positive P sur $(\Omega, \mathcal{A})$ par la formule

$$P(A) = \int_{\Omega \times E} \hat{P}(d\omega)\, \lambda(dt)\, 1_A(\theta_{-t}\omega)\, a(\theta_{-t}\omega, t) \qquad (A \in \mathcal{A})$$

(que nous avons rencontrée après le lemme II.8). Nous établirons d'abord que

$$\theta\,[P(d\omega)\, N(\omega, ds)] = \hat{P}(d\omega)\ \lambda(ds) \qquad \text{sur } \Omega \times E$$

si $\theta\,[(\omega, s)] = (\theta_s\omega, s)$.

A cet effet écrivons que pour toute fonction mesurable positive $f : \Omega \times E \to R_+$

$$\int_{\Omega \times E} P(d\omega)\, N(\omega, ds)\ f \circ \theta(\omega, s)$$

$$= \int_E \lambda(dt) \int_\Omega \hat{P}(d\omega)\, a(\theta_{-t}\omega, t) \int_E N(\theta_{-t}\omega, ds)\, f \circ \theta\,[(\theta_{-t}\omega, s)]$$

$$= \int_E \lambda(dt) \int_\Omega \hat{P}(d\omega)\, a(\theta_{-t}\omega, t) \int_E N(\omega, ds)\, f(\theta_s\omega, s+t)$$

par la stationnarité de N et maintenant par l'hypothèse d'invariance de $\hat{P}(d\omega)\, N(\omega, ds)$ par ρ

$$= \int_E \lambda(dt) \int_{\Omega \times E} \hat{P}(d\omega)\ N(\omega, ds)\, a(\theta_{s-t}\omega, t)\, f(\omega, -s+t)$$

$$= \int_E \lambda(dt) \int_{\Omega \times E} \hat{P}(d\omega)\, N(\omega, ds)\, a(\theta_{-t}\omega, t+s)\ f(\omega, t)$$

par l'invariance de λ par τ_s

$$= \int_E \lambda(dt) \int_\Omega \hat{P}(d\omega)\ f(\omega, t)$$

car par la stationnarité de N et le choix de la fonction a

$$\int_E N(\omega, ds)\, a(\theta_{-t}\omega, t+s) = \int_E N(\theta_{-t}\omega, ds)\, a(\theta_{-t}\omega, s) = 1$$

$\widehat{P}$ presque sûrement pour λ-presque tout $t \in E$.

La formule annoncée étant démontrée, nous en déduisons que pour tout $t \in E$

$$(\theta_t \times \tau_t)\left[P\,(d\omega)\; N\,(\omega,\, ds)\right] = P\,(d\omega)\; N\,(\omega,\, ds) \quad \text{sur } \Omega \times E$$

car d'une part d'après la démonstration du théorème 4

$$\theta_t \times \tau_t = \theta^{-1} \circ (i_\Omega \times \tau_s) \circ \theta$$

et car d'autre part la mesure $\widehat{P}\,(d\omega)\;\lambda\,(ds)$ est invariante par $i_\Omega \times \tau_s$. Cela entraîne que la mesure P qui par construction est portée par $\{N\,(E) \neq 0\}$ est invariante par le flot mesurable $(\theta_t\ ,\ t \in E)$. En effet pour un t fixé si $a : \Omega \times E \to R_+$ est une fonction mesurable strictement positive telle que $\int N\,(\omega,\, ds)\; a\,(\omega,\, s) = 1_{\{N(\omega,E)\neq 0\}}$ P et θ_t P p.s., la stationnarité de N entraîne aussi que

$$\int_E N\,(\omega,\, ds)\; a\,(\theta_t\,\omega,\, \tau_t\, s) = \int_E N\,(\theta_t\,\omega,\, ds)\; a\,(\theta_t\,\omega,\, s) = 1_{\{N(\omega,E)\neq 0\}}$$

P presque partout (compte tenu de ce que $N\,(\theta_t\,\omega,\, E) = N\,(\omega,\, E)$). Donc pour toute fonction mesurable positive $f :\ \Omega \to R_+$

$$\begin{aligned}\int_\Omega P\,(d\omega)\; f\,(\theta_t\,\omega) &= \int_{\Omega\times E} P\,(d\omega)\; N\,(\omega,\, ds)\; f\,(\theta_t\,\omega)\;\; a\,(\theta_t\,\omega,\, \tau_t\, s) \\ &= \int_{\Omega\times E} P\,(d\omega)\; N\,(\omega,\, ds)\; f\,(\omega)\; a\,(\omega,\, s) \\ &= \int_\Omega P\,(d\omega)\; f\,(\omega)\end{aligned}$$

ce qui montre bien que $\theta_t\, P = P$. Comme d'après la première formule démontrée, $\widehat{P}$ est la mesure de Palm de N par rapport à P, la proposition est établie.□

II.3. PROCESSUS PONCTUELS STATIONNAIRES SIMPLES SUR R

La structure ordonnée de la droite réelle permet de représenter un processus ponctuel sur R comme une suite croissante de points aléatoires. L'objet de ce paragraphe est d'étudier cette représentation pour les processus stationnaires et d'en déduire de nouveaux résultats sur les mesures de Palm de ces processus.

Commençons par démontrer un lemme simple de théorie ergodique qui va nous permettre de ne considérer que des processus "doublement infinis", c'est-à-dire des processus tels que $N(R_+) = N(R_-) = +\infty$.

Lemme II.12

Etant donné un flot réel $(\theta_t, t \in R)$ sur un espace de probabilité $(\Omega, \mathcal{A}, P)$, pour toute fonction mesurable positive $f : \Omega \to R_+$ localement intégrable sur les trajectoires du flot, l'ensemble

$$I_f = \{\omega : \int_0^\infty f(\theta_t \omega)\, dt = +\infty = \int_0^\infty f(\theta_{-t} \omega)\, dt\}$$

est invariant par le flot et $f = \int_{-\infty}^{+\infty} f \circ \theta_t\, dt = 0$ p.s. hors de I_f.

Démonstration

Pour tout réel a, la formule $f_a = \int_a^\infty f \circ \theta_t\, dt$ définit une fonction mesurable $f_a : \Omega \to [0,\infty]$ dont l'ensemble de finitude $J_+ = \{f_a < \infty\}$ ne dépend pas de a puisque par hypothèse $f(\theta_t \omega)$ est localement intégrable en t pour tout ω. Il est facile de vérifier que $f_a \circ \theta_s = f_{a+s}$ pour tous a, s $\in$ R. Ceci entraîne d'abord que $\theta_s^{-1}(\{f_a < \infty\}) = \{f_{a+s} < \infty\}$ et donc que J_+ est invariant par les θ_s (s $\in$ R). D'autre part pour tout $\varepsilon > 0$ fixé, les ensembles $J_+ \cap \{f_a \geq \varepsilon\}$ décroissent vers $\emptyset$ lorsque $a \uparrow \infty$ puisque $f_a \downarrow 0$ sur J_+ si $a \uparrow \infty$; or, d'après l'invariance de P par le flot, ces ensembles ont tous la même probabilité puisque pour tout s $\in$ R

$$P[J_+ \{f_a \geq \varepsilon\}] = \theta_s P [J_+ \{f_a \geq \varepsilon\}| = P(J_+ \{f_{a+s} \geq \varepsilon\})$$

grâce à l'invariance de J_+ et à la formule $f_{a+s} = f_a \circ \theta_s$. Donc $P(J_+ \{f_a \geq \varepsilon\}) = 0$ pour tout $a \in R$ et tout $\varepsilon > 0$ et en faisant tendre $a \to -\infty$ et $\varepsilon \downarrow 0$, nous trouvons que $\int_{-\infty}^{+\infty} f \circ \theta_t \, dt = 0$ p.s. sur J_+. Nous démontrerions de même que cette dernière intégrale est nulle p.s. sur l'ensemble invariant $J_- = \{\int_{-\infty}^{0} f \circ \theta_t \, dt < \infty\}$. Comme $I_f = (J_+ \cup J_-)^c$, le lemme est démontré à la dernière égalité près. Celle-ci résulte de ce que

$$0 = \int_{I_f^c} dP \int_{-\infty}^{+\infty} f \circ \theta_t \, dt = \int_{-\infty}^{+\infty} dt \, . \int_{I_f^c} f \, dP \quad . \square$$

Etant donné un processus ponctuel stationnaire N sur R, le lemme précédent s'applique en particulier à la variable aléatoire $N(]0, 1])$ puisque

$$\int_{-a}^{+a} N(\theta_t \omega,]0, 1]) \, dt = \int_{-a}^{+a} N(\omega,]t, t+1]) \, dt \leq N(\omega,]-a, a+1]) < \infty$$

pour tout $\omega \in \Omega$ et tout $a \in R_+$. Ce lemme montre donc que l'évènement $I = \{N(., R_+) = +\infty = N(., R_-)\}$ est invariant par le flot et que $N(., R) = 0$ p.s. hors de I, puisque

$$N(]0, \infty[) \leq \int_0^{\infty} N(]0,1]) \circ \theta_t \, dt = \int_0^{\infty} N(]t, t+1[) \, dt \leq N(]-1, +\infty[)$$

et $\int_{-\infty}^{+\infty} N(]0,1]) \circ \theta_t \, dt = N(R)$.

Quitte à nous restreindre à I, <u>nous ne considérerons plus dans la suite de ce paragraphe que des processus doublement infinis.</u>

Toute mesure ponctuelle m simple et doublement infinie sur R s'écrit sous la forme $m = \sum_{\mathbb{Z}} \varepsilon_{t_n}$ pour une suite $(t_n, n \in \mathbb{Z})$ strictement croissante de réels telle que $\lim_{n\to\pm\infty} t_n = \pm\infty$; de plus cette suite est unique si nous lui imposons de vérifier les inégalités $t_0 \leq 0 < t_1$. Par conséquent tout processus ponctuel sur R, soit $N : \Omega \to M_p(R)$, qui est simple et doublement infini s'écrit d'une et d'une seule manière sous la forme

(14) $$N(\omega, .) = \sum_{\mathbb{Z}} \varepsilon_{T_n(\omega)}$$

pour une suite de variables aléatoires réelles T_n $(n \in \mathbb{Z})$ telle que

$$\ldots \; T_{-1} < T_0 \leq 0 < T_1 < T_2 \; \ldots \text{ et que } \lim_{n\to\pm\infty} T_n = \pm\infty \quad ;$$

le caractère mesurable des fonctions $T_n : \Omega \to R$ découle immédiatement des deux formules $\{\omega : T_n(\omega) \leq t\} = \{\omega : N(\omega,]0, t]) \geq n\}$ et $\{\omega : T_n(\omega) \geq t\} = \{\omega : N(\omega,]t, 0]) \geq n\}$ valables respectivement lorsque n et t sont strictement positifs ou sont négatifs.

Pour toute fonction mesurable $S : \Omega \to R$, l'application $\theta_S : \Omega \to \Omega$ définie par $\theta_S(\omega) = \theta_s(\omega)$ si $S(\omega) = s$ est mesurable en tant que composée des deux applications mesurables $\omega \to (\omega, S(\omega))$ de Ω dans $\Omega \times R$ et $(\omega, s) \to \theta_s(\omega)$ de $\Omega \times R$ dans Ω. La composition de θ_S avec un processus ponctuel $N = \sum_{Z} \varepsilon_{T_n}$ translate ce processus ponctuel de S comme le montre la formule de stationnarité

$$N(\theta_S\,\omega, .) = \tau_{S(\omega)} \circ N(\omega, .) = \sum_{Z} \varepsilon_{T_n(\omega)-S(\omega)}$$

En particulier $N(\theta_{T_0}\,\omega, .)$ est la mesure ponctuelle obtenue en translatant l'origine au premier point de la mesure ponctuelle $N(\omega, .)$ qui soit à gauche de 0, c'est-à-dire à $T_0(\omega)$. L'objet des propositions suivantes est d'étudier les transformations θ_{T_n} $(n \in Z)$. Notons auparavant qu'à la différence de ce qui se passe lorsque S est constante sur Ω, les transformations θ_S en général ne préservent pas la mesure probabilité P et ne sont pas non plus inversibles.

Nous commencerons par traduire sur les variables T_n $(n \in Z)$ la propriété de stationnarité du processus ponctuel N. Cette propriété nous permet d'écrire pour tout $\omega \in \Omega$ et tout $s \in R$ que

$$\sum_{Z} \varepsilon_{T_n(\omega)} = N(\omega, .) = \tau_{-s}\,[N(\theta_s\,\omega, .)] = \sum_{Z} \varepsilon_{s+T_n(\theta_s\,\omega)} \cdot$$

Ceci montre que les deux suites croissantes $(T_n(\omega), n \in Z)$ et $(s + T_n(\theta_s\,\omega), n \in Z)$ dans R coïncident. Il ne faut pas en conclure que $T_n(\omega) = s + T_n(\theta_s\,\omega)$ en général car si $T_0(\omega)$ et $T_1(\omega)$ sont les points de la première suite qui encadrent 0, <u>les points</u> $s + T_0(\theta_s\,\omega)$ <u>et</u> $s + T_1(\theta_s\,\omega)$ <u>de mêmes indices</u> 0 <u>et</u> 1 <u>de la deuxième suite sont ceux qui encadrent</u> s. La conclusion correcte à tirer des égalités ci-dessus est en fait que

(15) $s + T_n(\theta_s\,\omega) = T_{m+n}(\omega)$ si $T_m(\omega) \leq s < T_{m+1}(\omega)$

($\omega \in \Omega$, $s \in R$, $n \in \mathbb{Z}$). En particulier la formule précédente montre que pour tous $m, n \in \mathbb{Z}$:

(16) $T_m + T_n \circ \theta_{T_m} = T_{m+n}$ sur Ω .

Proposition II.13

Soit $N = \sum_{\mathbb{Z}} \varepsilon_{T_n}$ un processus ponctuel stationnaire sur R, simple et doublement infini. Si $\widehat{\Omega}$ désigne le sous-ensemble mesurable de Ω défini par

(17) $\widehat{\Omega} = \{\omega : N(\omega, \{0\}) \neq 0\} = \{\omega : T_0(\omega) = 0\}$,

l'application mesurable θ_{T_0} applique Ω sur $\widehat{\Omega}$ et la mesure de Palm $\widehat{P}$ du processus N est liée à l'image de P par θ_{T_0} par la formule

(18) $\theta_{T_0} \circ P = T_1 \,.\, \widehat{P}$,

ce qui entraîne que cette mesure de Palm est portée par $\widehat{\Omega}$.

Réciproquement la probabilité P peut être récupérée à partir de la mesure de Palm $\widehat{P}$ par la formule

(19) $\int_\Omega P(d\omega)\, f(\omega) = \int_{\widehat{\Omega}} \widehat{P}(d\omega) \int_0^{T_1(\omega)} f(\theta_s\,\omega)\, ds$

où $f : \Omega \to R_+$ est une fonction mesurable positive arbitraire. D'ailleurs les deux formules précédentes ne sont que deux cas particuliers de la formule

(20) $\int_\Omega P(d\omega)\; g\,[\theta_{T_0}(\omega), -T_0(\omega)] = \int_{\widehat{\Omega}} \widehat{P}(d\omega) \int_0^{T_1(\omega)} g(\omega, s)\, ds$

valable pour toute fonction mesurable positive $g : \Omega \times R_+ \to R_+$.

Démonstration

D'après la formule qui précède la proposition, la mesure ponctuelle $N(\theta_{T_o}\omega, .)$ admet la suite de points $(T_n(\omega) - T_o(\omega), n \in \mathbb{Z})$ pour support et charge donc 0 ; il s'en suit que $\theta_{T_o}(\omega) \in \widehat{\Omega}$ pour tout $\omega \in \Omega$. D'autre part si $\omega \in \widehat{\Omega}$, il résulte de $T_o(\omega) = 0$ que $\theta_{T_o}(\omega) = \omega$; ainsi θ_{T_o} se réduit à l'identité sur $\widehat{\Omega}$ et il est donc clair que $\theta_{T_o}(\Omega) = \widehat{\Omega}$.

Pour démontrer la dernière formule de la proposition, nous utiliserons la fonction $a_o : \Omega \times R \to R_+$ définie par

(21) $$a_o(\omega, s) = 1_{\{T_o(\omega) \leq s < T_1(\omega)\}}$$

qui possède la propriété d'invariance : $a_o(\theta_s\omega, -s) = a_o(\omega, s)$ $(\omega \in \Omega, s \in R)$. En effet $a_o(\theta_s\omega, -s) = 1_{\{s+T_o(\theta_s\omega) \leq 0 < s+T_1(\theta_s\omega)\}}$ et nous avons remarqué avant la proposition 13 que $s + T_o(\theta_s\omega)$, respectivement $s + T_1(\theta_s\omega)$ est le premier point de la mesure ponctuelle $N(\omega, .)$ à gauche de s, resp. à droite de s ; la condition $s + T_o(\theta_s\omega) \leq 0$ équivaut donc à $s < T_1(\omega)$ et de même la condition $s + T_1(\theta_s\omega) > 0$ équivaut à $T_o(\omega) \leq s$, de sorte que $a_o(\theta_s\omega, -s) = a_o(\omega, s)$.

Appliquons alors la première formule du théorème II.4 à la fonction $f(\omega, t) = g(\omega, -t)\, a_o(\omega, -t)$; nous trouvons ainsi compte tenu de la remarque précédente que

$$\int_{\Omega\times R} P(d\omega)\, N(\omega, dt)\, a_o(\omega, t)\, g(\theta_t\omega, -t)$$

$$= \int_{\Omega\times R} \widehat{P}(d\omega)\, \lambda(dt)\, a_o(\omega, -t)\, g(\omega, -t)$$

Etant donné la forme de la fonction a_o et compte tenu de l'invariance de la mesure λ par $t \to -t$, nous avons obtenu que

$$\int_{\Omega} P(d\omega)\, g\left[\theta_{T_o}(\omega), - T_o(\omega)\right] = \int_{\Omega\times R} \widehat{P}(d\omega)\, \lambda(dt)\, 1_{\{T_o(\omega)<t<T_1(\omega)\}}\, g(\omega,t).$$

En particulier pour une fonction g ne dépendant que de ω, soit g (ω,t) = f (ω), la formule que nous venons d'obtenir s'écrit

$$\int_{\Omega} P\,(d\omega)\, f \circ \theta_{T_0}\,(\omega) = \int_{\Omega} \hat{P}\,(d\omega)\,[T_1\,(\omega) - T_0\,(\omega)]\; f\,(\omega)\;;$$

la mesure positive $(T_1 - T_0)$. $\hat{P}$ est donc égale à la mesure image $\theta_{T_0} \circ P$ et comme θ_{T_0} applique Ω sur $\hat{\Omega}$, cette mesure est portée par $\hat{\Omega}$. Nous retrouvons ainsi que la mesure de Palm $\hat{P}$ est portée par $\hat{\Omega}$ (lemme II.6) puisque la variable $T_1 - T_0$ est strictement positive, c'est-à-dire que $T_0 = 0$ $\hat{P}$ p.p.

Enfin si dans la formule établie ci-dessus nous prenons pour g une fonction de la forme g (ω, s) = f $(\theta_s\, \omega)$ sur $\Omega \times R$, nous trouvons que

$$\int_{\Omega} P\,(d\omega)\, f\,(\omega) = \int_{\hat{\Omega}} \hat{P}\,(d\omega) \int_0^{T_1(\omega)} f\,(\theta_s\, \omega)\, ds,$$

compte tenu de ce que $g\left[\theta_{T_0}\,(\omega),\, - T_0\,(\omega)\right] = f\,(\omega)$.□

La proposition précédente nous donne en particulier des indications sur la loi du couple (T_0, T_1) formé par les deux points aléatoires du processus ponctuel qui entourent 0.

<u>Corollaire II.14</u>

<u>La formule</u> $\hat{F} = \hat{P}\,(T_1 \in .)$ <u>définit une mesure positive sur</u> $]0, \infty[$ telle que $\int v\, \hat{F}\,(dv) = 1$ <u>et telle que la loi du couple de variables aléatoires positives</u> $(- T_0,\, T_1 - T_0)$ <u>pour la probabilité</u> P <u>soit donnée par</u>

$$P\,(u \leqslant - T_0 < u + du,\; v \leqslant T_1 - T_0 < v + dv) = 1_{\{0 \leqslant u < v\}}\, du\, \hat{F}\,(dv). \tag{22}$$

<u>En particulier</u>

$$P\,(v \leqslant T_1 - T_0 < v + dv) = v\, \hat{F}\,(dv),$$

$$P\,[u \leqslant - T_0 < u + du \,/\, T_1 - T_0 = v] = \frac{1}{v}\, 1_{\{0 \leqslant u < v\}}\, du.$$

<u>et</u>

$$P\,(u \leqslant - T_0 < u + du) = P\,(u \leqslant T_1 < u + du) = du\, f\,(u)$$

<u>si</u> $f\,(u) = \hat{F}\,([u, \infty[\,)$.

La dernière formule de ce corollaire montre donc que la variable T_1 d'un processus ponctuel stationnaire possède toujours une densité et que cette densité est une fonction décroissante sur R_+ .

Démonstration

Dans la dernière formule de la proposition précédente, prenons $g(\omega, s) = h(T_1(\omega), s)$ où $h : R_+^2 \to R_+$ est borélienne. Comme $T_1(\theta_{T_0}\omega) = (T_1 - T_0)(\omega)$, nous trouvons que

$$\int_\Omega h(T_1 - T_0, -T_0)\, dP = \int_{\hat{\Omega}} \left[\int_0^{T_1} h(T_1, s)\, ds\right] d\hat{P}$$
$$= \int_{R_+} \left[\int_0^v h(v, s)\, ds\right] \hat{F}(dv)$$

ce qui prouve la première formule du corollaire ; en particulier en prenant $h \equiv 1$ nous trouvons que $\int v\, \hat{F}(dv) = 1$ de sorte que $\hat{F}([u, \infty[) < \infty$ pour tout $u > 0$. Les autres formules se déduisent de la formule ci-dessus.□

Remarquons bien que la mesure $\hat{F} = \hat{P}(T_1 \in .)$ n'est pas la loi de $T_1 - T_0$ (ni celle de T_1) pour la probabilité P : c'est la mesure $v . \hat{F}(dv)$ qui donne cette loi. D'ailleurs la masse totale de la mesure $\hat{F}$ n'est pas égale à 1 en général puisque $\hat{F}(R_+) = \hat{P}(\Omega)$; la mesure $\hat{F}$ peut même être infinie.

La formule suivante

$$\text{(23)} \quad \begin{aligned} &P\left[u \leq -T_0 < u + du \; ; \; v_k \leq T_{k+1} - T_k < v_k + dv_k \quad (|k| \leq n)\right] \\ &= 1_{\{0 \leq u < v_0\}}\, du\, \hat{P}\left[v_k \leq T_{k+1} - T_k < v_k + dv_k \qquad (|k| \leq n)\right] \end{aligned}$$

où $n \in N$ et u, v_k $(|k| \leq n)$ appartiennent à R_+ , se démontre comme la formule (22) ci-dessus (prendre pour $g(\omega, s)$ une fonction de dépendant de ω que par l'intermédiaire de $T_{k+1} - T_k$ $(|k| \leq n)$). Cette formule exprime la loi du vecteur $(T_{-n}, T_{-n+1}, \dots T_{n+1})$ pour la probabilité P en fonction de la loi de ce même vecteur pour la mesure de Palm $\hat{P}$, compte tenu de ce que

$T_o = 0$ pour $\widehat{P}$; nous reviendrons sur les implications de cette formule lorsque nous étudierons à la fin du paragraphe les processus de renouvellement.

Les formules (22) et (23) qui précèdent ne sont enfin que des cas particuliers du résultat plus général explicité dans le corollaire suivant de la proposition 13.

Corollaire II.15

La loi conditionnelle de la variable $-T_o$ par rapport à la tribu $\theta_{T_o}^{-1}(\mathcal{C})$ est la loi uniforme sur $[0, T_1 - T_o]$.

Remarquons que la formule $\theta_{T_o} \circ P = T_1 . \widehat{P}$ de la proposition 13 détermine P sur la tribu $\theta_{T_o}^{-1}(\mathcal{C})$ à partir de $\widehat{P}$: $P\left[\theta_{T_o}^{-1}(A)\right] = \int_A T_1 \, d\widehat{P}$, mais que d'après le corollaire que nous venons d'énoncer cette sous-tribu $\theta_{T_o}^{-1}(\mathcal{C})$ de $\mathcal{C}$ ne peut être égale à $\mathcal{C}$ puisqu'elle ne peut rendre T_o mesurable.

Démonstration

Notons d'abord que $T_1 - T_o$ est une variable mesurable par rapport à la tribu $\theta_{T_o}^{-1}(\mathcal{C})$ puisqu'elle coïncide avec $T_1 \circ \theta_{T_o}$.

La propriété énoncée dans le corollaire revient à dire que la formule

$$E^{\theta_{T_0}^{-1}(\mathcal{A})}\left[h(-T_0)\right] = \frac{1}{T_1-T_0}\int_0^{T_1-T_0} h(s)\,ds$$

est valable pour toute fonction borélienne positive $h : R_+ \to R_+$, ce qui se déduit d'une double application de la dernière formule de la proposition en écrivant que

$$E\left[g \circ \theta_{T_0}\ h(-T_0)\right] = \int_{\hat{\Omega}} d\hat{P}\ g \int_0^{T_1} h(s)\,ds$$

$$= \int_{\hat{\Omega}} d\hat{P}\ g\ \frac{1}{T_1}\int_0^{T_1} h(s)\,ds\ .\ T_1$$

$$= E\left[g \circ \theta_{T_0}\ \left(\frac{1}{T_1}\int_0^{T_1} h(s)\,ds\right) \circ \theta_{T_0}\right]$$

pour toute fonction mesurable positive $g : \Omega \to R_+$.□

Remarquons ensuite que comme la loi conditionnelle de $-T_0$ par rapport à la tribu $\theta_{T_0}^{-1}(\mathcal{A})$ ne dépend que de $T_1 - T_0 = T_1 \circ \theta_{T_0}$, cette loi est aussi la loi conditionnelle de $-T_0$ par rapport à $T_1 - T_0$ et nous retrouvons ainsi la dernière partie du corollaire II.14 . Notons enfin que les variables aléatoires $S_n = T_{n+1} - T_n$ $(n \in Z)$ sont mesurables par rapport à la tribu $\theta_{T_0}^{-1}(\mathcal{A})$ car elles sont invariantes par θ_{T_0} . Par conséquent la loi de la suite $(S_n,\ n \in Z)$ pour la probabilité P, c'est-à-dire l'image $\sigma \circ P$ de P par l'application $\sigma(\omega) = (S_n(\omega),\ n \in N)$ de Ω dans R_+^Z est donnée par la formule

$$\sigma \circ P\,(ds) = s_0 .\ \sigma \circ \hat{P}\,(ds)$$

où s parcourt R_+^Z et s_0 désigne la coordonnée d'indice 0 de s. Quant à la loi conditionnelle de T_0 par rapport à $(S_n,\ n \in Z)$ elle est égale à la loi uniforme sur $[0, S_0[$ puisque la tribu engendrée par les S_n $(n \in Z)$ est comprise entre $\theta_{T_0}^{-1}(\mathcal{A})$ et celle engendrée par $S_0 = T_1 - T_0$].

La proposition II.10 qui permet de passer de la mesure de Palm d'un processus ponctuel stationnaire N_1 à celle d'un second entraîne pour les processus réels le résultat particulièrement simple suivant.

Proposition II.16

Soient $N_1 = \sum_{\mathbb{Z}} \varepsilon_{S_n}$ et $N_2 = \sum_{\mathbb{Z}} \varepsilon_{T_n}$ deux processus ponctuels stationnaires sur R, simples et doublement infinis. La mesure de Palm $\hat{P}_2$ du deuxième processus peut alors se déduire de celle du premier processus par la formule

$$(24) \quad \int_{\hat{\Omega}_2} f(\omega)\, \hat{P}_2(d\omega) = \int_{\hat{\Omega}_1} \left[\sum_{n:0<T_n(\omega)\leqslant S_1(\omega)} f(\theta_{T_n}\omega) \right] \hat{P}_1(d\omega)$$

En particulier, la formule suivante est valable

$$\theta_{S_0} \circ \hat{P}_2 = N_2([0, S_1[) \,.\, \hat{P}_1$$

Dans le deuxième membre de la 1ère formule, on peut écrire indifféremment $0 < T_n(\omega) \leqslant S_1(\omega)$ ou $0 \leqslant T_n(\omega) < S_1(\omega)$. La démonstration suivante que nous faisons dans le premier cas s'applique aussi bien au second cas avec quelques permutations de $<$ en $\leqslant$.

Démonstration

Considérons la fonction $a_0(\omega, s) = 1_{\{S_0(\omega)<s\leqslant S_1(\omega)\}}$ dont nous avons vu dans la démonstration de la proposition II.13 qu'elle était invariante par la transformation $(\omega, s) \to (\theta_s\omega, -s)$ de $\Omega \times R$. Comme en outre $\int_R N_1(\omega, ds)\, a_0(\omega, s) = 1$ pour tout $\omega \in \Omega$, la formule de la proposition II.10 implique que pour toute fonction mesurable positive $f : \Omega \to R_+$ nous avons

$$\int_\Omega \hat{P}_2(d\omega) f(\omega) = \int_{\Omega\times R} \hat{P}_2(d\omega)\, N_1(\omega, ds)\, f(\omega)\, a_0(\omega, s)$$

$$= \int_{\Omega\times R} \hat{P}_1(d\omega)\, N_2(\omega, ds)\, f(\theta_s\omega)\, a_0(\omega, s)$$

en vertu de l'invariance de a et donc puisque $S_o = 0$ $\hat{P}_1$ p.s.

$$= \int_\Omega \hat{P}_1 (d\omega) \sum_{\mathbb{Z}} f(\theta_{T_n} \omega) 1_{(0<T_n(\omega)\leq S_1(\omega))}$$

La première formule de la proposition est démontrée. La seconde formule s'en déduit en prenant pour f une fonction de la forme $g \circ \theta_{T_o}$ et en notant que $T_n + S_o \circ \theta_{T_n} = S_o$ et donc que $\theta_{S_o} \circ \theta_{T_n} = \theta_{S_o}$, lorsque $0 \leq T_n < S_1$; ceci montre en effet que

$$\sum_{n:\, 0\leq T_n<S_1} f \circ \theta_{T_n} = \sum_{n:0\leq T_n<S_1} g \circ \theta_{S_o} = g \circ \theta_{S_o} . N_2([0, S_1[).\square$$

En particulier si les points des processus N_1 et N_2 alternent sur R, c'est-à-dire si $N_1([T_n, T_{n+1}[) = 1$ sur Ω pour tout $n \in \mathbb{Z}$ la formule précédente se réduit simplement à

(25) $$\hat{P}_2 = \theta_{T_1} \circ \hat{P}_1 .$$

En particulier si $N_1 = N_2$, cette formule établit une propriété d'invariance des mesures de Palm dont la proposition suivante montre qu'elle caractérise les mesures de Palm.

Proposition II.17

Soit $N = \sum_{\mathbb{Z}} \varepsilon_{T_n}$ un processus ponctuel réel, simple et doublement infini, stationnaire par le flot mesurable réel $(\theta_t, t \in R)$ défini comme N sur l'espace mesurable $(\Omega, \mathcal{A})$. Si $\hat{\Omega} = \{N(\{o\}) = 0\} = \{T_o = 0\}$, les applications θ_{T_n} sont des surjections mesurables de Ω sur $\hat{\Omega}$; les restrictions $\hat{\theta}_n$ des θ_{T_n} à $\hat{\Omega}$ forment un groupe à un paramètre entier de transformations mesurables sur $\hat{\Omega}$, $\hat{\theta}_o$ se réduisant à l'identité.

Pour toute probabilité P sur $(\Omega, \mathcal{A})$, la mesure de Palm $\hat{P}$ de N relativement à P est invariante par les transformations $\hat{\theta}_n$ $(n \in \mathbb{Z})$.

Réciproquement si $\hat{P}$ est une mesure positive sur $\hat{\Omega}$ telle que $\int_{\hat{\Omega}} T_1 \, d\hat{P} = 1$ et si $\hat{P}$ est invariante par $\hat{\theta}_1$ il existe une et une seule probabilité P sur $(\Omega, \mathcal{A})$ invariante par le flot telle que $\hat{P}$ soit la mesure de Palm de N pour P. Elle est donnée par la formule

$$(19) \qquad \int_{\Omega} P\,(d\omega)\, f\,(\omega) = \int_{\hat{\Omega}} \hat{P}\,(d\omega) \int_0^{T_1(\omega)} f\,(\theta_s\, \omega)\, ds$$

valable pour toute fonction mesurable positive $f : \Omega \to R_+$.

Puisque les variables $\tau_n = T_{n+1} - T_n$ $(n \in Z)$ sont telles que $\tau_n \circ \theta_k = {}_{n+k}$ $(n, k \in Z)$, la première partie de la proposition montre que la mesure image $\hat{P}\,(\tau_n \in .)$ sur R_+ ne dépend pas de n ; il en est de même plus généralement des mesures positives $\hat{P}\,[(\tau_n, \tau_{n+1}, \ldots \tau_{n+k}) \in .]$ sur R_+^{k+1} quel que soit $k \in \mathbb{N}$.

La signification probabiliste de la dernière formule de la proposition est la suivante. En utilisant la probabilité $T_1 . \hat{P}$, choisissons un point $\hat{\omega}$ au hasard dans $\hat{\Omega}$ et donc une mesure ponctuelle changeant l'origine $N\,(\hat{\omega}, .)$. Modifions ensuite ce point $\hat{\omega}$ en $\omega = \theta_\sigma\,(\hat{\omega})$, ce qui translate la mesure ponctuelle $N\,(\hat{\omega}, .)$ de σ, en choisissant σ au hasard dans l'intervalle réel $[0, T_1\,(\hat{\omega})]$ suivant la densité uniforme. Le point $\omega = \theta_\sigma\, \hat{\omega}$ ainsi obtenu est réparti sur l'espace Ω suivant la probabilité P puisque d'après la formule ci-dessus

$$P\,(A) = \int_{\hat{\Omega}} (T_1 . \hat{P})\,(d\hat{\omega}) . \frac{1}{T_1(\hat{\omega})} \int_0^{T_1(\hat{\omega})} 1_A\,(\theta_s\, \hat{\omega})\, ds$$

pour tout $A \in \mathcal{A}$.

Démonstration

Les applications mesurables θ_{T_m} $(m \in Z)$ appliquent Ω dans $\hat{\Omega}$ car

$$N\,(\theta_{T_m}\, \omega, \{o\}) = \sum_{Z} \varepsilon_{T_n(\omega) - T_m(\omega)}\,(\{o\}) = 1.$$

Elles vérifient les relations $\theta_{T_n} \circ \theta_{T_m} = \theta_{T_{m+n}}$ sur Ω $(m, n \in Z)$ car

$$\theta_{T_n}\,(\theta_{T_m}\, \omega) = \theta_{T_n(\theta_{T_m}\omega)} \circ \theta_{T_m(\omega)}\, (\omega) = \theta_{T_m(\omega) + T_n(\theta_{T_m}\omega)}\, (\omega)$$

et car d'après le début de ce paragraphe $T_m + T_n \circ \theta_{T_m} = T_{m+n}$. Les applications $\hat{\theta}_n : \hat{\Omega} \to \hat{\Omega}$ définies par restriction des θ_{T_n} à $\hat{\Omega}$ sont donc telles que $\hat{\theta}_n \circ \hat{\theta}_m = \hat{\theta}_{n+m}$ $(m, n \in \mathbb{Z})$. Comme $\theta_{T_0}(\omega) = \omega$ si $\omega \in \hat{\Omega}$, la transformation $\hat{\theta}_0$ se réduit à la transformation identique.

D'après la formule qui précède la proposition $\hat{\theta}_1 \circ \hat{P} = \hat{P}$ et par conséquent $\hat{\theta}_n \circ \hat{P} = \hat{P}$ pour tout $n \in \mathbb{Z}$.

Réciproquement si $\hat{P}$ est une mesure positive sur $\hat{\Omega}$ invariante par les $\hat{\theta}_n$ $(n \in \mathbb{Z})$, montrons que la mesure $\hat{P}(d\omega)\, N(\omega, ds)$ est invariante par la transformation $\rho(\omega, s) = (\theta_s\, \omega, -s)$ de $\Omega \times R$ pour pouvoir appliquer la proposition II.11. Pour toute fonction mesurable positive $f : \hat{\Omega} \times R \to R_+$

$$\int_{\hat{\Omega}\times R} \hat{P}(d\omega)\, N(\omega, ds)\, f(\theta_s\, \omega, -s) = \int_{\hat{\Omega}} \hat{P}(d\omega) \sum_{\mathbb{Z}} f\left[\hat{\theta}_n(\omega), -T_n(\omega)\right]$$

$$= \sum_{\mathbb{Z}} \int_{\hat{\Omega}} \hat{\theta}_n \circ \hat{P}(d\omega)\ f\ \left[\omega, -T_n \circ \hat{\theta}_{-n}(\omega)\right]$$

Mais de la formule $T_{-n} + T_n \circ \theta_{T_{-n}} = T_0$ valable sur Ω il résulte que $T_{-n} + T_n \circ \hat{\theta}_{-n} = 0$ sur $\hat{\Omega}$; compte tenu de l'hypothèse $\hat{\theta}_n \circ \hat{P} = \hat{P}$, les expressions précédentes sont encore égales à

$$= \sum_{\mathbb{Z}} \int_{\hat{\Omega}} \hat{P}(d\omega)\ f\left[\omega, T_{-n}(\omega)\right]$$

$$= \int_{\hat{\Omega}\times R} \hat{P}(d\omega)\, N(\omega, ds)\, f(\omega, s).$$

La proposition II.11 citée donne alors l'existence d'une mesure P invariante par le flot $(\theta_t, t \in R)$ relativement à laquelle $\hat{P}$ est la mesure de Palm de N. La proposition II.13 achève alors de démontrer notre proposition.□

Nous terminerons ce paragraphe par l'étude d'un exemple important.

Proposition II.19

Soit μ une probabilité définie sur $]0, \infty[$ et d'espérance c finie. Un processus ponctuel stationnaire $N = \sum_{Z} \varepsilon_{T_n}$ sur R est dit de renouvellement de loi μ s'il vérifie l'une des propriétés équivalentes suivantes :

a) le vecteur $(-T_o, T_1 - T_o)$ et les variables $\tau_n = T_{n+1} - T_n$ $(n \in Z, n \neq 0)$ sont indépendants entre eux, le premier suit la loi

$c^{-1} 1_{\{0 \leq u < v\}} du\, \mu(dv)$ sur R_+^2 tandis que les secondes suivent chacune la loi μ sur R_+ ;

a') les variables $T_1, T_2 - T_1, T_3 - T_2, \ldots$ sont indépendantes entre elles et μ est la loi de $T_2 - T_1$;

b) $\widehat{P}(\Omega) = c^{-1}$ et les variables $\tau_n = T_{n+1} - T_n$ $(n \in Z)$ sont indépendantes et équidistribuées suivant μ pour la probabilité $c\,\widehat{P}(.)$;

b') $\widehat{P}(\Omega) < \infty$, les variables τ_n $(n \geq 0)$ sont indépendantes pour la probabilité $\frac{1}{\widehat{P}(\Omega)} \widehat{P}(.)$ et μ est la loi de τ_o pour cette probabilité.

Notons que les propriétés a et b sont des propriétés fortes qui déterminent la loi de N pour P, resp. pour $\widehat{P}$; les propriétés a' et b' par contre sont beaucoup plus faibles a priori bien que d'après la proposition elles soient équivalentes aux premières.

Démonstration

Nous établirons que a $\Longrightarrow$ a' $\Longrightarrow$ b' $\Longrightarrow$ b $\Longrightarrow$ a. La première implication est évidente. Pour démontrer la seconde de ces implications, partons de la formule (23) :

$$P\left[u \leq -T_o < u+du\, ,\ v \leq T_1 - T_o < v + dv\ ;\ v_k \leq \tau_k < v_k + dv_k\ (1 \leq k \leq n)\right]$$

$$= 1_{\{0 \leq u < v\}} du\, \widehat{P}\left[v \leq T_1 < v + dv\ ;\ v_k \leq \tau_k < v_k + dv_k \qquad (1 \leq k \leq n)\right]$$

qui implique en toute généralité que sur R_+^{n+1}

$$P\left[w \leqslant T_1 < w + dw \,;\, v_k \leqslant \tau_k \qquad (1 \leqslant k \leqslant n)\right]$$

$$= dw\ \hat{P}\left[T_1 \geqslant w \,;\, v_k \leqslant \tau_k \qquad (1 \leqslant k \leqslant n)\right].$$

Sous l'hypothèse a', le premier membre de cette dernière égalité vaut encore

$$P\,(w \leqslant T_1 < w + dw)\ \prod_1^n P\,(v_k \leqslant \tau_k)\ ;$$

Comme $P\,(w \leqslant T_1 < w + dw) = dw\,\hat{P}\,(T_1 \geqslant w)$, nous avons trouvé que

$$\hat{P}\ \left[\tau_o \geqslant w \,;\, \tau_k \geqslant v_k\ (1 \leqslant k \leqslant n)\right] = \hat{P}\,(\tau_o \geqslant w)\ \prod_1^n P\,(\tau_k \geqslant v_k)$$

pour presque tout w et donc pour tout $w > 0$ par continuité à gauche en w.

Faisons alors tendre w et les v_k $(1 \leqslant k \leqslant n)$ sauf l'un d'eux vers zéro pour obtenir que $\hat{P}\,(\tau_k \geqslant v) = \hat{P}\,(\Omega)\ P\,(\tau_k \geqslant v)$ $(k \geqslant 1)$. Le premier membre ne dépend pas de k et est fini pour tout $v > 0$; cela entraîne que $\hat{P}\,(\Omega) < \infty$ et que si μ désigne la loi de probabilité de τ_o pour $\frac{1}{\hat{P}\,(\Omega)}\ \hat{P}$, les variables τ_k $(k \geqslant 1)$ suivent cette loi μ relativement à P.

En revenant à la dernière formule du paragraphe précédent, nous obtenons ainsi que

$$\hat{P}\left[\tau_o \geqslant w \,;\, \tau_k \geqslant v_k\ (1 \leqslant k \leqslant n)\right] = \hat{P}\,(\Omega)\ \ \mu\ (\left[w, \infty\right[\ )\ \prod_1^n \mu\,(\left[v_k, \infty\right[\ )$$

pour tout $n \in N^*$, ce qui démontre b'.

L'implication b' $\Longrightarrow$ b résulte essentiellement de l'invariance de $\hat{P}$ par les $\hat{\theta}_k$ $(k \in Z)$ puisque $\tau_k = \tau_o \circ \hat{\theta}_k$ pour tout $k \in Z$. Cette invariance entraîne en effet d'abord que la loi de probabilité $\frac{1}{\hat{P}\,(\Omega)}\ \hat{P}\,(\tau_k \in .)$ ne dépend pas de k $(k \in Z)$; elle entraîne ensuite que si la suite $(\tau_n\ , n \geqslant 0)$ est indépendante, il en est de même des suites $(\tau_{n-k}\ , n \geqslant 0)$ $(k \in \mathbb{N})$; la double suite $(\tau_n\ , n \in Z)$ est donc aussi indépendante. Enfin comme $\int T_1\, d\hat{P} = 1$, nécessairement $c\,\hat{P}\,(\Omega) = 1$.

Il reste à démontrer que b $\Longrightarrow$ a. Or la formule 23 qui sous l'hypothèse b s'écrit

$$P\left[u \leqslant -T_0 < u + du,\ v_k \leqslant \tau_k < v_k + dv_k\ (|k| \leqslant n)\right]$$

$$= 1_{\{0 \leqslant u < v_0\}}\ du\ \widehat{P}(\Omega) \prod_{-n}^{+n} \mu(dv_k)$$

établit bien a.□

Faisons quelques remarques. D'abord pour un processus de renouvellement stationnaire, la loi pour P de la variable $\tau_0 = T_1 - T_0$ n'est pas la même que celle des τ_n $(n \neq 0)$ que dans le seul cas où cette loi commune μ des τ_n est dégénérée, soit $\mu = \varepsilon_c$; en effet la loi de τ_0 est donnée par $c^{-1} x\, \mu(dx)$ en fonction de μ. La différence d'énoncé des propriétés a et b est donc tout à fait essentielle.

Ensuite dans la propriété a', les variables $T_2 - T_1$, $T_3 - T_2$; ... ont nécessairement la même loi (puisque a' $\Longrightarrow$ a) ; par contre cette loi commune μ n'est aussi celle de T_1 que dans le seul cas exponentiel, soit $\mu = \theta\, e^{-\theta t}\, dt$ pour $\theta = \frac{1}{c}$, c'est-à-dire lorsque N est un processus de Poisson.

Enfin la propriété d'équidistribution pour P des variables τ_n $(n \neq 0)$ est rarement vérifiée en dehors des processus de renouvellement. En effet soit à partir de la formule (23) soit directement à partir de la proposition 13, en remarquant que $\tau_k \circ \theta_{T_0} = \tau_k$, il est facile de voir que

$$P(\tau_k \in F) = \int 1_{\{\tau_k \in F\}}\ \tau_0\ d\widehat{P} \qquad (F \in \mathcal{R}_+,\ k \in Z)$$

Or le second membre dépend en général de k $(k \neq 0)$ puisque la mesure $\tau_0 \widehat{P}$ n'est pas invariante par translation, et il en est donc de même du premier ; néanmoins si τ_0 et τ_k sont indépendantes pour $\frac{1}{\widehat{P}(\Omega)} \widehat{P}$ le second membre se réduit à $\frac{1}{\widehat{P}(\Omega)} \widehat{P}(\tau_k \in F)$ car $\int \tau_0\, d\widehat{P} = 1$ et ne dépend donc pas de k.

La proposition II.17 permet de construire très simplement un processus de renouvellement stationnaire sur R de loi μ. Prenons pour espace $(\Omega, \mathcal{Q})$ le

sous-espace mesurable de $[M_p(R), \mathcal{M}_p(R)]$ formé par les mesures ponctuelles simples doublement infinies et pour flot $(\theta_t, t \in R)$ la restriction à l'espace invariant Ω du flot des translations sur $M_p(R)$; notons $N = \sum_Z \varepsilon_{T_n}$ le processus ponctuel canonique égal à l'application identique de Ω dans $M_p(R)$.
Le sous-espace $\widehat{\Omega} = \{T_o = 0\}$ de Ω peut alors être regardé comme l'image du sous-espace S de $]0, \infty[^Z$ formé par les suites $(s_n, n \in Z)$ telles que $\sum_{\mathbb{N}} s_n = +\infty = \sum_{\mathbb{N}} s_{-n}$, par l'application bijective $\nu : S \to \widehat{\Omega}$ définie par

$$\nu[(s_n, n \in Z)] = \sum_Z \varepsilon_{t_n}$$

où les t_n sont définis par les relations $t_o = 0$, $t_{n+1} - t_n = s_n$ $(n \in Z)$. Cette bijection $\nu : S \to \widehat{\Omega}$ est mesurable ainsi que son inverse ; de plus si $(\varphi_p, p \in Z)$ désigne les translations de coordonnées de $]0, \infty[^Z$ par lesquelles S est d'ailleurs stable, nous avons la relation

$$\nu \circ \varphi_p = \widehat{\theta}_p \circ \nu \quad \text{sur } S.$$

Il résulte alors de la proposition 16 que les mesures de Palm du processus ponctuel stationnnaire $N : \Omega \to M_p(R)$ sont exactement sur $\widehat{\Omega}$ les images par la bijection ν des mesures positives Q sur $]0, \infty[^Z$, portées par S, invariantes par $(\varphi_p, p \in Z)$ et telles que $\int s_o \, Q(ds) = 1$.

En particulier si μ est une probabilité sur $]0, \infty[$ d'espérance $c = \int t \, d\mu(t)$ finie, la mesure $Q_\mu = c^{-1} \mu^{Z \otimes}$ sur $]0, \infty[^Z$ est une mesure du type précédent et son image $\widehat{P_\mu} = \nu \circ Q_\mu$ est donc la mesure de Palm de N relativement à une probabilité P_μ sur Ω. D'après la proposition II.19, il est clair que N est pour cette probabilité P_μ un processus de renouvellement stationnaire de loi μ.

COMPLEMENT. LE THEOREME D'AMBROSE-KAKUTANI

Ce théorème d'existence, dont nous ne nous servirons pas dans la suite, est célèbre en théorie ergodique car il ramène essentiellement l'étude d'un flot réel $(\theta_t\ ,\ t \in R)$ à celle d'une seule transformation $\hat{\theta}$ sur une section de l'espace. Nous nous contenterons de l'énoncer dans le cas ergodique bien que l'extension au cas général ne présente pas de difficulté.

Proposition II.18

Soit $(\theta_t\ ,\ t \in R)$ un flot réel ergodique défini sur un espace de probabilité $(\Omega, \mathcal{A}, P)$ non dégénéré (c'est-à-dire tel que $0 < P(A) < 1$ pour un $A \in \mathcal{A}$ au moins). Il existe alors au moins un processus ponctuel simple sur R stationnaire par le flot et presque sûrement non nul.

La proposition précédente peut être étendue à des flots non réels en s'appuyant sur un article de D.A. Lind (cf bibliographie)

Démonstration

L'espace $L^2(\Omega, \mathcal{A}, P)$ contient par hypothèse des fonctions qui ne sont pas p.s. constantes ; l'algèbre de fonctions H du lemme II.3 contient donc au moins une fonction h non p.s. constante. Les fonctions h_a $(a > 0)$ définies par $h_a(\omega) = \sup_{s \in [0,a]} h(\theta_s\,\omega)$ appartiennent encore à H (h_a est mesurable car dans la borne supérieure précédente s peut être restreint aux rationnels, sans rien changer) ; comme $\lim_{a \to 0} h_a = h$ sur Ω, pour un a assez petit que nous fixerons, la fonction h_a n'est pas p.s. constante et nous pouvons donc trouver une constante b telle que $0 < P(h_a > b) < 1$.

Pour tout $\omega \in \Omega$, le sous-ensemble de R

$$G(\omega) = \{t : t \in R,\ h_a(\theta_t\,\omega) > b\}$$

est ouvert. Comme tout ouvert de R, $G(\omega)$ est la réunion d'intervalles ouverts deux à deux disjoints ; la longueur de chacun de ces intervalles dépasse a comme il est facile de le voir sur l'équivalence

$$t \in G(\omega) \iff h(\theta_u \omega) > b \quad \text{pour un} \quad u \in [t, t+a]$$

qui résulte de la définition de h_a. D'autre part, d'après le lemme II.12, l'ensemble I des ω tels que les deux intégrales sur R_+ et R_-

$$\int_{R_\pm} 1_{G(\omega)}(t)\, \lambda(dt) = \int_{R_\pm} 1_{\{h_a > b\}}(\theta_t \omega)\, \lambda(dt)$$

soient infinies, est mesurable, invariant par le flot et contient p.s. l'ensemble $\{h_a > b\}$; comme le flot est ergodique, P (I) vaut 0 ou 1 et alors P (I) = 1 puisque $\{h_a > b\}$ n'est pas négligeable. De la même façon, puisque $\{h_a \leq b\}$ n'est pas négligeable, l'ensemble I' des ω tels que $\int_{R_\pm} 1_{G(\omega)^c}(t)\, \lambda(dt) = +\infty$ est mesurable, invariant et de probabilité 1. Ainsi pour tout $\omega \in I \cap I'$, donc pour presque tout ω, l'ouvert G (ω) et son complémentaire $G(\omega)^c$ contiennent tous les deux des points t arbitrairement proches de + ∞ et des points t' arbitrairement proches de - ∞.

Ces propriétés des G (ω) nous permettent de définir pour tout $\omega \in I \cap I'$ une mesure ponctuelle N (ω, .) en affectant de masses unités les bornes droites des intervalles ouverts constituant G (ω) : les points de N (ω, .) sont distances 2 à 2 de a au moins et ils sont en nombre infini dans R_+ et dans R_-. Lorsque $\omega \notin I \cap I'$, nous poserons N (ω, .) = 0. Il est clair que $G(\theta_s \omega) = \tau_s^{-1} [G(\omega)]$ et par conséquent que $N(\theta_s \omega, .) = \tau_s [N(\omega, .)]$ pour tout $s \in R$ et tout $\omega \in I \cap I'$. Le processus ponctuel N est donc stationnaire ; il reste à montrer qu'il est mesurable.

A cet effet, considérons la fonction $T : I \cap I' \to R_-$ définie par $T(\omega) = \sup [t : t \in R_- \cap G(\omega)]$; comme pour tout s < 0,

$$\{T > s\} = \bigcup_{\substack{t \in Q \\ \delta < t \leq 0}} \{h_a \circ \theta_t > b\}$$

, cette fonction T est mesurable. Il en est alors de même de la mesure aléatoire $\nu = \{\nu(\omega, .), \omega \in I \cap I'\}$ définie pour chaque ω comme l'image de la mesure de Lebesgue λ par l'application

$t \to t + T(\theta_t\, \omega)$. Or, il est facile de voir que $\nu(\theta_t\, \omega, .) = \tau_t\, [\nu(\omega, .)]$ et que

$$\nu(\omega, dt) = 1_{G(\omega)}(t)\, \lambda(dt) + \xi(\theta_t\, \omega)\, N(\omega, dt)$$

si $\xi(\omega) = \nu(\omega, \{o\})$, pour tout ω ; la quantité $\xi(\theta_t\, \omega) = \nu(\omega, \{t\})$ est nulle sauf si t est la borne droite de l'un des intervalles constituant $G(\omega)$, auquel cas $\xi(\theta_t\, \omega)$ est la distance à t de $G(\omega) \cap [t, +\infty[$

Par conséquent la fonction η définie par $\eta(\omega) = 1 / \xi(\omega)$ si $\xi(\omega) > 0$, $= 0$ sinon est une fonction mesurable et comme

$$N(\omega, dt) = \eta(\theta_t\, \omega)\, \nu(\omega, dt),$$

le processus ponctuel N est mesurable.□

Remarquons que si $\hat{\Omega} = \{N(\{o\}) \neq 0\}$ dans Ω et si $\hat{P}$ désigne la mesure de Palm de N, l'application $\phi(\omega) = (\theta_{T_o}(\omega), -T_o(\omega))$ établit une bijection du sous-ensemble invariant $I \cap I'$ de mesure un de Ω sur le sous-ensemble $\tilde{\Omega} = \{(\hat{\omega}, s) ; 0 \leqslant s \leqslant T_1(\hat{\omega})\}$ de $\hat{\Omega} \times R$. Cette bijection est mesurable ; son inverse qui est donnée par $(\hat{\omega}, s) \to \theta_s\, \hat{\omega}$ l'est aussi. Il est facile de vérifier en se servant de la formule (15), que l'image par ϕ du flot θ, soit $(\tilde{\theta}_t, t \in R_+)$, est donnée par

$$\tilde{\theta}_t\, [(\hat{\omega}, s)] = (\hat{\theta}_n\, \hat{\omega}, t - s - T_n(\hat{\omega})) \text{ si } s + T_n(\hat{\omega}) \leqslant t < s + T_{n+1}(\hat{\omega}).$$

De plus la formule (19) montre que l'image de P par ϕ est égale à la restriction à $\tilde{\Omega}$ de la mesure $\hat{P} \otimes \lambda$ définie sur $\hat{\Omega} \times R$. La représentation $(\tilde{\theta}_t, t \in R)$ du flot $(\theta_t, t \in R)$ sur $(\Omega, \mathcal{A}, P)$ que nous avons ainsi obtenue sur $\tilde{\Omega}$ se dessine comme ci-dessous :

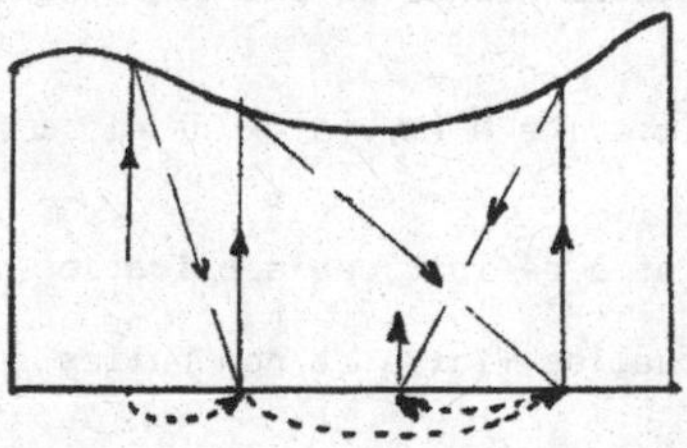

II.4. PROCESSUS PONCTUELS STATIONNAIRES MARQUES ; AGGLOMERATS

Souvent les points d'un processus ponctuel sont marqués par une marque, un point d'un espace auxiliaire ; par exemple les instants d'arrivée des clients dans un organe de service peuvent être marqués par le type ou la durée du service requis par chaque client. Evidemment un processus ponctuel sur E marqué par les points d'un espace M peut être considéré comme un processus ponctuel sur E x M, du moins s'il est simple ; aussi n'est-il intéressant d'étudier la notion de processus ponctuel marqué qu'en présence d'une structure supplémentaire comme celle par exemple qu'introduit la stationnarité.

Une courte étude préliminaire est nécessaire pour arriver à la bonne définition des processus ponctuels stationnaires marqués. Etant donné un sous-groupe fermé E de R^d, un espace de probabilité $(\Omega, \mathcal{A}, P)$ muni d'un flot $(\theta_t, t \in E)$ indexé par E et un espace mesurable auxiliaire, soit $\overline{N} : \Omega \to M_p(E \times M)$ un processus ponctuel sur E x M jouissant des deux propriétés suivantes :

a) le processus marginal $N(\omega, .) = \overline{N}[\omega, . \times M]$ est un processus ponctuel sur E, de sorte que, la mesure $\overline{N}(\omega, \{s\} \times .)$ est finie sur M quels que soient $\omega \in \Omega$ et $s \in E$,

b) $\overline{N}(\theta_t \omega, .) = (\tau_t \times i_M) \overline{N}(\omega, .)$ sur E x M, pour tout $\omega \in \Omega$ et tout $t \in E$, si i_M désigne la transformation identique de M.

Un processus ponctuel stationnaire sur E marqué par M pourrait être défini s'il est simple comme un processus de ce type avec en plus la condition que le processus ponctuel marginal N est simple.

Etant donné un processus $\overline{N}$ du type précédent, le processus ponctuel marginal N qui lui est associé est évidemment stationnaire, ce qui nous amène à poser

$$\hat{\Omega} = \{\omega : N(\omega, \{o\}) \neq 0\} \quad \text{et} \quad \Delta(\omega, .) = \overline{N}(\omega, \{o\} \times .) \text{ si } \omega \in \hat{\Omega} ;$$

alors $\hat{\Omega}$ est une partie mesurable de Ω et Δ définit une application mesurable de $\hat{\Omega}$ dans l'ensemble des mesures ponctuelles finies et non nulles sur M.

De plus grâce à la propriété (b) de $\overline{N}$ si $N(\omega, \{s\}) > 0$ pour un $s \in E$, alors $\theta_s\omega \in \hat{\Omega}$ car $N(\theta_s\omega, \{o\}) = N(\omega, \{s\})$ et $\overline{N}[\omega, \{s\} \times .] = \Delta(\theta_s\omega, .)$ sur M ; ceci nous donne la représentation suivante de $\overline{N}$ en fonction de N et de Δ

$$\overline{N}(\omega ; ds\, dx) = N(\omega, ds) \frac{1}{N(\theta_s\omega, \{o\})} \Delta(\theta_s\omega, dx) \text{ sur } E \times M \ (\omega \in \Omega)$$

(la définition de Δ hors de $\hat{\Omega}$ est sans importance puisque $N(\omega, ds)$-presque sûrement sur E, $\theta_s\omega \in \hat{\Omega}$ d'après ce qui précède). Lorsque le processus marginal N est simple, $\Delta(\omega, M) = N(\omega, \{o\}) = 1$ pour tout $\omega \in \hat{\Omega}$ et il existe par suite une application mesurable unique $\delta : \hat{\Omega} \to M$ telle que $\Delta(\omega, .) = \varepsilon_{\delta(\omega)}$ du moins si nous supposons que la tribu $\mathcal{M}$ sépare les points de M et est de type dénombrable, ce qui est le cas habituel.

Nous sommes ainsi conduits à la définition suivante dans laquelle nous renversons l'ordre des définitions de $\overline{N}$ et de N.

Définition II.20

Etant donné un processus ponctuel stationnaire $N : \Omega \to M_p(E)$ défini sur un espace de probabilité $(\Omega, \mathcal{A}, P)$ muni d'un flot $(\theta_t, t \in E)$, une marque de N est une application mesurable $\delta : \hat{\Omega} \to M$ du sous-ensemble $\hat{\Omega} = \{N(\{o\}) \neq 0\}$ de Ω dans un espace mesurable auxiliaire M. Le processus ponctuel sur $E \times M$ défini par

(27) $\overline{N}(\omega ; ds\, dx) = N(\omega, ds)\ \varepsilon_{\delta(\theta_s\omega)}(dx)$

est alors appelé un processus ponctuel stationnaire marqué.

Le lemme suivant s'applique en particulier aux processus marqués ; la forme générale sous laquelle nous l'énonçons nous sera utile ultérieurement.

Lemme II.21

Si $N : \Omega \to M_p(E)$ est un processus stationnaire défini sur $[\Omega, \mathcal{A}, P ; (\theta_t, t \in E)]$ et si $\chi : \hat{\Omega} \equiv \{N(\{o\}) \neq 0\} \to \mathbb{N}$ est une application mesurable,

<u>la formule</u>

(28) $\tilde{N}(\omega, ds) = X(\theta_s \omega) N(\omega, ds)$

<u>définit un processus ponctuel stationnaire sur</u> E <u>dont la mesure de Palm est égale à</u> $X . \hat{P}$ <u>si</u> $\hat{P}$ <u>désigne la mesure de Palm de</u> N.

<u>Démonstration</u>

Le processus ponctuel $\tilde{N}$ est stationnaire car pout tout $t \in E$:

$$\tilde{N}(\theta_t \omega, ds) = X(\theta_{s+t} \omega) N(\theta_t \omega, ds) = \tau_t \left[X(\theta_s \omega) . N(\omega, ds)\right]$$

Le calcul suivant où $f : \hat{\Omega} \to R_+$ et $g : E \to R_+$ sont deux fonctions mesurables positives et où $\int g \, d\lambda = 1$, montre que $X . \hat{P}$ est la mesure de Palm de $\tilde{N}$

$$\int_{\hat{\Omega}} f X \, d\hat{P} = \int_{\Omega \times E} P(d\omega) N(\omega, dt) f(\theta_t \omega) X(\theta_t \omega) g(t)$$

$$= \int_{\Omega \times E} P(d\omega) \tilde{N}(\omega, dt) f(\theta_t \omega) g(t) = \int_{\Omega} \hat{P}_{\tilde{N}}(d\omega) f(\omega). \square$$

Pour le processus $\overline{N}$ obtenu en marquant le processus ponctuel stationnaire N par la marque δ comme dans la définition 17 ci-dessus, il résulte du lemme précédent que pour tout $B \in \mathcal{M}$, le processus ponctuel $\overline{N}_B$ sur E défini par

$$\overline{N}_B(\omega ; ds) = \overline{N}(\omega ; ds \times B) = 1_B \circ \delta(\theta_s \omega) \; N(\omega, ds)$$

est stationnaire et admet la mesure de Palm $1_B \circ \delta . \hat{P}$. Il s'en suit immédiatement que si θ désigne comme précédemment la transformation $\theta[(\omega, t)] = (\theta_t \omega, t)$ de $\Omega \times E$, nous avons sur $\Omega \times E \times M$

$$(\theta \times i_M)\left[P(d\omega) N(\omega, dt) \; \varepsilon_{\delta(\theta_t \omega)}(dx)\right] = \hat{P}(d\omega) \lambda(dt) \varepsilon_{\delta(\omega)}(dx)$$

ce qui par une extension naturelle des définitions, nous conduit à dire que la <u>mesure positive</u> $\hat{P}(d\omega) \quad \varepsilon_{\delta(\omega)}(dx)$ <u>sur</u> $\hat{\Omega} \times M$ <u>est la mesure de Palm du processus marqué</u> $\overline{N}$.

Exemple. La marque : multiplicité

L'étude de tout processus ponctuel (stationnaire) sur E dont les points ont une multiplicité aléatoire quelconque se ramène à celle d'un processus ponctuel (stationnaire) simple sur E marqué par $\mathbb{N}^* = \{1, 2, \dots\}$ en utilisant l'application de $M_p(E)$ dans $M_p(E \times N^*)$ définie par

$$\phi\left(\sum_I n_i \ \varepsilon_{t_i}\right) = \sum_I \varepsilon_{(t_i, n_i)} \ ;$$

nous désignons ici par $(t_i, i \in I)$ une famille localement finie de points deux à deux distincts de E et par n_i des entiers ≥ 1 quelconques $(i \in I)$. La formule

$$\phi(m)\left[F \times \{n\}\right] = \frac{1}{n} \int_F 1_{(m(\{t\})=n)} \ m(dt) \qquad (m \in M_p(E))$$

où $F \in \mathcal{E}$ et $n \in \mathbb{N}^*$ permet de démontrer facilement que l'application ϕ est mesurable et il est clair que $\phi(\tau_t m) = (\tau_t \times i_{N^*})\,\phi(m)$ pour tout $m \in M_p(E)$ et tout $t \in E$. Notons aussi que la mesure marginale $\phi\, m(. \times \mathbb{N}^*)$ de ϕm sur E vaut $\sum_I \varepsilon_{t_i}$ si $m = \sum_I n_i \ \varepsilon_{t_i}$, c'est-à-dire est l'unique mesure ponctuelle simple de même support que m.

Si $\tilde{N} : \Omega \to M_p(E)$ est alors un processus ponctuel sur E, associons lui le processus ponctuel $\overline{N}$ sur $E \times \mathbb{N}^*$ défini par $\overline{N}(\omega, .) = \phi \circ \tilde{N}(\omega, .)$ ou plus explicitement par

$$\overline{N}(\omega\, ; F \times \{n\}) = \frac{1}{n} \int_F 1_{(\tilde{N}(\omega,\{t\})=n)} \ \tilde{N}(\omega, dt) \qquad (F \in \mathcal{E}, \ n \in \mathbb{N}^*)$$

ainsi que le processus marginal N de $\overline{N}$ sur E qui vaut

$$N(\omega, F) = \overline{N}(\omega, F \times \mathbb{N}^*) = \int_F \frac{1}{\tilde{N}(\omega, \{t\})} \ \tilde{N}(\omega, dt) \qquad (F \in \mathcal{E}).$$

Ce processus N est l'unique processus ponctuel simple sur E tel que $N(\omega, .)$ et $\tilde{N}(\omega, .)$ aient le même support dans E pour tout $\omega \in \Omega$.

De plus si $\tilde{N}$ est stationnaire sur E, N l'est évidemment aussi et le processus ponctuel $\overline{N}$ sur $E \times \mathbb{N}^*$ dont la donnée est équivalente à celle de $\tilde{N}$ est le processus ponctuel obtenu en marquant le processus simple N par la marque

$\delta : \hat{\Omega} \to \mathbb{N}^*$ que définit la multiplicité, soit :

$$\delta(\omega) = \tilde{N}(\omega, \{o\}) \quad \text{si} \quad \omega \in \hat{\Omega} = \{N(\{o\}) \neq 0\} \equiv \{\tilde{N}(\{o\}) \neq 0\}$$

Comme $\delta(\theta_t \omega) = \tilde{N}(\theta_t \omega, \{o\}) = \tilde{N}(\omega, \{t\})$ les formules ci-dessus s'écrivent bien sur $E \times \mathbb{N}^*$ et E respectivement

(30) $\bar{N}(\omega ; dt \times \{n\}) = N(\omega, dt) \quad \varepsilon_{\delta(\theta_t \omega)}(n)$, $\tilde{N}(\omega, dt) = \delta(\theta_t \omega) N(\omega, dt)$

pour tout $\omega \in \Omega$. Les résultats antérieurs de ce paragraphe montrent aussi que si $\hat{P}$ est la mesure de Palm de N, $\hat{P}(d\omega) \quad \varepsilon_{\delta(\omega)}(n)$ sur $\hat{\Omega} \times \mathbb{N}^*$ et $\delta(\omega) . \hat{P}$ (sur $\hat{\Omega}$ sont les mesures de Palm de $\bar{N}$ et $\tilde{N}$. Alternativement puisque $\delta \geqslant 1$ sur $\hat{\Omega}$, <u>la mesure de Palm du processus stationnaire simple associé à un processus stationnaire quelconque</u> $\tilde{N}$ <u>s'obtient en divisant la mesure de Palm de</u> $\tilde{N}$ <u>par sa multiplici</u>

Les points d'un processus ponctuel dit <u>primaire</u> sont parfois susceptibles de déclencher chacun un processus ponctuel dit <u>secondaire</u> et défini sur le même espace ou sur un espace différent. Nous allons procéder à une étude un peu formel de ces schémas, de manière à lever toute ambiguité dans leur définition.

D'un premier point de vue le processus secondaire peut être considéré comme une marque attachée à chaque point du processus primaire. Soit donc $N : \Omega \to M_p(E)$ un processus ponctuel stationnaire sur E, dit primaire et pour tout $\omega \in \hat{\Omega}$, c'est-à-dire chaque fois que $N(\omega, .)$ charge l'origine, soit $\Delta(\omega, .)$ le processus ponctuel sur E' déclenché par le point 0 du processus primaire ; bien entendu l'application $\omega \to \Delta(\omega, .)$ de $\hat{\Omega}$ dans $M_p(E')$ est supposée mesurable. Le processus ponctuel $\bar{N}$ sur $E \times M_p(E')$ obtenu en marquant N par Δ est donné par

$$\bar{N}(\omega ; ds\, dm') = N(\omega ; ds) \quad \varepsilon_{\Delta(\theta_s \omega)}(dm') ;$$

il est stationnaire relativement aux deux flots $(\theta_t , t \in E)$ sur Ω et $(\tau_t \times i_{M_p}(E') , t \in E)$ sur $E \times M_p(E')$ et sa mesure de Palm est égale à

$\hat{P}(d\omega)\quad \varepsilon_{\Delta(\omega)}(dm')$ sur $\hat{\Omega}\times M_p(E')$ si $\hat{P}$ est la mesure de Palm de N.

D'un second point de vue les deux processus primaire et secondaire ensemble peuvent être considérés comme un processus sur $E \times E'$. Considérons en effet l'application ϕ de l'espace $M_p\left[E \times M_p(E')\right]$ des mesures ponctuelles sur $E \times M_p(E')$ finies sur tout $F \times M_p(E')$ $(F \in \mathcal{E}_c)$ dans l'espace plus simple $M_p(E \times E')$ définie par

$$\phi\left[\sum_I \varepsilon_{(t_i,m_i')}\right] = \sum_I \varepsilon_{t_i} \otimes m_i' .$$

[Cette application est injective quand on la restreint aux mesures ponctuelles sur $E \times M_p(E')$ dont la marge sur E est simple] . L'application ϕ est mesurable et telle évidemment que

$$\phi \circ \left[\tau_t \times i_{M_p(E')}\right] = (\tau_t \times i_{E'}) \circ \phi \qquad (t \in E).$$

Donc l'image par ϕ du processus marqué $\overline{N}$ ci-dessus que nous désignerons par $\tilde{N}$ et que nous pouvons écrire

(31) $\tilde{N}(\omega\,;\,ds\,dx) = N(\omega, ds)\,\Delta(\theta_s\,\omega, dx)$ sur $E \times E'$ $(\omega \in \Omega)$,

est un processus ponctuel sur $E \times E'$ relativement aux deux flots $(\theta_t\ ,\ t \in E)$ sur Ω et $(\tau_t \times i_{E'}\ ,\ t \in E)$ sur $E \times E'$. Pour tout borélien relativement compact F' de E' , $\tilde{N}(\omega, dt \times F')$ est donc un processus ponctuel stationnaire sur E dont la mesure de Palm vaut d'après le lemme 20 : $\hat{P}(d\omega)\,\Delta(\omega, F')$; il s'en suit (comme dans l'exemple des processus à multiplicité) que

(32) $(\theta \times i_{E'})\left[P(d\omega)\,\tilde{N}(\omega\,;\,ds\,dx)\right] = \hat{P}(d\omega)\,\lambda(ds)\,\Delta(\omega, dx)$

sur $\Omega \times E \times E'$ si $\theta(\omega, t) = (\theta_t\,\omega, t)$ sur $\Omega \times E$ et nous appellerons la mesure positive $\hat{P}(d\omega)\,\Delta(\omega, dx)$ sur $\hat{\Omega} \times E'$ la mesure de Palm de $\tilde{N}$.

Enfin lorsque E' = E, il peut être intéressant de considérer <u>l'agglomérat</u> formé par la superposition des processus secondaires respectivement translatés par les points du processus primaire dont ils sont les marques, si du moins cette superposition donne encore un processus ponctuel localement fini.

Considérons donc l'addition s, t $\to$ s + t sur E et l'application ψ qu'elle induit de l'espace M_p (E × E) dans l'espace des mesures ponctuelles sur E non nécessairement de Radon. Comme

$$\psi \left(\sum_I \varepsilon_{t_i} \otimes m'_i \right) = \sum_I \varepsilon_{t_i} * m'_i$$

l'opérateur composé de ϕ et ψ applique la mesure ponctuelle $\sum_I \varepsilon_{t_i}$ marquée par les mesures m'_i ($i \in I$) sur la mesure $\sum_I \varepsilon_{t_i} * m'_i$ obtenue par superposition des mesures m'_i translatées de $- t_i$. Le processus ponctuel N^* sur E image de $\tilde{N}$ par ψ s'écrit

$$N^* (\omega, F) = \int_{E^2} N(\omega, dt) \, \Delta(\theta_t \omega, dx) \, 1_F (t+x) \qquad (F \in \mathcal{E})$$

à ceci près que les mesures positives $N^*(\omega, .)$ ne sont pas nécessairement de Radon. Il est stationnaire parce que l'application ψ vérifie la relation $\psi \circ (\tau_t \times i_E) = \tau_t \circ \psi$ sur E × E pour tout $t \in E$. Sa mesure de Palm est calculée dans le lemme suivant.

<u>Lemme II.22</u>

<u>Si</u> $N : \Omega \to M_p(E)$ <u>est un processus ponctuel stationnaire sur</u> E <u>et si</u> $\Delta : \hat{\Omega} \to M_p(E)$ <u>est une application mesurable définie sur</u> $\hat{\Omega} = \{N(\{o\}) \neq 0\}$ <u>à valeurs dans</u> $M_p(E)$, <u>le processus ponctuel</u> N^* <u>sur</u> E <u>image de</u> $N(\omega, ds) \; \Delta(\theta_s \omega, du)$ <u>par l'addition</u> s, u $\to$ s + u <u>est un processus ponctuel stationnaire sur</u> E <u>qui admet pour mesure de Palm l'image de la mesure</u> $\hat{P}(d\omega) \, \Delta(\omega, du)$ <u>par l'application</u> $(\omega, u) \to \theta_u \omega$, <u>du moins s'il est presque sûrement localement fini. Il en sera bien ainsi en particulier si</u> $\int_{\hat{\Omega}} \hat{P}(d\omega) \, \Delta(\omega, E) < \infty$; <u>en tous les cas cette intégrale donne l'intensité du processus ponctuel</u> N^*.

<u>Démonstration</u>

La stationnarité de N^* a déjà été démontrée. Le calcul suivant où $f : \Omega \times E \to R_+$ est une fonction mesurable positive arbitraire donne la mesure de Palm de N^*

$$\int_{\Omega\times E} P\,(d\omega)\; N^*\;(\omega,\,ds)\; f\,(\theta_s\,\omega,\,s)$$

$$= \int_{\Omega\times E} P\,(d\omega)\; N\,(\omega,\,ds) \int_E \Delta\,(\theta_s\,\omega,\,du)\; f\,(\theta_{s+u}\,\omega,\,s+u)$$

$$= \int_{\Omega\times E} \widehat{P}\,(d\omega)\; \lambda\,(ds) \int_E \Delta\,(\omega,\,du)\; f\,(\theta_u\,\omega,\,s+u)$$

$$= \int_{\Omega\times E} \widehat{P}\,(d\omega)\; \Delta\,(\omega,\,du) \int_E \lambda\,(ds)\; f\,(\theta_u\,\omega,\,s).$$

En particulier pour $f = 1 \otimes 1_F$ ce calcul montre que pour tout $F \in \mathcal{E}_c$

$$E\,[N^*\,(F)] = \int_{\Omega\times E} \widehat{P}\,(d\omega)\; \Delta\,(\omega,\,du)\,.\,\lambda\,(F) < +\infty$$

sous la dernière hypothèse du lemme ; alors N^* (ω, .) est bien une mesure de Radon pour presque tout ω.□

II.5. ERGODISME ET ANALYSE SPECTRALE DES PROCESSUS PONCTUELS STATIONNAIRES.

A. Donnons-nous un flot $(\theta_t\ , t \in E)$ sur un espace de probabilité $(\Omega, \mathcal{A}, P)$ et désignons par $\mathcal{J}$ la sous-tribu de $\mathcal{A}$ formée des évènements invariants par les θ_t , soit $\mathcal{J} = \{I\ ;\ I \in \mathcal{A}\ ,\ \theta_t^{-1}(I) = I$ pour tout $t \in E\}$; rappelons qu'une application $f : \Omega \to R$ est $\mathcal{J}$ -mesurable si et seulement si elle est $\mathcal{A}$ -mesurable et invariante par les θ_t $(t \in E)$. Le flot est dit ergodique si P ne prend que les valeurs 0 ou 1 sur $\mathcal{J}$.

Lemme II.23

Etant donné un processus ponctuel $N : \Omega \to M_p(E)$ stationnaire pour le flot $(\theta_t\ , t \in E)$, il existe une fonction $\mathcal{J}$ -mesurable, soit $i : \Omega \to \overline{R}_+$ telle que

a) $E^{\mathcal{J}}\ [N(g)] = \lambda(g)\, i$ P p.s. pour toute fonction borélienne positive λ -intégrable $g : \Omega \to R_+$

b) $\hat{P} = i\ .\ P$ sur $\mathcal{J}$, si $\hat{P}$ est la mesure de Palm de N. En particulier si le flot est ergodique : $i = \hat{P}(\Omega) \leqslant \infty$.

Démonstration

La définition de la mesure de Palm $\hat{P}$ montre que pour toute fonction borélienne positive $g : E \to R_+$ et tout $I \in \mathcal{J}$:

$$\int_{\Omega\times E} \hat{P}(d\omega)\ \lambda(dt)\ 1_I(\omega)\ g(t) = \int_{\Omega\times E} P(d\omega)\ N(\omega, dt)\ 1_I(\omega)\ g(t)$$

ce qui peut encore s'écrire en prenant une espérance conditionnelle dans le second membre

$$\hat{P}(I)\ \lambda(g) = \int_I N(g)\ dP = \int_I E^{\mathcal{J}}\ [N(g)]\ dP.$$

Pour une fonction g_o d'intégrale $\lambda(g_o)$ ni nulle , ni infinie posons alors

$i = \frac{1}{\lambda(g_o)}\ E^{\mathcal{J}}\ [N(g_o)]$; la formule précédente montre d'abord que la fonction $\mathcal{J}$ -mesurable de Ω dans $\overline{R}_+$ ainsi définie vérifie $\hat{P}(I) = \int_I i\ dP$ pour tout $I \in \mathcal{J}$;

elle montre ensuite que $E^{\mathcal{J}}[N(g)] = \lambda(g)\, i$ pour toute fonction borélienne positive λ-intégrable g. Enfin si le flot est ergodique, c'est-à-dire si P = 0 ou 1 sur $\mathcal{J}$, la fonction i est p.s. constante et ne peut donc que valoir $\hat{P}(\Omega)$ p.p.□

Il résulte alors du théorème ergodique ponctuel (théorème de Birkhoff si E = Z ou R ; théorème de Dunford-Schwartz dans le cas général) que pour toute fonction borélienne positive intégrable $g : E \to R_+$, l'ensemble invariant où $s \to N(g) \circ \theta_s$ est localement intégrable sur E par rapport à λ et où

(33) $$\frac{1}{\lambda(\{s : 0 \leq s \leq t\})} \int_{\{s: 0 \leq s \leq t\}} N(g) \circ \theta_s \, dP \to \lambda(g)\, i$$

lorsque $t \uparrow \infty$ est de mesure pleine pour P et donc aussi d'après le lemme précédent pour $\hat{P}$.

Remarquons d'autre part que pour un processus ponctuel stationnaire simple et doublement infini défini sur R, la tribu $\mathcal{J}$ peut encore être décrite à partir de la tribu $\hat{\mathcal{J}}$ des parties de $\hat{\Omega}$ invariantes par les $\hat{\theta}_n$ $(n \in Z)$: ces deux tribus sont en effet liées par les deux relations

(34) $$\hat{\mathcal{J}} = \hat{\Omega} \cap \mathcal{J} \quad , \quad \mathcal{J} = \theta_{T_0}^{-1}(\hat{\mathcal{J}}) ,$$

si bien qu'une fonction $f : \Omega \to R_+$ $\mathcal{A}$-mesurable est invariante par le flot $(\theta_t, t \in R)$ si et seulement si elle est de la forme $g \circ \theta_{T_0}$ pour une fonction $g : \hat{\Omega} \to R_+$ mesurable par rapport à $\hat{\Omega} \cap \mathcal{A}$ et invariante par les $\hat{\theta}_n$ $(n \in Z)$, qui est d'ailleurs nécessairement égale à la restriction de f à $\hat{\Omega}$.

La démonstration des deux relations ci-dessus est facile. Si $\theta_t^{-1}(I) = I$ pour tout $t \in R$, alors $\theta_{T_n}^{-1}(I) = I$ et par suite $\hat{\theta}_n^{-1}(I \cap \hat{\Omega}) = I \cap \hat{\Omega}$ pour tout $n \in Z$ puisque $\hat{\theta}_n$ est la restriction à $\hat{\Omega}$ de θ_{T_n}. Inversement comme $\theta_{T_0} \circ \theta_t = \theta_{T_n} = \hat{\theta}_n \circ \theta_{T_0}$ sur $\{T_n \leq t < T_{n+1}\}$, si $\hat{I} \in \hat{\mathcal{J}}$ et si $I = \theta_{T_-}^{-1}(\hat{I})$, nous avons

$$\theta_t^{-1} I \cap \{T_n \leqslant t < T_{n+1}\} = I \cap \{T_n \leqslant t < T_{n+1}\}$$

pour tout $n \in \mathbb{Z}$ et par conséquent en sommant sur n nous trouvons que $I \in \mathcal{J}$.

B. Pour qu'un processus ponctuel stationnaire N sur E soit de carré intégrable, c'est-à-dire pour que $E[N(F)^2] < \infty$ pour tout $F \in \mathcal{E}_c$, (il faut et) il suffit que $E[N(G)^2]$ soit fini pour un seul ouvert relativement compact non vide puisque, grâce à la stationnarité du processus, $E[N(\tau_t G)^2]$ ne dépend pas de t et puisque tout $F \in \mathcal{E}_c$ est recouvert par un nombre fini de translatés de G. La proposition suivante qui pourrait être démontrée directement, est encore une conséquence presque immédiate de l'existence de la mesure de Palm.

Proposition II.24

Pour qu'un processus ponctuel stationnaire N sur E soit de carré intégrable, il faut et il suffit que la mesure positive et symétrique σ sur E définie à partir de la mesure de Palm $\hat{P}$ de N par

$$(35) \qquad \sigma(F) = \int_\Omega \hat{P}(d\omega)\, N(\omega, F) \qquad (F \in \mathcal{E})$$

soit une mesure de Radon sur E. S'il en est ainsi, $\sup_{t \in E} \sigma(F+t)$ est fini pour tout $F \in \mathcal{E}_c$ et pour tout couple de fonctions boréliennes, soit positives, soit bornées à supports compacts $f, g : E \to R$

$$(36) \qquad E[N(f)\, N(g)] = \int_E f * \check{g}\, d\sigma$$

où $f * \check{g}(t) = \int_E f(s+t)\, g(s)\, \lambda(ds)$ désigne le produit de convolution sur E de la fonction f et de la fonction retournée $\check{g}(s) = g(-s)$.

Cette mesure σ est appelée la mesure spectrale de N.

Démonstration

Il résulte de la formule

$$P(d\omega)\, N(\omega, ds)\, N(\omega, dt) = \bar{\theta}^{-1}[\hat{P}(d\omega)\, N(\omega, ds)\, \lambda(dt)]$$

où $\bar{\theta}(\omega, s, t) = (\theta_t\, \omega, \tau_t\, s, t)$ sur $\Omega \times E^2$, qui est la clé de la démonstration de la proposition II.10 que

$$\int_{\Omega\times E^2} P\,(d\omega)\ N\,(\omega,\ ds)\ N\,(\omega,\ dt)\ f\,(s)\ g\,(t)$$

$$=\int_{\Omega\times E^2} \hat{P}\,(d\omega)\ N\,(\omega,\ ds)\ \lambda\,(dt)\ f\,(s+t)\ g\,(t)$$

pour tout couple de fonctions boréliennes positives $f,\ g : E \to R_+$, ce qui s'écrit encore

$$E\left[N\,(f)\ N\,(g)\right] = \int_{E^2} \sigma\,(ds)\ \lambda\,(dt)\ f\,(s+t)\ g\,(t).$$

Si σ est une mesure de Radon, le second membre est fini lorsque f et g sont bornées à supports compacts et cela montre que $E\left[N\,(F)^2\right] < \infty$ pour tout $F \in \mathscr{E}_c$. Inversement pour tout compact K de E il existe au moins une fonction $f \in C_k^+\,(E)$ telle que $f * \check{f} \geqslant 1_K$, ce qui nous permet d'écrire pour tout $u \in E$ que

$$\begin{aligned}\tau_u \circ \sigma\,(K) \leqslant \tau_u \circ \sigma\,(f * \check{f}) &= \sigma\,(f \circ \tau_u * \check{f})\\ &= E\left[N\,(f \circ \tau_u)\ N\,(f)\right] \leqslant E\left[N\,(f)^2\right] < \infty\end{aligned}$$

grâce à l'inégalité de Schwarz, à la stationnarité de N et à l'hypothèse que N est de carré intégrable ; ainsi sous cette hypothèse $\sup_u \sigma\,(K + u) < \infty$. Enfin l'invariance de la mesure $\hat{P}\,(d\omega)\ N\,(\omega,\ dt)$ par la transformation $(\omega,\ t) \to (\theta_t\,\omega,\ -t)$ entraîne l'invariance de σ par la symétrie $t \to -\,t$. □

Il résulte de la proposition précédente que la mesure μ_2 moment d'ordre deux d'un processus ponctuel stationnaire est égale à l'image de la mesure $\sigma\,(ds)\ \lambda\,(dt)$ par l'application $(s,\ t) \to (s + t,\ t)$.

Pour faire ensuite l'analyse spectrale de N, considérons d'abord la mesure $\tau = \sigma - c^2\ \lambda\ \left[c = \hat{P}\,(\Omega)\right]$ qui donne la variance du processus ponctuel puisque

$$\begin{aligned}\mathrm{Var}\left[N\,(f)\right] &= E\left[N\,(f)^2\right] - \left(E\left[N\,(f)\right]\right)^2\\ &= \sigma\,(f * \check{f}) - \left[c\ \lambda\,(f)\right]^2 = (\sigma - c^2\,\lambda)\,(f * \check{f})\end{aligned}$$

La mesure positive σ étant tempérée (puisque $\sup_t \sigma(t + K) < \infty$), la mesure τ l'est aussi ; de plus cette mesure est de type positif car elle est symétrique et car $\tau(f * \check{f}) \geq 0$ d'après le calcul qui précède au moins pour les fonctions f de l'espace S de Schwartz. La transformée de Fourier au sens de la théorie des distributions de τ est alors une mesure positive tempérée, soit $\hat{\tau}$, qui vérifie par définition $\hat{\tau}(f) = \tau(\hat{f})$ pour tout $f \in S$.

D'autre part comme

$$E\left[\int N(.,dx)\frac{1}{1+|x|^{d+\varepsilon}}\right] = c\int \lambda(dx)\frac{1}{1+|x|^{d+\varepsilon}} < \infty \quad \text{dès que } \varepsilon > 0,$$

pour presque tout ω, la mesure positive $N(\omega, .)$ est aussi tempérée et nous pouvons en prendre la transformée de Fourier qui est une distribution, soit $\hat{N}(\omega, .)$. Considérons enfin l'application U qui à toute fonction $f \in S$ fait correspondre une variable aléatoire suivant la formule

$$U(f)[\omega] = \hat{N}(\omega, f) - c\, f(o) = N(\omega, \hat{f}) - c\,\lambda(\hat{f}).$$

Alors

$$E\left[U(f)^2\right] = \mathrm{Var}\left[N(\hat{f})\right] = \tau\left[\hat{f} * \hat{f}^{\vee}\right] = \int |f|^2\, d\hat{\tau}$$

pour tout $f \in S$ et puisque S est dense dans $L^2(\hat{\tau})$, l'application U se prolonge en une isométrie de $L^2(\hat{\tau})$ dans l'espace $L^2(\Omega, \mathcal{A}, P)$ qui s'appelle <u>la transformée de Fourier du processus ponctuel centré</u> $N - c\lambda$.

<u>Exemples</u>

1) Si N est une progression arithmétique de pas 1 stationnaire, donc si

$N(\omega, .) = \sum_{Z} \varepsilon_{n-a(\omega)}$ où $a(\omega) = -T_o(\omega)$ est une variable à valeurs dans $[0, 1[$ uniformément répartie sur cet intervalle, il résulte de la formule de Poisson de la théorie de l'intégrale de Fourier que

$$\hat{N}(\omega, ds) = \exp\left[2\pi i s\, a(\omega)\right] . \sum_{Z} \varepsilon_n(ds).$$

2) Pour un processus de renouvellement $N = \sum_{\mathbb{Z}} \varepsilon_{T_n}$ sur $\mathbb{R}$ associé à une probabilité μ d'espérance finie, la mesure σ vaut

$$\sigma = \sum_{\mathbb{Z}} \int_\Omega \hat{P}(d\omega)\, \varepsilon_{T_n(\omega)} = c \left\{ \varepsilon_o + \sum_{N^*} \mu^{n*} + \sum_{N^*} (\mu^{n*})^{\vee} \right\}$$

où c est l'inverse de l'espérance de μ. Un calcul de transformation de Fourier montre alors que

$$\hat{\tau} = \frac{1}{c}\, \varphi(t)\, \lambda(dt) \qquad \text{si } \varphi(t) = \frac{1 - |\hat{\mu}(t)|^2}{|1 - \hat{\mu}(t)|^2}\, .\square$$

II.6. GEOMETRIE ALEATOIRE

La géométrie aléatoire dont l'objet est d'étudier des choix au hasard d'éléments géométriques, des droites dans un plan par exemple, sera abordée dans ce paragraphe sous l'angle des processus ponctuels définis sur un espace homogène. Nous commencerons par quelques rappels de théorie des groupes sans démonstration.

A. Préliminaires

Soient G un groupe localement compact à base dénombrable et H un sous-groupe fermé de G ; soient $E = G/H$ l'espace homogène des classes à droite suivant H et $i : G \to E$ la surjection canonique. Le groupe G opère continûment à droite sur E ; nous noterons indifféremment par τ_s ($s \in G$) les translations à droite sur G définies par $\tau_t(s) = s\,t^{-1}$ ($s, t \in G$) ou leurs images sur E si bien que avec ces notations $i \circ \tau_s = \tau_s \circ i$ sur G pour tout $s \in G$. Nous désignerons d'autre part par σ_s ($s \in G$) les translations à gauche sur G définies par $\sigma_s(t) = s^{-1}t$ ($s, t \in G$).

Nous noterons λ_G une mesure de Haar invariante à droite sur G (que nous normaliserons par la condition $\mu_G(G) = 1$ lorsque G est compact) et Δ_G la fonction modulaire de G définie de sorte que

$$\tau_s \circ \lambda_G = \lambda_G \quad , \quad \sigma_s \circ \lambda_G = \Delta_G(s)\,\lambda_G \quad \text{sur } G.$$

Nous adopterons des notations analogues sur H.

La formule

(37) $$\Lambda f \circ i = \int_H f(t.)\,\lambda_H(dt)$$

définit alors une application linéaire, positive et surjective de l'espace $C_k(G)$ des fonctions continues $f : G \to R$ à supports compacts sur l'espace $C_k(E)$ qui est telle que

a) $\Lambda(f \circ \tau_s) = \Lambda f \circ \tau_s$ pour tout $s \in G$,

b) $\Lambda(f \circ \sigma_t) = \Delta_H(t)\,\Lambda f$ pour tout $t \in H$,

c) $\Lambda(f \,.\, g \circ i) = \Lambda f \,.\, g$ quels que soient $f \in C_k(G)$ et $g \in C_k(E)$,

d) il existe au moins une fonction continue positive h sur G telle que l'application linéaire positive $g \to h \, . \, g \circ i$ envoie C_k (E) dans C_k (G) et inverse à droite Λ au sens où Λ (h . g o i) = g pour tout $g \in C_k$ (E) ; lorsque H est compact, bien entendu h = 1 convient.

L'application transposée Λ' qui applique l'espace M (E) des mesures de Radon sur E dans M (G) en plus d'être injective jouit des propriétés importantes suivantes :

1°) Une mesure de Radon positive sur G, soit $\nu \in M_+$ (G), appartient à l'image $\Lambda' \, [M_+ (E)]$ du cône des mesures de Radon positives sur E par Λ' si et seulement si elle vérifie la condition de relative invariance par H

$$\tau_t \circ \nu = \Delta_H (t) \, \nu \qquad \text{quel que soit } t \in H.$$

2°) Soit χ un caractère réel sur G, la fonction 1 éventuellement. Pour qu'il existe une mesure de Radon positive non nulle μ sur E vérifiant la condition de relative invariance $\tau_s \circ \mu = \chi (s) \, \mu$ pour tout $s \in G$, il est nécessaire et suffisant que $\chi \, \Delta_G = \Delta_H$ sur H. Sous cette condition, la mesure μ qui est alors unique à une constante multiplicative près est caractérisée par la propriété $\Lambda' \mu = c \, \chi \, . \, \lambda_G$ (c constante > 0). En particulier si $\Delta_G = \Delta_H$ sur H,ce qui sera toujours le cas si H est compact, il existe une mesure de Radon positive unique sur E que nous noterons λ_E , telle que $\Lambda' \lambda_E = \lambda_G$ et les multiples de λ_E sont alors les mesures de Radon sur E invariantes par l'action de G, soit $\tau_s \circ \lambda_E = \lambda_E$ pour tout $s \in G$.

3°) Pour qu'une mesure de Radon positive μ sur E soit quasi-invariante par l'action à gauche de G, c'est-à-dire pour que les mesures $\tau_s \circ \mu$ et μ soient équivalentes quel que soit $s \in G$, il est nécessaire et suffisant que l'image $\Lambda' \mu$ de μ par Λ' soit équivalente à la mesure de Haar λ_G de G ; de plus si cette condition est remplie et si $\Lambda' \mu = f \, . \, \lambda_G$ sur G alors $\tau_s \circ \mu = g_s \, . \, \mu$ sur E pour une densité de Radon-Nikodym donnée par la formule $g_s \circ i = f \circ \tau_s^{-1} / f$ $(s \in G)$. D'ailleurs il existe toujours au moins une mesure de Radon positive non nulle

μ sur E quasi-invariante par G, les autres mesures de ce type étant alors les mesures équivalentes à μ sur E.

B. Processus ponctuels stationnaires sur un espace homogène

Une famille $(\theta_t,\ t \in G)$ de transformations d'un espace de probabilité $(\Omega, \mathcal{A}, P)$ est appelée un G-flot mesurable préservant la mesure, en abrégé G-flot, si elle vérifie les conditions de la définition II.1 avec G au lieu de E et st au lieu de s + t (donc $\theta_s\ \theta_t = \theta_{st}$). Un processus ponctuel stationnaire sur l'espace homogène E = G/H sera alors défini comme une application mesurable N appliquant un espace de probabilité $(\Omega, \mathcal{A}, P)$ muni d'un G-flot $(\theta_t,\ t \in G)$ dans l'espace $M_p\ (E)$ de telle manière que

$$N\ (\theta_s\ \omega,\ .) = \tau_s\ [N\ (\omega,\ .)] \qquad \text{pour tous } s \in G \text{ et } \omega \in \Omega\ .$$

La loi de probabilité P_N d'un tel processus étant définie sur $M_p\ (E)$ comme l'image de P par N, est invariante par l'action de G sur $M_p\ (E)$, soit $\tau_s \circ P_N = P_N$ pour tout $s \in G$.

Existe-t-il un processus de Poisson stationnaire non trivial sur E ? Pour qu'un processus de Poisson sur E soit stationnaire en loi, c'est-à-dire pour que $\tau_s \circ P_N = P_N$ $(s \in G)$, il faut et il suffit, comme il est facile de le voir, que son intensité μ qui est une mesure de Radon positive sur E, soit invariante par l'action à droite de G. D'après le paragraphe A (alinéa 2°), pour qu'une telle mesure $\mu \neq 0$ existe, il faut et il suffit que $\Delta_G = \Delta_H$ sur H , ce que nous supposerons désormais. [D'ailleurs si l'on ne fait pas cette hypothèse, il n'existe aucun processus ponctuel stationnaire d'intensité de Radon sur E].

Il n'est pas difficile de définir une mesure de Palm pour un processus ponctuel stationnaire sur E bien que cet espace E ne soit plus un groupe. La transformation Λ' du paragraphe A nous permet en effet de remonter tout processus ponctuel stationnaire N sur E en une mesure aléatoire stationnaire G en posant

$$\tilde{N}\ (\omega,\ .) = \Lambda'\ [N\ (\omega,\ .)] \qquad (\ \omega \in \Omega\ )\ ;$$

la mesurabilité de $\tilde{N}$ découle de ce que $\tilde{N}$ (f) = N (Λ f) pour toute fonction $f \in C_k$ (G) et sa stationnarité résulte de la propriété d'invariance (a) de Λ. Désignons alors par $\hat{P}$ la mesure de Palm de $\tilde{N}$ (voir la remarque suivant le théorème II.4) ; nous avons donc sur $\Omega \times G$

$$\theta\left[P(d\omega)\,\tilde{N}(\omega, ds)\right] = \hat{P}(d\omega)\,\lambda_G(ds) \quad \text{si} \quad \theta\left[(\omega, s)\right] = (\theta_s\,\omega, s).$$

En traduisant cette formule sur le processus ponctuel N, nous obtiendrons le résultat suivant.

Proposition II.25

Soit N un processus ponctuel sur E stationnaire par le flot (θ_t, $t \in G$) définis sur l'espace de probabilité (Ω, $\mathcal{A}$, P). Il existe alors une mesure positive σ-finie unique $\hat{P}$ sur (Ω, $\mathcal{A}$) telle que toute fonction mesurable positive $f : \Omega \times G \to R_+$ vérifie l'égalité

$$\int_{\Omega\times E} P(d\omega)\,N(\omega, dx)\,\Lambda f(\omega, x) = \int_{\Omega\times G} \hat{P}(d\omega)\,\lambda_G(ds)\,f(\theta_{s^{-1}}\,\omega, s)$$

(si $\Lambda f(\omega, x) = \Lambda\left[f(\omega, .)\right](x)$). Cette mesure $\hat{P}$ est en outre invariante par le sous-flot (θ_t, $t \in H$).

Si h est une fonction ayant la propriété (d) du paragraphe A et si $\ominus_x^{\pm}$ sont les opérateurs définis par

$$(38) \quad \ominus_x^{+} g(\omega) = \Lambda\left[h\, g(\theta_.\,\omega)\right](x)\,, \quad \ominus_x^{-} g(\omega) = \Lambda\left[h\, g(\theta_.^{-1}\,\omega)\right](x)$$

sur les fonctions mesurables positives $g : \Omega \to R_+$, la formule précédente permet d'écrire encore pour toute fonction mesurable positive $g : \Omega \times E \to R_+$ que

$$(39) \quad \begin{aligned} \int_{\Omega\times E} P(d\omega)\,N(\omega, dx)\,g(\omega, x) &= \int_{\Omega\times E} \hat{P}(d\omega)\,\lambda_E(dx)\,\ominus^{-} g(\omega, x). \\ \int_{\Omega\times E} \hat{P}(d\omega)\,\lambda_E(dx)\,g(\omega, x) &= \int_{\Omega\times E} P(d\omega)\,N(\omega, dx)\,\ominus^{+} g(\omega, x). \end{aligned}$$

en convenant que $\ominus^{\pm} g(\omega, x) = \left[\ominus_x^{\pm}\, g(., x)\right](\omega)$.

Démonstration

La première formule résulte immédiatement de ce que

$$\int_{\Omega\times G} P\,(d\omega)\ \tilde{N}\,(\omega,\ ds)\ f\,(\omega,\ s) = \int_{\Omega\times G} \hat{P}\,(d\omega)\ \lambda_G\,(ds)\ f\,(\theta_{s-1}\ \omega,\ s)$$

par la définition de la mesure de Palm de $\tilde{N}$. Pour en déduire la deuxième formule, posons $f\,(\omega,\ s) = h\,(s)\ g\,(\omega,\ i\,(s))$ de sorte que $\Lambda\, f\,(\omega,\ x) = g\,(\omega,\ x)$ [propriété (d) de Λ] et que

$$\int_G \lambda_G\,(ds)\ f\,(\theta_{s-1}\ \omega,\ s) = \int_E \lambda_E\,(dx) \int_G \Lambda\,(x,\ ds)\ f\,(\theta_{s-1}\ \omega,\ s)$$

$$= \int_E \lambda_E\,(dx)\ \ominus^-\ g\,(\omega,\ x)$$

d'après la définition de λ_E [alinéa 2° de Λ] et celle de $\ominus^-$. Ensuite pour déduire la troisième formule de la première, posons $f\,(\omega,\ s) = h\,(s)\ g\,(\theta_s\ \omega,\ i(s))$ de sorte que $\Lambda\, f\,(\omega,\ x) = \ominus^+\ g\,(\omega,\ x)$ par définition de $\ominus^+$ et

$$\int_G \lambda_G\,(ds)\ f\,(\theta_{s-1}\ \omega,\ s) = \int_E \lambda_E\,(dx) \int_G \Lambda\,(x,\ ds)\ h\,(s)\ g\,[\omega,\ i\,(s)]$$

$$= \int_E \lambda_E\,(dx)\ g\,(\omega,\ x).$$

Il reste à montrer que la mesure de Palm $\hat{P}$ est invariante sur Ω par les θ_t $(t \in H)$. D'après la propriété (b) de l'application Λ, nous avons $\sigma_t \circ \tilde{N}\,(\omega,\ .) = \Delta_H\,(t)\ \tilde{N}\,(\omega,\ .)$ sur G et par suite

$$(i_\Omega \times \sigma_t)\ [P\,(d\omega)\ \tilde{N}\,(\omega,\ ds)] = \Delta_H\,(t)\ .\ P\,(d\omega)\ \tilde{N}\,(\omega,\ ds)\quad \text{sur } \Omega \times G$$

pour tout $t \in H$. En passant à la mesure de Palm, nous en déduisons que

$$(\theta_t \times \sigma_t)\ \left[\hat{P}\,(d\omega)\ \lambda_G\,(ds)\right] = \Delta_H\,(t)\ .\ \hat{P}\,(d\omega)\ \lambda_G\,(ds) \qquad (t \in H).$$

Or $\sigma_t \circ \lambda_G = \Delta_G\,(t)\ \lambda_G$ et par hypothèse $\Delta_G = \Delta_H$ sur H ; il s'en suit bien que $\theta_t \circ \hat{P} = \hat{P}$ sur Ω .□

Remarque. Il résulte des 2 dernières formules de la proposition que la mesure $\hat{P} \otimes \lambda_E$ sur $\Omega \times E$ est invariante par l'opérateur $\textcircled{H}^- \textcircled{H}^+$ mais ce résultat est contenu dans l'invariance de $\hat{P}$ par les θ_t $(t \in H)$ car le calcul de cet opérateur produit montre que $\textcircled{H}^-_x \textcircled{H}^+_x$ g est une moyenne sur H des fonctions $g \circ \theta_t$ $(t \in H)$ [lorsque H est compact et h = 1, $\textcircled{H}^-_x \textcircled{H}^+_x g = \int_H \lambda_H(dt)\, g(\theta_t\, .)$ pour tout $x \in E$].

COMPLEMENTS ET EXERCICES

1. On doit à Palm non pas l'introduction de la mesure qui porte son nom mais celle des "fonctions de Palm" définies pour un processus ponctuel stationnaire N sur R par

$$\phi_k(t) = \hat{P}\,[N_t = k] \quad \text{où } N_t = N(]0,t[) \qquad (t \in R_+ \,,\ k \in N).$$

1°) Etablir que si le processus N est simple

$$P(N_t > k) = \int_0^t \phi_k(s)\,ds \quad \text{pour tous } t \quad R_+ \,,\ k \quad N$$

après avoir remarqué que

$$1_{\{N_t > k\}} = \int_0^t N(ds)\ 1_{\{N_t - N_s = k\}} \,.$$

Soit en utilisant la formule précédente, soit en procédant directement, démontrer les formules suivantes sur les fonctions génératrices et les moments factoriels des variables N_t $(t \in R_+)$:

$$1 - E(z^{N_t}) = (1-z)\int_0^t \hat{E}(z^{N_s})\,ds, \quad E\left[\binom{N_t}{k}\right] = \int_0^t \hat{E}\left[\binom{N_s}{k-1}\right] ds$$

$(z \in [0,1]$ dans R, $k \in N^*)$ où $\hat{E}$ désigne l'intégrale par rapport à $\hat{P}$.

2°) Montrer que sur R_+ la fonction ϕ_o est décroissante, continue à droite et finie sauf peut-être en 0 où elle vaut $\hat{P}(\Omega)$.

3°) Que valent les fonctions de Palm d'un processus de renouvellement ?

2. a) Pour toute probabilité μ sur $]0, \infty[$, montrer qu'il existe une <u>progression arithmétique aléatoire</u>, c'est-à-dire un processus ponctuel simple et doublement infini $N = \sum_Z \varepsilon_{T_n}$ tel que les variables $T_{n+1} - T_n$ $(n \in Z)$ soient toutes égales, qui soit stationnaire et dont le pas, c'est-à-dire les variables $T_{n+1} - T_n$, suive la loi μ . Montrer que la mesure de Palm $\hat{P}$ de ce processus a une masse totale égale à $\int v^{-1}\, d\mu(v) \leq +\infty$.

b) Soit $(a_1, a_2, \ldots a_k)$ une suite finie de réels > 0. Soient α_j $(j = 1, 2, \ldots k)$ les mesures ponctuelles périodiques de période $a = \sum_1^k a_j$ dont les restrictions à $[0, a[$ valent respectivement

$$\varepsilon_o + \varepsilon_{a_j} + \varepsilon_{a_j+a_{j+1}} + \ldots + \varepsilon_{a_j+\ldots+a_k+a_1+\ldots a_{j-2}}.$$

Montrer qu'il existe une mesure de Palm unique sur M_p (R) qui soit portée par $\{\alpha_1, \alpha_2, \ldots \alpha_k\}$. Montrer qu'il existe un processus ponctuel stationnaire sur R dont la loi est unique, de la forme $N = \sum_Z \varepsilon_{T_n}$ avec $T_o \leq 0 < T_1$ et $(T_{n+1} - T_n, n \in Z)$ presque sûrement égale à l'une des suites périodiques $(a_{n+j(\text{mod } k)}, n \in Z)$ $(j = 1, \ldots k)$; deux constructions sont possibles : soit à partir de leur mesure de Palm en utilisant la proposition 17 (vérifier dans ce cas l'interprétation probabiliste de cette proposition), soit comme agglomérat construit à partir d'une progression arithmétique aléatoire de pas déterministe égal à a.

Etablir que les mesures de Palm de processus ponctuels simples sur R qui ont un support fini dans M_p (R) sont nécessairement de la forme précédente.

3. a) Soit P la loi de probabilité sur M_p (E) d'un processus de Poisson d'intensité $c\,\lambda$ $(c \in R_+)$. Montrer que la mesure de Palm correspondante vaut c fois l'image de P par l'application $m \to m + \varepsilon_o$ de M_p (E) dans lui-même et en déduire que la mesure σ du paragraphe 5 vaut $c^2 \lambda + c\,\varepsilon_o$ pour ce processus.

b) Si P est la loi sur M_p (E) d'un processus poissonnien de nuages sur E associé à la mesure Q [voir I.3.C] , montrer que P est invariante par le flot des translations de E si et seulement si Q l'est. Sous cette condition, comment la mesure de Palm de P s'exprime-t-elle à l'aide de celle de $\hat{Q}$.

En déduire que si le processus est de carré intégrable

$$\sigma = c^2 \lambda + \int m\, \hat{Q}\,(dm)$$

si c désigne la masse totale de $\hat{Q}$.

Comment les résultats de (a) se déduisent-ils des précédents ? Examiner le cas où Q est concentrée sur les mesures ponctuelles de masse 2 et où Q peut donc s'écrire comme l'image de $\lambda \otimes q$ par $s, t \to \varepsilon_s + \varepsilon_{s+t}$ pour une mesure positive et finie q sur $]0, \infty[$.

4. Soit N un processus ponctuel stationnaire sur E de mesure de Palm $\hat{P}$.

Si $S : \Omega \to E$ est mesurable, montrer que les images de $N(\omega, .)$ par $t \to t + S(\theta_t\, \omega)$ définissent un nouveau processus ponctuel stationnaire dont la mesure de Palm vaut $\theta_S \circ \hat{P}$. En déduire que le processus N est invariant par la transformation précédente si et seulement si $\theta_S \circ \hat{P} = \hat{P}$.

5. Pour tout processus ponctuel $N : \Omega \to M_p(E)$ montrer qu'il existe au moins une fonction continue strictement positive $f : E \to R_+$ telle que $N(f) < \infty$ p.p. sur Ω. Si le processus ponctuel N est stationnaire et si $\hat{P}$ désigne sa mesure de Palm, établir que pour toute fonction mesurable positive $g : \Omega \to R_+$

$$\int_{\{N \neq 0\}} g(\omega)\, P(d\omega) = \int_{\hat{\Omega}} \left[\frac{1}{N(\omega, f)} \int_E g(\theta_t\, \omega)\, f(t)\, \lambda(dt) \right] \hat{P}(d\omega)$$

après avoir remarqué que $N(., f) > 0$ p.p. pour $\hat{P}$.

6. Soit $N = \sum_{Z} \varepsilon_{T_n}$ un processus ponctuel stationnaire, simple et doublement infini sur R. Montrer que pour toute fonction mesurable positive $f : \Omega \to R$

$$E^{\theta_{T_0}^{-1}(\mathcal{A})}(f) = F \circ \theta_{T_0} \quad \text{si} \quad F = \frac{1}{T_1} \int_0^{T_1} f \circ \theta_s\, ds.$$

Montrer que la tribu $\mathcal{A}$ est engendrée par la tribu $\theta_{T_0}^{-1}(\mathcal{A})$ et la variable T_0.

7. <u>Superposition de processus stationnaires indépendants</u>. Soient $N_i : [\Omega_i, \mathcal{A}_i, P_i ; (\theta_t^i, t \in E)] \to M_p(E)$ $(i = 1, \dots k)$ k processus ponctuels stationnaires sur E définis sur des espaces de probabilité différents. On forme l'espace produit

$$[\Omega, \mathcal{A}, P ; (\theta_t, t \in E)] = \prod_{i=1}^{k} [\Omega_i, \mathcal{A}_i, P_i ; (\theta_t^i, t \in E)]$$

ainsi que le processus ponctuel $N : \Omega \to M_p(E)$

$$N(\omega, .) = \sum_1^k N_i(\omega_i, .) \text{ sur } E \text{ où } \omega = (\omega_1, \dots \omega_k) \in \Omega.$$

Montrer que N est stationnaire et trouver sa mesure de Palm en fonction des P_i et des $\hat{P}_i$ $(i = 1, \dots k)$.

8. Soit $N : \Omega \to M_p(E^2)$ un processus ponctuel sur E^2 tel que pour le flot $(\theta_t, t \in E)$ donné sur $(\Omega, \mathcal{A}, P)$: $N(\theta_t \omega, .) = [\tau_t \times \tau_t] N(\omega, .)$
Etablir qu'il existe alors une mesure positive σ-finie Q sur $\Omega \times E$ telle que $\overline{\theta} [P(d\omega) N(\omega, ds\, dt)] = Q(d\omega\, ds) \lambda(dt)$ si $\overline{\theta}(\omega, s, t) = (\theta_t \omega, \tau_t s, t)$.
Montrer que si pour presque tout ω la mesure ponctuelle $N(\omega, .)$ est symétrique sur E^2, la mesure Q est invariante par l'involution $(\omega, s) \to (\theta_s \omega, -s)$ de E et réciproquement. Si N est la seconde puissance d'un processus ponctuel stationnaire N_o sur E, exprimer la mesure Q en fonction de la mesure de Palm du processus N_o.

9. <u>Une mesure aléatoire stationnaire sur un groupe</u> E est une application mesurable $A : (\Omega, \mathcal{A}, P) \to (M_+(E), \mathcal{M}_+(E))$ telle que $A(\theta_t \omega, .) = \tau_t A(\omega, .)$ $(t \in E)$ pour un flot $(\theta_t, t \in E)$ sur $(\Omega, \mathcal{A}, P)$ et les translations τ_t de E.
Etendre à ces mesures aléatoires stationnaires la notion de mesure de Palm.

1°) Si $a : \Omega \to R_+$ est une fonction mesurable positive localement intégrable sur les trajectoires du flot, montrer que $A(\omega, ds) = a(\theta_s \omega) \lambda_E(ds)$ est une mesure aléatoire stationnaire de mesure de Palm a.P. Si A est une mesure aléatoire stationnaire telle que $A(\omega, .) \leqslant \lambda$ resp. $A(\omega, .) << \lambda$ pour presque tout ω, montrer que sa mesure de Palm vérifie $\hat{P} \leqslant P$, resp. $\hat{P} << P$ et en déduire que si $a = d\hat{P} / dP$ la mesure stationnaire est donnée par
$A(\omega, .) = a(\theta . \omega) \lambda_E$ p.p. sur Ω .

Montrer que la mesure de Palm de A est étrangère à P si et seulement si pour presque tout ω, A $(\omega, .)$ et λ_E sont étrangères sur E [Si P (B) = 0 = $\widehat{P}$ (B^c) pour un B $\in \mathcal{A}$, montrer que pour presque tout ω A $(\omega, .)$ est portée par l'ensemble λ-négligeable $\{t : \theta_t \quad \omega \in B\}$. D'autre part si $\widehat{P} \geq a \, . \, P$, presque sûrement A $(\omega, .) \geq a \, (\theta_{.} \, \omega) \, \lambda$] . Quel lien y-a-t-il entre la décomposition de Lebesgue de $\widehat{P}$ par rapport à P et celle des mesures A $(\omega, .)$ par rapport à λ ?

2°) Si $\Delta : \Omega \to M_+ (E)$ est une mesure aléatoire sur E telle que $\int P(d\omega) \, \Delta(\omega, E) < \infty$, montrer que la formule

$$N(\omega, .) = \int_E \lambda(ds) \, \tau_{-s} \circ \Delta(\theta_s \, \omega, .)$$

définit une mesure aléatoire stationnaire sur E dont la mesure de Palm est l'image de la mesure P $(d\omega)$ Δ (ω, ds) sur $\Omega \times E$ par l'application $\omega, s \to \theta_s \, \omega$ de $\Omega \times E$ dans Ω .

Montrer qu'un résultat analogue reste valable si dans la formule ci-dessus on remplace la mesure λ par une mesure aléatoire stationnaire A (ω, ds) à condition de remplacer la condition d'intégrabilité par $\int \widehat{P}(d\omega) \, \Delta(\omega, E) < \infty$ où $\widehat{P}$ désigne la mesure de Palm de A. En particulier si $\Delta(\omega, .) = \varepsilon_{S(\omega)}$ pour une fonction mesurable $S : \Omega \to E$, trouver N ainsi que sa mesure de Palm.

10. Soit $(\Omega, \mathcal{A}, P)$ un espace de probabilité muni d'un flot $(\theta_t, t \in E)$ et soit $T : \Omega \to E$ une application mesurable.

a) Pour tout processus ponctuel simple stationnaire $N : \Omega \to M_p(E)$ de mesure de Palm $\widehat{P}_N$ montrer que P [N ({T}) = 1] = $\int \xi_T \, d\widehat{P}_N$ si $\xi_T(\omega) = \lambda(\{t : t + T(\theta_t \, \omega) = 0\})$ pour tout ω.

b) Montrer que l'image de la mesure λ par $t \to t + T(\theta_t \, \omega)$ définit une mesure aléatoire stationnaire $\nu_T(\omega, .)$ sur E de mesure de Palm égale à $\theta_T \circ P$. De plus s'il existe un processus ponctuel stationnaire simple N sur E tel que P (N ({T}) = 1), établir que $\nu_T(\omega, ds) = \xi_T(\theta_s \, \omega) \, N(\omega, ds)$ et que la formule

$$N_T(\omega, ds) = \xi'_T(\theta_s\,\omega)\,\nu_T(\omega, ds) \text{ où } \xi'_T = \frac{1}{T}\,1_{\{\xi_T>0\}}$$

définit un processus ponctuel stationnaire simple sur E tel que $N_T(\{T\}) = 1$ sur Ω et minimal au sens où $N_T \leqslant N$ p.s. pour tout autre processus ponctuel stationnaire simple tel que $P\,[N(\{T\}) = 1]$.

c) Si S et T sont deux applications mesurables de Ω dans E telles que $N(\{S\}) = N(\{T\}) = 1$ p.s. pour un processus ponctuel stationnaire simple au moins, montrer que les conditions suivantes de disjonction sont équivalentes :

1) $N_S(\{T\}) = 0$ P p.s. ,

2) $N_S(\omega, .)$ et $N_T(\omega, .)$ n'ont pour presque tout ω, aucun point commun,

3) $\widehat{P}_S(\xi_T > 0) = 0$ si $\widehat{P}_S$ désigne la mesure de Palm de N_S ,

4) $\xi_S\,\xi_T = 0$ $\widehat{P}$ p.s. si $\widehat{P}$ est la mesure de Palm d'un processus N tel que $N(\{S\}) = 1 = N(\{T\})$ p.s.

d) Pour tout processus ponctuel stationnaire simple N sur R^d, montrer que $N = N_T$ si pour tout ω, $T(\omega)$ désigne le point de $N(\omega, .)$ le plus proche de 0.

11. Inhibition. Soient $\nu = \sum_{\mathbb{Z}} \varepsilon_{S_n}$ et $N = \sum_{\mathbb{Z}} \varepsilon_{T_n}$ deux processus ponctuels stationnaires sur E, simples et doublement infinis définis sur le même espace de probabilité. Tout point de $N(\omega, .)$ qui suit immédiatement un point de $\nu(\omega, .)$ sera dit inhibé.

Montrer que les points de N qui sont inhibés sont les points qui pour un $n \in \mathbb{Z}$ au moins sont égaux à $T^*_n = S_n + T_1 \circ \theta_{S_n}$. Etablir que le processus ponctuel $N^* = \sum_{Z} \varepsilon_{T^*_n}$ qui en général n'est pas simple a les deux propriétés suivantes :

a) $N^*(\omega, .)$ est l'image de $\nu(\omega, .)$ par $t \to t + T_1(\theta_t\,\omega)$

b) $N^*(\omega, dt = \xi(\theta_t\,\omega)\,N(\omega, dt)$ si $\xi : \widehat{\Omega}_N \to \mathbb{N}$ est défini par $\xi(\omega) = \nu(\omega, [-T_1(\omega), 0[\,)$.

Montrer ensuite que le processus ponctuel N^- formé par les <u>différents</u> points inhibés de N est donné par la formule $N^-(\omega, dt) = 1_{\{\xi(\theta_t\omega)>0\}} N(\omega, dt)$. En déduire que le rapport des intensités des processus ponctuels N^- et N vaut $\widehat{P}_N(S_1 < T_1) / \widehat{P}_N(\Omega)$; en particulier si les processus ν et N sont indépendants, montrer que ce rapport vaut $\int_0^\infty f(u)\, g(u)\, du / \int_0^\infty f(u)\, du$ si $f(u) = \widehat{P}_\nu(S_1 > u) / \widehat{P}_\nu(\Omega)$ et $g(u) = \widehat{P}_N(T_1 > u) / \widehat{P}_N(\Omega)$. Calculer aussi la probabilité $P[T_1$ est inhibé$]$.

12. <u>Construction d'un processus ponctuel stationnaire marqué sur R.</u>

Soit $N : \Omega \to M_p(R)$ un processus ponctuel stationnaire, simple et doublement infini, donc de la forme $N = \sum_Z \varepsilon_{T_n}$ sur R, défini sur l'espace de probabilité $(\Omega, \mathcal{A}, P)$ muni du flot réel $(\theta_t, t \in R)$. Si $(M, \mathcal{M})$ est un espace mesurable auxiliaire, soit W l'espace M^Z muni de la tribu produit et des translations de coordonnées ϕ_n $(n \in Z)$; formons l'espace produit mesurable

$$(\Omega^\bullet, \mathcal{A}^\bullet) = (\Omega, \mathcal{A}) \times (W, \mathcal{W})$$

et le flot $(\theta_t^\bullet, t \in R)$ défini par

$$\theta_t^\bullet[(\omega, w)] = (\theta_t\omega, \phi_n(w)) \text{ si } T_n(\omega) \leq t < T_{n+1}(\omega).$$

Ensuite, soit $\Pi = [\Pi(\omega, dw) ; \omega \in \widehat{\Omega}, w \in W]$ une probabilité de transition de $\widehat{\Omega}$ vers W telle que $\Pi(\widehat{\theta}_n\omega, .) = \phi_n \circ \Pi(\omega, .)$ sur W et si $\widehat{P}$ désigne la mesure de Palm de N sur $\widehat{\Omega} \subset \Omega$, considérons la mesure

$$\widehat{P}^\bullet(d\omega\, dw) = \widehat{P}(d\omega)\, \Pi(\omega, dw) \quad \text{sur } \widehat{\Omega}^\bullet = \widehat{\Omega} \times W$$

Montrer que la formule

$$P^\bullet(A^\bullet) = \int_{\Omega^\bullet} \widehat{P}^\bullet(d\omega\, dw) \int_0^{T_1(\omega)} 1_A \circ \theta_t^\bullet[(\omega, w)]\, dt$$

définit une probabilité sur $(\Omega^\bullet, \mathcal{A}^\bullet)$ invariante par le flot $(\theta_t^\bullet, t \in R)$ et que la mesure $\widehat{P}^\bullet$ est la mesure de Palm du processus ponctuel stationnaire $N^\bullet$ défini sur $\Omega^\bullet$ par $N^\bullet[(\omega, w), .] = N(\omega, .)$ [Utiliser la proposition 16]

La formule $\Delta[(\omega, w)] = w_o$ où w_o désigne la première coordonnée de w, définit alors une marque sur $\hat{\Omega}^{\cdot}$. Pour tout $\omega^{\cdot} = (\omega, w)$ de $\Omega^{\cdot}$, $\{\Delta \circ \theta^{\cdot}_{T_n} [(\omega, w)], n \in Z\} = w$ ce qui montre que w est la suite des marques attachées aux points $T_n(\omega)$ $(n \in Z)$ de $N(\omega, .)$ par notre construction.

13. Soient P_n $(n \in N)$ et P des probabilités sur $M_p(E)$ invariantes par le flot des translations et dont les mesures de Palm $\hat{P}_n$ et $\hat{P}$ sont supposées finies. Etablir que les deux conditions suivantes sont équivalentes

a) $P_n \to P$ au sens de la convergence étroite sur $M_p(E)$ lorsque $n \uparrow \infty$ et de plus $\lim_{n\to\infty} \hat{P}_n(\Omega) = \hat{P}(\Omega)$,

b) $\hat{P}_n \to \hat{P}$ au sens de la convergence étroite sur $M_p(E)$ lorsque $n \uparrow \infty$

[Pour établir que (a) $\Longrightarrow$ (b) montrer que pour toute fonction $f \in C_b^+[M_p(E)]$, $\int f(m)\hat{P}(dm) = \int g(m) P(dm)$ où $g(m)$ est la fonction continue positive majorée par $||f||$ $m(a)$ définie par $g(m) = \int_E m(dt)\, a(t)\, f(\theta_t m)$, la fonction a étant une fonction arbitraire de $C_k^+(E)$ telle que $\lambda(a) = 1$. Procéder de manière analogue pour montrer que (b) $\Longrightarrow$ (a)].

14. Sur un espace homogène $E = G/H$, montrer qu'il ne peut exister de processus ponctuel stationnaire $N : \Omega \to M_p(E)$ tel que $N(E) = n$ p.s. pour un entier fini n (ou plus généralement tel que $P[N(E) < \infty] \neq 0$) que si E possède une probabilité invariante par les τ_t $(t \in G)$.

CHAPITRE III

METHODE DES MARTINGALES

Sommaire

L'auteur n'a pu rédiger que le début de ce chapitre mais il espère pouvoir compléter son travail dans un proche avenir. Les paragraphes rédigés s'appuient fortement sur les traités de C. Dellacherie et P.A. Meyer.

III.1. MESURABILITES OPTIONNELLE ET PREVISIBLE DES PROCESSUS ET DES MESURES ALEATOIRES.

Dans tout ce paragraphe, nous nous fixerons un espace mesurable $(\Omega, \mathfrak{F})$ et une famille croissante, continue à droite $(\mathfrak{F}_t, t \in R_+)$ de sous-tribus de $\mathfrak{F}$; par définition $\mathfrak{F}_s \subset \mathfrak{F}_t$ si $s < t$ dans R_+ et $\bigcap_{t>s} \mathfrak{F}_t = \mathfrak{F}_s$ pour tout $s \in R_+$.

Un processus sur R_+, c'est-à-dire une application X de $\Omega \times R_+$ dans un espace mesurable arbitraire est dit adapté à la famille $(\mathfrak{F}_t, t \in R_+)$ si pour tout t fixé dans R_+, l'application $\omega \to X(\omega, t)$ qui sera notée X_t, est $\mathfrak{F}_t$-mesurable sur Ω. Il est immédiat de vérifier que la classe de tous les processus adaptés réels (c'est-à-dire à valeurs dans R) est stable par les opérations habituelles : addition, multiplication, enveloppes supérieures et inférieures dénombrables.

La condition d'adaptation que nous venons d'introduire sur un processus n'impose aucune régularité aux trajectoires de ce processus, c'est-à-dire aux applications $t \to X(t, \omega)$ définies sur R_+ pour chaque $\omega \in \Omega$. Aussi peut-il sembler nécessaire de la renforcer en imposant à ces trajectoires d'être continues à droite et limitées à gauche sur R_+, en abrégé cadlag, compte tenu de ce qu'une large majorité des processus étudiés jusqu'à présent satisfont cette hypothèse. Mais si la classe des processus réels adaptés et cadlag est bien une algèbre de fonctions réelles sur $\Omega \times R_+$, cette classe n'est malheureusement pas stable par enveloppes dénombrables ; en stabilisant les classes des processus adaptés et cadlag pour les opérations d'enveloppes supérieure et inférieure dénombrables, nous obtiendrons exactement la classe des processus optionnels de la définition suivante.

Définition III.1

La tribu optionnelle (ou bien-mesurable) associée à la famille $(\mathfrak{F}_t, t \in R_+)$ soit $\mathcal{O}$, est définie comme la tribu de parties de $\Omega \times R_+$ engendrée par les processus réels $X : \Omega \times R_+ \to R$ adaptés et cadlag. Les parties de $\Omega \times R_+$ appartenant à cette tribu seront dites optionnelles et les processus $\mathcal{O}$-mesurables sur $\Omega \times R_+$ seront appelés optionnels.

La tribu optionnelle est intimement liée aux temps d'arrêt qui pour cette raison seront plutôt appelés ici temps optionnels. Un temps optionnel relativement à la famille $(\mathcal{F}_t\ ,\ t \in R_+)$ est défini comme une application $T : \Omega \to \overline{R}_+ = [0, \infty]$ telle que

(1) $$\{T \leqslant t\} \in \mathcal{F}_t \qquad \text{pour tout } t \in R_+ .$$

Si de manière générale pour deux applications $S, T : \Omega \to \overline{R}_+$ quelconques, l'intervalle stochastique $[\![S, T[\![$ est défini comme le sous-ensemble $\{(\omega, t) : S(\omega) \leqslant t < T(\omega)\}$ de $\Omega \times R_+$, dire que T est un temps optionnel équivaut à dire que le processus réel cadlag $1_{[\![T, \infty[\![}$ est adapté. De cette remarque il résulte facilement que la classe des temps optionnels est stable par enveloppes supérieures dénombrables et enveloppes inférieures finies, puisque $[\![\sup_N T_n , \infty[\![= \inf_N [\![T_n , \infty[\![$ et $[\![\min_{1 \leqslant k \leqslant n} T_k , \infty[\![= \sup_{1 \leqslant k \leqslant n} [\![T_k , \infty[\![$

La tribu $\mathcal{F}_T$ des évènements antérieurs à un temps optionnel T est définie sur Ω par

(2) $$\mathcal{F}_T = \{F \ ;\ F \in \mathcal{F}\ ,\ F\ \{T \leqslant t\} \in \mathcal{F}_t \ \text{ pour tout } t \in R_+\} \ ;$$

cette classe est une tribu car elle contient $\emptyset$, car elle est manifestement stable par réunion dénombrable et car d'autre part, parce que T est optionnel, elle est stable par complémentation. Tout temps constant sur Ω, soit $T \equiv t_o$, est optionnel et pour un tel temps $\mathcal{F}_T$ coïncide bien avec $\mathcal{F}_{t_o}$, ce qui justifie nos notations. Inversement toute propriété intéressante qui fait intervenir un temps constant peut le plus souvent être étendue au cas d'un temps optionnel et une telle extension s'avèrera fréquemment fructueuse.

Une variable aléatoire réelle $\mathcal{F}$- mesurable $Z : \Omega \to R$ est mesurable par rapport à la sous-tribu $\mathcal{F}_T$ de $\mathcal{F}$ si et seulement si le processus réel cadlag $Z 1_{[\![T,\infty[\![}$ qui vaut $Z(\omega) 1_{(T(\omega) \leq t)}$ en tout point (ω, t) de $\Omega \times R_+$ est adapté ; cela se vérifie immédiatement.

Voici alors le double lien qui unit la tribu optionnelle $\mathcal{O}$ à la famille des tribus $\mathfrak{F}_T$ et qui va nous permettre le plus souvent dans la suite de ne travailler que sur la seule tribu $\mathcal{O}$ (sur $\Omega \times R_+$) plutôt que sur toute la famille des tribus $\mathfrak{F}_T$ (sur Ω).

Proposition III.2

a) Pour tout temps optionnel T et tout processus réel optionnel X, le processus réel $X 1_{[\![0,T]\!]}$ est mesurable sur $\Omega \times R_+$ relativement à la tribu produit $\mathfrak{F}_T \otimes \mathcal{R}_+$ et la variable X_T est $\mathfrak{F}_T$ - mesurable sur $\{T < \infty\} \subset \Omega$.

b) La tribu optionnelle $\mathcal{O}$ est engendrée par la famille des intervalles stochastiques $[\![T, \infty[\![$ obtenue lorsque T parcourt l'ensemble des temps optionnels. Pour tout processus réel $Y : \Omega \times R_+ \to R$ mesurable par rapport à la tribu $\mathfrak{F}_T \otimes R_+$, le processus réel $Y 1_{[\![T,\infty[\![}$ est optionnel.

La deuxième partie de (a) pour les temps constants montre que tout processus optionnel est adapté. Compte tenu de la définition de la tribu optionnelle, nous voyons ainsi qu'un processus réel cadlag est optionnel si et seulement s'il est adapté. Nous en déduirons que

1°) T est un temps optionnel si et seulement si $[\![T, \infty[\![\in \mathcal{O}$,

2°) si T est un temps optionnel et si Z est une variable réelle $\mathfrak{F}$-mesurable sur Ω , Z est $\mathfrak{F}_T$-mesurable si et seulement si $Z 1_{[\![T,\infty[\![}$ est $\mathcal{O}$-mesurable ; en particulier $F \in \mathfrak{F}$ appartient à $\mathfrak{F}_T$ si et seulement si $(F \times R_+) \cap [\![T, \infty[\![\in \mathcal{O}$.

Voici une autre application de cette proposition, de l'alinéa (b) cette fois : si T est un temps optionnel, toute variable réelle $\mathfrak{F}_T$-mesurable S telle que $S \geq T$ sur Ω est encore un temps optionnel. En effet, par hypothèse $[\![S, \infty[\![$ appartient à $\mathfrak{F}_T \otimes \mathcal{R}_+$ et est contenu dans $[\![T, \infty[\![$; ainsi

$1_{[\![S,\infty[\![} = 1_{[\![S,\infty[\![} \; 1_{[\![T,\infty[\![}$ est optionnel d'après (b) et S est donc un temps optionnel.

Démonstration

a) L'argument des classes monotones nous permet de ne démontrer la première partie de la proposition que lorsque X est un processus réel adapté et cadlag. Or tout processus réel cadlag X vérifie la formule suivante pour toute application T : $\Omega \to \overline{R}_+$:

$$X(\omega, s)\, 1_{(s<T(\omega))}$$

$$= \lim_{n\uparrow\infty} \sum_{k\in N} X\left[\omega, (k+1)\, 2^{-n}\right] 1_{((k+1)2^{-n}<T(\omega))} \cdot 1_{(k\, 2^{-n}\leqslant s<(k+1)\, 2^{-n})}$$

Si X est adapté et si T est un temps optionnel, chaque terme de la série du second membre est une fonction $\mathcal{F}_T \otimes \mathcal{R}_+$ mesurable de (ω, t), ce qui entraîne que $X\; 1_{[\![0,T[\![}$ est mesurable par rapport à cette tribu.

Ce premier résultat vaut en particulier si T est un temps constant : $X\, 1_{[\![0,t[\![}$ est $\mathcal{F}_t \otimes \mathcal{R}_+$ mesurable pour tout $t \in R_+$. Mais pour un temps constant t, comme X est adapté, le processus $X\, 1_{[\![t]\!]}$ qui vaut $X(\omega, t)\, 1_{\{t\}}(s)$ au point (ω, s), est manifestement aussi $\mathcal{F}_t \otimes \mathcal{R}_+$ mesurable ; par addition nous trouvons que $X\, 1_{[\![0,t]\!]}$ est $\mathcal{F}_t \otimes \mathcal{R}_+$ mesurable pour tout $t \in R_+$. D'autre part pour tout temps optionnel T, l'application $\omega \to (\omega, T(\omega) \wedge t)$ de $(\Omega, \mathcal{F}_t)$ dans $(\Omega \times R_+, \mathcal{F}_t \otimes \mathcal{R}_+)$ est mesurable parce que $T \wedge t$ est $\mathcal{F}_t$ - mesurable sur Ω ; en composant ce résultat de mesurabilité avec le précédent, nous trouvons que $X_{T\wedge t}$ est $\mathcal{F}_t$ - mesurable sur Ω pour tout $t \in R_+$. Par définition de la tribu $\mathcal{F}_T$, ceci montre que X_T est $\mathcal{F}_T$ - mesurable sur $\{T < \infty\}$. Pour achever la démonstration de la première partie de la proposition, il suffit alors de remarquer que le processus $X\, 1_{[\![T]\!]}$ qui vaut $X_T(\omega)\, 1_{[\![T]\!]}(\omega, s)$ au point (ω, s), est $\mathcal{F}_T \otimes \mathcal{R}_+$ mesurable puisque X_T est $\mathcal{F}_T$ - mesurable et $(\omega, s) \to T(\omega) - s$ est $\mathcal{F}_T \otimes \mathcal{R}_+$ mesurable (pour voir que T est $\mathcal{F}_T$ - mesurable prendre $X(\omega, t) = t$). Par conséquent $X 1_{[\![0,T]\!]} = X 1_{[\![0,T[\![} + X 1_{[\![T]\!]}$ est aussi $\mathcal{F}_T \times \mathcal{R}_+$ mesurable.

b) Si T est un temps optionnel, le processus réel $1_{[\![T,\infty[\![}$ est cadlag et adapté ; par conséquent la tribu $\mathcal{C}$ de parties de $\Omega \times R_+$ est engendrée par les intervalles stochastiques $[\![T, \infty[\![$ obtenus lorsque T parcourt les temps optionnels, est contenue dans la tribu optionnelle $\mathcal{O}$.

D'un autre côté pour tout processus réel $\mathcal{F}_T \otimes \mathcal{R}_+$ mesurable Y sur $\Omega \times R_+$, le processus $Y 1_{[\![T,\infty[\![}$ est $\mathcal{C}$-mesurable sur cet espace, quel que soit le temps optionnel T. Il suffit en effet par les arguments habituels de le démontrer lorsque Y est de la forme $1_{F \times [t,\infty[}$ pour un $F \in \mathcal{F}_T$ et un $t \in R_+$. Or, dans ce cas $Y 1_{[\![T,\infty[\![} = 1_{[\![T',\infty[\![}$ si l'application $T' : \Omega \to \overline{R}_+$ est définie par $T' = T \vee t$ sur F, $= +\infty$ sur F^c ; mais il est facile de vérifier que ce T' est un temps optionnel et donc que $[\![T', \infty[\![\in \mathcal{C}$.

Pour achever la démonstration de la proposition, il reste à établir l'inclusion $\mathcal{O} \subset \mathcal{C}$. Nous aurons besoin à cet effet de l'un des deux résultats suivants.

<u>Lemme</u> III.3

<u>Soit</u> Z <u>un processus réel adapté et cadlag.</u> <u>Pour tout ouvert</u> G <u>de</u> R, <u>la formule</u> $T(\omega) = \inf(t : Z_t(\omega) \in G)$ <u>définit un temps optionnel grâce à la continuité à droite de la famille</u> $(\mathcal{F}_t, t \in R_+)$. <u>Par contre sans utiliser cette hypothèse</u>, <u>pour tout fermé</u> $\overline{F}$ <u>de</u> R <u>la formule</u> $T'(\omega) = \inf(t : Z_t(\omega)$ <u>ou</u> $Z_{t-}(\omega) \in F]$ <u>définit un temps optionnel</u>.

Ce lemme découle des formules suivantes faciles à vérifier

$$\{T < t\} = \bigcup_{r \in Q_+,\, r<t} \{Z_r \in G\} \quad , \quad \{T' \leq t\} = \bigcap_n \bigcup_{\substack{r \in Q_+,\, r<t \\ \text{ou } r=t}} \{Z_r \in G_n\}$$

où $(G_n, n \in \mathbb{N})$ désigne une suite d'ouverts décroissant vers F.

Soit alors X un processus réel adapté et cadlag. Pour tout $\varepsilon > 0$, définissons par récurrence sur n une suite croissante $(T_n^\varepsilon, n \in N)$ de temps optionnels en partant de $T_0^\varepsilon = 0$ et en posant

$$T_{n+1}^\varepsilon = \inf (t : |Z_t^{\varepsilon,n}| \text{ ou } |Z_{t-}^{\varepsilon,n}| \geqslant \varepsilon) \text{ où } Z_t^{\varepsilon,n} = (X_t - X_{T_n^\varepsilon}) 1_{(t \geqslant T_n^\varepsilon)}$$

Parce que X est cadlag, $\lim_n \uparrow T_n^\varepsilon = +\infty$; par suite le processus réel

$$X^\varepsilon = \sum_N X_{T_n^\varepsilon} 1_{[\![T_n^\varepsilon, T_{n+1}^\varepsilon [\![}$$

vérifie l'inégalité $|X - X^\varepsilon| \leqslant \varepsilon$ sur $\Omega \times R_+$. Mais d'après ce qui précède ce processus X^ε est $\mathcal{C}$-mesurable sur $\Omega \times R_+$; il en est donc de même de X et ceci établit bien que $\mathcal{O} \subset \mathcal{C}$. □

La proposition précédente entraîne que tout intervalle stochastique $||S, T||$ qu'il soit fermé, semi-fermé ou ouvert, est un ensemble optionnel dès que S et T sont des temps optionnels. En effet d'une part d'après le paragraphe (b) de la proposition, $[\![S, T[\![= [\![S, \infty[\![\cap [\![T, \infty[\![^c$ appartient à $\mathcal{O}$; d'autre part si T est un temps optionnel, il en est de même des T + 1/n (qui sont $\mathcal{F}_T$ - mesurables et majorent T) pour tout $n \in N^*$ et alors $[\![T]\!] = \lim_n \downarrow [\![T, T + \frac{1}{n}[\![\in \mathcal{O}$.

La définition suivante introduit les mesures aléatoires optionnelles.

Définition III.4

Une mesure aléatoire positive sur R_+, c'est-à-dire une application $\mathcal{F}$-mesurable N de Ω dans $M_+(R_+)$ est dite optionnelle si elle vérifie l'une des deux conditions équivalentes suivantes :

a) le processus réel croissant et cadlag défini sur $\Omega \times R_+$ comme la primitive de N, soit

(3) $\quad A(\omega, t) = N(\omega, [0, t])$

est adapté (ou, ce qui est équivalent, optionnel) ;

b) pour tout processus réel positif optionnel X sur R_+, le processus réel positif Y défini par

$$Y(\omega, t) = \int_{[0,t]} X(\omega, s)\, N(\omega, ds)$$

sur $\Omega \times R_+$ est optionnel.

L'équivalence de ces deux conditions est quasiment évidente. La condition (b) se réduit à (a) pour $X \equiv 1$. Inversement sous la condition (a), pour tout t fixé dans R_+, $N(., I)$ est $\mathfrak{F}_t$ - mesurable pour tout intervalle $[0, s]$ $(0 \leqslant s \leqslant t)$; cela entraîne par les arguments habituels que $\int_{[0,t]} X(\omega, s)\, N(\omega, ds)$ est $\mathfrak{F}_t$ - mesurable sur Ω, si X est un processus réel positif $\mathfrak{F}_t \otimes \mathcal{R}_+$ mesurable au moins sur $[\![0, t]\!]$. Nous voyons ainsi que si X est un processus réel positif optionnel, le processus Y est adapté ; mais ce processus Y est aussi croissant et cadlag : il est donc optionnel.

L'équivalence des deux conditions (a) et (b) ci-dessus a la conséquence suivante : si N est une mesure optionnelle positive et si X est un processus réel optionnel positif, localement intégrable par rapport à N, c'est-à-dire tel que le processus Y de la condition (b) soit fini sur $\Omega \times R_+$, alors la mesure aléatoire $X \cdot N$ de densité X par rapport à N est encore optionnelle puisque Y est précisément égal à sa primitive.

La primitive de la mesure aléatoire positive $N = \varepsilon_T$ ($N(\ ,\ .) = 0$ si $T(\omega) = +\infty$) définie à partir d'une application $\mathfrak{F}$- mesurable $T : \Omega \to \overline{R}_+$ est égal à $1_{[\![T,\infty[\![}$. Cette mesure aléatoire N est donc optionnelle si et seulement si T est un temps optionnel ; ainsi, par l'intermédiaire de la correspondance $T \to \varepsilon_T$, la notion de mesure optionnelle étend celle de temps optionnel.

Dans le cas plus général d'une mesure aléatoire positive discrète sur R_+ que nous écrirons avec la convention $\varepsilon_{+\infty} = 0$ sous la forme

$$N(\omega, .) = \sum_{\mathbb{N}^*} Z_n(\omega)\, \varepsilon_{T_n(\omega)} \qquad (\omega \in \Omega)$$

pour une suite strictement croissante d'applications $T_n : \Omega \to [0, \infty]$ et une suite d'applications $Z_n : \Omega \to R_+$ telles que $\{Z_n = 0\} = \{T_n = \infty\}$, il n'est pas difficile de vérifier que cette mesure aléatoire est optionnelle si et seulement si

a) les T_n sont des temps optionnels $(n \in \mathbb{N}^*)$,

b) les Z_n sont des variables $\mathfrak{F}_{T_n}$ - mesurables $(n \in \mathbb{N}^*)$.

En effet cette double condition est suffisante puisque d'après la proposition 3.2 elle entraîne que le processus $A = \sum_{\mathbb{N}^*} Z_n \, 1_{[\![T_n,\infty[\![}$ est optionnel. Pour voir qu'elle est aussi nécessaire il suffit en raisonnant par récurrence sur n de remarquer que $[\![T_{n+1}, \infty[\![= \{A > \sum_1^n Z_k \, 1_{[\![T_k,\infty[\![}\}$ et que $Z_{n+1} = A_{T_{n+1}} - \sum_1^n Z_k$ sur $\{T_{n+1} < \infty\}$.

A côté de la tribu optionnelle une seconde tribu de parties de $\Omega \times R_+$, la tribu prévisible, qui est intimement reliée aux tribus $\mathfrak{F}^-$ définies ci-dessous, joue un rôle important dans l'étude des processus et des mesures aléatoires sur R_+ en relation avec une famille $(\mathfrak{F}_t, t \in R_+)$.

Pour tout $t \in R_+$ strictement positif, la tribu $\mathfrak{F}_t^- = \bigvee_{s<t} \mathfrak{F}_s$ engendrée par les évènements strictement antérieurs à t est contenue dans $\mathfrak{F}_t$ mais peut être distincte de $\mathfrak{F}_t$; pour t = 0, nous poserons $\mathfrak{F}_o^- = \mathfrak{F}_o$. Plus généralement pour tout temps optionnel T la tribu $\mathfrak{F}_T^-$ des évènements strictement antérieurs à T est définie comme la tribu de parties de Ω engendrée par $\mathfrak{F}_o$ et par les évènements $F\{t < T\}$ obtenus lorsque F parcourt $\mathfrak{F}_t$ et t parcourt R_+. Il est immédiat de vérifier sur cette définition que $\mathfrak{F}_T^- \subset \mathfrak{F}_T$ mais même si $\mathfrak{F}_t^- = \mathfrak{F}_t$ pour tout $t \in R_+$, il existe généralement des temps optionnels pour lesquels $\mathfrak{F}_T^- \neq \mathfrak{F}_T$.

Comme va nous le montrer la proposition 3.6 ci-dessous, la tribu prévisible que nous allons définir permet de mieux manipuler les tribus $\mathfrak{F}^-$.

Définition III.5

La tribu prévisible $\mathcal{P}$ associée à la famille $(\mathfrak{F}_t, t \in R_+)$ est la tribu de parties de $\Omega \times R_+$ engendrée par les processus réels $X : \Omega \times R_+ \to R$

adaptés et continus à gauche.

Les parties de $\Omega \times R_+$ dans cette tribu $\mathcal{P}$ seront dites prévisibles et les processus $\mathcal{P}$- mesurables définis sur $\Omega \times R_+$ seront dits prévisibles.

Une application $\mathcal{F}$- mesurable $S : \Omega \to \overline{R}_+$ est appelée un temps prévisible lorsque $[\![T, \infty[\![\in \mathcal{P}$.

La proposition suivante pourra alors être comparée pour ses analogies et ses différences, à la proposition 3.2.

Proposition III.6

a) Si X est un processus réel prévisible et si T est un temps optionnel, le processus $X\, 1_{[\![0, T[\![}$ est mesurable sur $\Omega \times R_+$ par rapport à la tribu $\mathcal{F}_T^- \otimes R_+$ et la variable X_T est $\mathcal{F}_T^-$ - mesurable sur $\{T < \infty\}$.

b) La tribu prévisible $\mathcal{P}$ est engendrée par les parties de $\Omega \times R_+$ de la forme $F \times \{o\}$ où $F \in \mathcal{F}_o$ et par celle de la forme $F \times]t, \infty[$ où $F \in \mathcal{F}_t$ et $t \in R_+$. Elle est aussi engendrée par la famille des intervalles stochastiques $]\!]T, \infty[\![$ obtenue lorsque T parcourt les temps optionnels ainsi que par les parties $F \times \{o\}$ $(F \in \mathcal{F}_o)$ de $\Omega \times R_+$. Pour tout processus réel $Y : \Omega \times R_+ \to R$ mesurable par rapport à la tribu $\mathcal{F}_T \otimes \mathcal{R}_+$, le processus réel $Y\, 1_{]\!]T, \infty[\![}$ est prévisible.

c) La tribu $\mathcal{P}$ est engendrée également par la classe des parties de $\Omega \times R_+$ de la forme $F \times [t, \infty[$ où $F \in \mathcal{F}_t^-$ et $t \in R_+$; elle est encore engendrée par la famille des intervalles stochastiques $[\![S, \infty[\![$ obtenue lorsque S parcourt les temps prévisibles. Pour tout temps prévisible S et tout processus réel $Y : \Omega \times R_+ \to R$ mesurable par rapport à la tribu $\mathcal{F}_S^- \otimes \mathcal{R}_+$, le processus réel $Y\, 1_{[\![S, \infty[\![}$ est prévisible.

Démonstration

a) Un argument de classes monotones nous permet de ne démontrer la première partie de la proposition que lorsque X est un processus réel adapté et continu à gauche. Or, pour un tel processus et pour tout temps optionnel T, nous pouvons écrire que

$$X\, 1_{[\![0,\, T]\!]}(\omega, s)$$

$$= \lim_{n\uparrow\infty} X_o(\omega)\, 1_{\{o\}}(s) + \sum_{k\in\mathbb{N}} X_{k2^{-n}}(\omega)\, 1_{(k\, 2^{-n} < T(\omega))} \cdot 1_{]k2^{-n},(k+1)2^{-n}]}(s);$$

comme les différents termes de la somme du second membre sont $\mathcal{F}_T^- \otimes \mathcal{R}_+$ mesurables, il en est de même du processus réel $X\, 1_{[\![0,\, T]\!]}$. Puisque T est $\mathcal{F}_T^-$ -mesurable, l'application $\omega \to (\omega, T(\omega))$ de $\{T < \infty\}$ dans $\Omega \times R_+$ est mesurable pour les tribus respectives $\mathcal{F}_T^-$ et $\mathcal{F}_T^- \otimes \mathcal{R}_+$; par composition avec le processus réel $X\, 1_{[\![0,\, T]\!]}$, nous trouvons que X_T est $\mathcal{F}_T^-$ - mesurable sur son domaine de définition $\{T < \infty\}$. La première partie de la proposition est ainsi démontrée.

b) Pour tout temps optionnel T, le processus réel continu à gauche $1_{]\!]T,\, \infty[\![}$ est adapté car pour tout $t \in R_+$, $\{T < t\} \in \mathcal{F}_t$; ce processus est donc prévisible. La tribu $\mathcal{P}^*$ de parties de $\Omega \times R_+$ engendrée par les intervalles stochastiques $]\!]T,\, \infty[\![$ obtenus lorsque T parcourt les temps optionnels ainsi que par les ensembles $F \times \{0\}$ où $F \in \mathcal{F}_o$ est donc contenue dans $\mathcal{P}$; notons en effet que $1_F \times \{o\}$ est aussi un processus adapté continu à gauche si $F \in \mathcal{F}_o$. Plus généralement, quel que soit le processus réel Y mesurable par rapport à la tribu $\mathcal{F}_T \otimes \mathcal{R}_+$, le processus $Y\, 1_{]\!]T,\, \infty[\![}$ est $\mathcal{P}^*$ - mesurable. Par les arguments habituels, il suffit pour l'établir de considérer des Y de l'une des deux formes $Y = 1_{F\times\{o\}}$ où $F \in \mathcal{F}_T$, $Y = 1_{F\times]t,\infty[}$ où $F \in \mathcal{F}_T$ et $t \in R_+$; or, dans le premier cas le processus $Y\, 1_{]\!]T,\, \infty[\![}$ est identiquement nul et dans le second, il vaut $1_{]\!]T',\, \infty[\![}$ où $T' = T \vee t$ sur F, $= +\infty$ sur F^c ; mais cette dernière formule définit un T' optionnel, par exemple car $1_{[\![T',\, \infty[\![} = Y\, 1_{[\![T,\, \infty[\![}$ est optionnel, et il s'ensuit que $]\!]T',\, \infty[\![\in \mathcal{P}^*$.

Soit $\mathcal{P}'$ la tribu de parties de $\Omega \times R_+$ engendrées par celles de la forme $F \times \{o\}$ $(F \in \mathcal{F}_o)$ ou $F \times]t, \infty[$ $(F \in \mathcal{F}_t,\ t \in R_+)$. Pour achever la démonstration de l'alinéa (b) de la proposition, il nous reste à établir que $\mathcal{P} \subset \mathcal{P}' \subset \mathcal{P}^*$. Or, si nous prenons $T \equiv \infty$ dans la première formule de la démonstration, cette formule nous montre que tout processus adapté et continu à gauche est $\mathcal{P}'$ - mesurable ; donc $\mathcal{P} \subset \mathcal{P}'$. D'autre part $\mathcal{P}' \subset \mathcal{P}^*$ car pour tout $F \in \mathcal{F}_t$ et tout $t \in R_+$, $F \times]t, \infty[$ coïncide avec $]\!]T, \infty[\![$ si T désigne le temps optionnel égal à t sur F et à $+\infty$ sur F^c.

c) Si S est un temps prévisible, pour démontrer que $Y 1_{[\![S, \infty[\![}$ est prévisible dès que le processus réel Y est $\mathcal{F}_S^- \otimes \mathcal{R}_+$ mesurable, il nous suffira, par les arguments habituels, de considérer le cas où $Y = 1_{G \times I}$ pour un G de l'une des deux formes $F\{u < S\}$ $(F \in \mathcal{F}_u, u \in R_+)$ ou F $(F \in \mathcal{F}_0)$ et pour un I égal soit à un intervalle $]t, \infty[$ $(t \in R_+)$, soit à R_+ tout entier. Or, si nous nous plaçons par exemple dans les premières des deux alternatives, nous trouvons que Y est égal au processus réel adapté continu à gauche $1_{F \times]u \vee t, \infty[}$ sur $[\![S, \infty[\![$, ce qui entraîne que $Y 1_{[\![S, \infty[\![}$ est prévisible puisque $[\![S, \infty[\![\in \mathcal{P}$ par hypothèse. Les trois autres cas se traitent de la même manière.

La tribu $\mathcal{P}^{**}$ engendrée par les intervalles stochastiques $[\![S, \infty[\![$ obtenus lorsque S parcourt les temps prévisibles est évidemment contenue dans $\mathcal{P}$. Elle contient d'autre part la tribu $\mathcal{P}''$ engendrée par les parties $F \times [t, \infty[$ de $\Omega \times R_+$ correspondant à des $F \in \mathcal{F}_t^-$ et $t \in R_+$; en effet tout temps constant est prévisible car $1_{[t, \infty[} \equiv 1$ si $t = 0$, $= \lim_{n/\infty} 1_{]t - \frac{1}{n}, \infty[}$ si $t > 0$ et les deuxièmes membres sont prévisibles ; puisque $F \in \mathcal{F}_t^-$, d'après ce que nous avons démontré à l'alinéa précédent, le processus $1_{F \times [t, \infty[}$ est prévisible et comme il est égal à $1_{[\![S, \infty[\![}$ si $S = t$ sur F, $= +\infty$ sur F^c, le temps S est prévisible et finalement $F \times [t, \infty[\in \mathcal{P}^{**}$. Nous savons donc maintenant que $\mathcal{P}'' \subset \mathcal{P}^{**} \subset \mathcal{P}$. Mais la tribu $\mathcal{P}''$ contient aussi la tribu $\mathcal{P}'$ du paragraphe (b) de la démonstration ; en effet si $t \in R_+$ et $F \in \mathcal{F}_t$,
$F \times]t, \infty[= \lim_n \nearrow F \times [t + \frac{1}{n}, \infty[\in \mathcal{P}''$ et pour tout $F \in \mathcal{F}_0$,

$F \times \{0\} = F \times [0, \infty[\; - F \times]0, \infty[\; \in \mathcal{P}''$. Comme nous avons établi antérieurement que $\mathcal{P}' = \mathcal{P}$, nous trouvons maintenant que $\mathcal{P}'' = \mathcal{P}^{**} = \mathcal{P}$ et ceci termine la démonstration. □

D'après la première partie de la proposition que nous venons de démontrer, tout processus réel prévisible est adapté à la famille $(\mathcal{F}_t^-, t \in R_+)$ et donc a fortiori à la famille $(\mathcal{F}_t, t \in R_+)$. Nous voyons alors qu'un processus réel <u>continu à gauche</u> est prévisible si et seulement s'il est adapté.

Le second paragraphe de la proposition entraîne que $\mathcal{P} \subset \mathcal{O}$ et donc que tout temps prévisible est optionnel ; en effet les intervalles stochastiques $]\!]T, \infty[\![$ appartiennent à $\mathcal{O}$ pour tout temps optionnel d'après le début de ce paragraphe. D'autre part d'après leur définition les temps prévisibles forment une classe stable par enveloppes supérieures dénombrables et enveloppes inférieures finies. Un temps optionnel n'est pas en général prévisible ; par contre un exemple fondamental de temps prévisible est donné par les limites strictement croissantes de temps optionnels ; de manière précise, si $(T_n, n \in N)$ est une suite croissante de temps optionnels dont la limite $T = \lim_n \uparrow T_n$ vérifie l'inégalité $T_n < T$ pour tout $n \in N$ sur $\{0 < T < \infty\}$, cette limite T est un temps prévisible car les hypothèses faites permettent d'écrire que

$$[\![T, \infty[\![\; = \{T = 0\} \times \{0\} + \lim_n \downarrow \;]\!]T_n, \infty[\![\; \in \mathcal{P}.$$

D'après la dernière partie de la proposition, si S est un temps prévisible une variable réelle $\mathcal{F}$- mesurable Z est $\mathcal{F}_S^-$ - mesurable si et seulement si $Z 1_{[\![S, \infty[\![}$ est un processus prévisible ; en particulier, pour qu'une partie F de Ω dans $\mathcal{F}$ appartienne à $\mathcal{F}_S^-$ il faut et il suffit que $(F \times R_+) \cap [\![S, \infty[\![\; \in \mathcal{P}$.

Les tribus $\mathcal{F}_.$ et $\mathcal{F}_.^-$ sont liées par des relations d'inclusion qu'il est facile d'obtenir et commode de retrouver en se servant des propositions 3.2 et 3.6. Ces relations, sauf celles liées aux temps prévisibles peuvent d'ailleurs se démontrer directement à partir des définitions.

Corollaire III.7

Si S et T sont deux temps optionnels alors :

(4) $\{S < T\} \in \mathcal{F}_S$ et $\mathcal{F}_S \cap \{S < T\} \subset \mathcal{F}_T^-$,

$\{S \leq T\} \in \mathcal{F}_S^-$ et $\mathcal{F}_S \cap \{S \leq T\} \subset \mathcal{F}_T$

En outre si S est prévisible, quel que soit T optionnel :

(5) $\mathcal{F}_S^- \cap \{S \leq T\} \subset \mathcal{F}_T^-$

Enfin $\{S < T\} \in \mathcal{F}_S^-$ si T est prévisible, quel que soit S optionnel.
En particulier en prenant Ω pour ensemble de $\mathcal{F}_S$ ou de $\mathcal{F}_S^-$ dans les inclusions de droite, nous trouvons que $\{S < T\} \in \mathcal{F}_T^-$, ...

Démonstration

D'après les propositions 3.2 et 3.6 (alinéas b) si S est un temps optionnel et si $F \in \mathcal{F}_S$, les processus $1_F \, 1_{[\![S,\infty[\![}$ et $1_F \, 1_{]\!]S,\infty[\![}$ sont respectivement optionnel et prévisible. D'après les alinéas (a) de ces mêmes propositions, les valeurs en un temps optionnel T de ces processus sont respectivement $\mathcal{F}_T$ et $\mathcal{F}_T^-$ mesurables ; d'où $F\{S \leq T\} \in \mathcal{F}_T$ et $F\{S < T\} \in \mathcal{F}_T^-$.
En prenant $F = \Omega$ et en échangeant le rôle de S et T nous trouvons en outre que $\{S < T\} = \{T \leq S\}^c \in \mathcal{F}_S$ et que $\{S \leq T\} = \{T < S\}^c \in \mathcal{F}_S^-$. Les relations (4) sont ainsi établies.

D'après la proposition III.6 (alinéa c), si S est un temps prévisible et si $F \in \mathcal{F}_S^-$, le processus $1_F \, 1_{[\![S,\infty[\![}$ est prévisible ; donc d'après l'alinéa (a) sa valeur en un temps optionnel est $\mathcal{F}_T^-$ mesurable, soit $F\{S \leq T\} \in \mathcal{F}_T^-$, ce qui prouve (5). Enfin $\{S < T\} = \{T \leq S\}^c \in \mathcal{F}_S^-$ si T est prévisible.□

La condition de quasi-continuité à gauche définie ci-dessous est souvent satisfaite dans les applications. Cette condition est en général strictement plus forte que la simple condition $\mathcal{F}_s^- = \mathcal{F}_s$ pour tout s > 0 de R.

Définition III.8

La famille de tribus ($\mathfrak{F}_t$, $t \in R_+$) est dite quasi-continue à gauche si pour tout temps prévisible S les deux tribus $\mathfrak{F}_S^-$ et $\mathfrak{F}_S$ coïncident.

Nous terminerons ce paragraphe par l'introduction des mesures aléatoires prévisibles.

Définition III.9

Une mesure aléatoire positive N sur R_+ est dite prévisible si elle vérifie l'une des deux conditions équivalentes suivantes :

a) la primitive de N, c'est-à-dire le processus réel croissant et cadlag

$A(\omega, t) = N(\omega, [0, t])$ est prévisible sur $\Omega \times R_+$,

b) pour tout processus réel positif prévisible X sur R_+ , le processus réel positif Y défini par $Y(\omega, t) = \int_{[0, t]} X(\omega, s)\, N(\omega, ds)$ est prévisible sur $\Omega \times R_+$.

L'équivalence de ces deux conditions s'établit facilement. D'abord (b) se réduit à (a) lorsque $X \equiv 1$. Pour établir ensuite l'implication (a) $\Longrightarrow$ (b), il nous suffira par les arguments habituels de considérer les deux cas où $X = 1_{F \times R_+}$ $(F \in \mathfrak{F}_o)$ et où $X = 1_{F \times]u,\infty[}$ $(F \in \mathfrak{F}_u$ et $u \in R_+)$; or dans le premier cas $Y_t = 1_F \cdot A_t$ et Y est donc prévisible (1_F est adapté et constant en t, donc prévisible) ; dans le second, $Y_t = 1_F (A_t - A_u)$ si $t \geq u$, = 0 sinon. Mais pour tout processus réel adapté et continu à gauche Z , le processus Z' défini par $Z'_t = 1_F (Z_t - Z_u)$ si $t \geq u$, = 0 sinon est encore adapté et continu à gaoche ; par les arguments habituels, il s'en suit plus généralement que Z' est prévisible si Z l'est ; ceci montre que le processus Y ci-dessus est prévisible.

L'équivalence des conditions (a) et (b) entraînent comme dans le cas optionnel que si X est un processus réel prévisible positif, localement intégrable par rapport à une mesure prévisible et positive N, la mesure aléatoire X . N de densité X par rapport à N est encore prévisible.

La première partie de la définition montre qu'une mesure prévisible est optionnelle. Par contre, comme la mesure aléatoire $N = \varepsilon_T$ est prévisible si et seulement si T est prévisible (puisque la primitive de N vaut $1_{[\![T, \infty [\![}$), il suffit qu'il existe un temps optionnel non prévisible T pour que la mesure aléatoire ε_T soit optionnelle mais non prévisible.

Une mesure aléatoire positive N est <u>diffuse</u>, c'est-à-dire telle que $N(\omega, \{t\}) = 0$ pour tout $(\omega, t) \in \Omega \times R_+$ si et seulement si sa primitive A est continue. Dans ce cas la mesure aléatoire N est prévisible si et seulement si A est adapté et ceci fournit l'exemple le plus important de mesures prévisibles ; notons aussi qu'une mesure aléatoire diffuse est prévisible dès qu'elle est optionnelle. Enfin une mesure aléatoire positive discrète

$N = \sum_{\mathbb{N}^*} Z_n \; \varepsilon_{T_n} \quad (0 \leq T_1 < T_2 < \dots ; \lim_n \uparrow T_n = +\infty ; \{Z_n = 0\} = \{T_n = \infty\})$

est prévisible si et seulement si les T_n sont des temps prévisibles et si pour chaque n, Z_n est une variable $\mathcal{F}^-_{T_n}$ mesurable ; la démonstration est analogue à celle que nous avons donnée dans le cas optionnel.

<u>Etude d'un exemple</u>

Nous commencerons par un exemple simple dont les résultats pourront être démontrés par le lecteur à titre d'exercice ; cet exemple ne sera d'ailleurs qu'un cas très particulier de celui que nous développerons ensuite et c'est pourquoi il peut être intéressant de l'approfondir d'abord.

Etant donnée une application T d'un espace Ω dans $]0, \infty]$, désignons par $\mathcal{T}$ la tribu triviale $\{\emptyset, \Omega\}$ de Ω et par $\mathcal{F}$ la tribu engendrée par T. Alors la famille de sous-tribus $\mathcal{F}_t = (F : F \in \mathcal{F}, F$ ou F^c contient $\{T > t\})$ $(t \in R_+)$ de $\mathcal{F}$ est croissante, continue à droite et T est un temps optionnel. Les tribus optionnelles et prévisibles pour cette famille valent respectivement

$$\mathcal{O} = (\mathcal{T} \otimes \mathcal{R}_+) \cap [\![0, T [\![\; + \; (\mathcal{F} \otimes \mathcal{R}_+) \cap [\![T, \infty [\![\; ,$$

$$\mathcal{P} = (\mathcal{T} \otimes \mathcal{R}_+) \cap [\![0, T]\!] \; + \; (\mathcal{F} \otimes \mathcal{R}_+) \cap]\!] T, \infty [\![\; .$$

Il en résulte qu'une application $\mathcal{F}$-mesurable $S : \Omega \to \overline{R}_+$ est un temps optionnel resp. prévisible si et seulement s'il existe une constante $a \in \overline{R}_+$ telle que

$T > a \Longrightarrow S = a$ et $T \leqslant a \Longrightarrow S \geqslant T$ et $T < a \Longrightarrow S > T)$

(resp. $T \geqslant a \Longrightarrow S = a$

en particulier un temps prévisible ne peut avoir plus d'une valeur en commun avec T et à moins que l'application T ne soit constante, le temps T est optionnel mais n'est pas prévisible.

L'exemple plus élaboré que nous allons étudier maintenant a comme nous le verrons de nombreuses applications. Soit $(\mathcal{B}_n\ , n \in N)$ une suite croissante de sous-tribus de $\mathcal{F}$ dans un espace mesurable $(\Omega, \mathcal{F})$ donné et soit $(T_n\ , n \in N)$ une suite strictement croissante de variables réelles positives, adaptée à la suite $(\mathcal{B}_n\ , n \in N)$ et telle que $T_o = 0$, $\lim_n \uparrow T_n = +\infty$. La formule

$$\mathcal{F}_t = \{F : F \in \mathcal{F} \text{ et } \forall n \in \mathbb{N},\ F\{t < T_{n+1}\} \in \mathcal{B}_n \cap \{t < T_{n+1}\}\}$$

où $t \in R_+$ définit alors comme il est facile de le vérifier, une famille croissante et continue à droite de sous-tribus de $\mathcal{F}$. Notons aussi que $\mathcal{F}_o = \mathcal{B}_o$.

En particulier supposons que pour chaque $n \in N^*$, la tribu $\mathcal{B}_n$ soit la tribu engendrée par $\mathcal{B}_o$ et par les variables $T_m\ , \xi_m\ (1 \leqslant m \leqslant n)$ où les "marques" ξ_m sont des variables $\mathcal{F}$-mesurables à valeurs dans un espace mesurable auxiliaire E. Chacune des tribus $\mathcal{F}_t$ ci-dessus est alors la plus petite tribu contenant $\mathcal{B}_o$ et rendant mesurable le processus ponctuel $1_{[0,t]} \cdot N$ si N désigne le processus ponctuel marqué $\sum_{\mathbb{N}^*} \varepsilon_{(T_m, \xi_m)}$. La démonstration est laissée au lecteur.

Identifions pour commencer les tribus optionnelles et prévisibles.

<u>Proposition III.10</u>

<u>Un processus réel</u> X <u>sur</u> R_+ <u>est optionnel, resp. prévisible, par rapport à la famille</u> $(\mathcal{F}_t\ , t \in R_+)$ <u>précédente si et seulement si pour tout</u> $n \in \mathbb{N}$, il <u>existe un processus réel</u> $X^{(n)}$ <u>mesurable sur</u> $\Omega \times R_+$ <u>par rapport à la tribu</u>

$\mathcal{B}_n \otimes \mathcal{R}_+$ <u>et tel que</u>

(6) $X = X^{(n)}$ <u>sur</u> $[\![0, T_{n+1} [\![$ <u>resp. sur</u> $]\!] 0, T_{n+1}]\!]$

<u>et si de plus dans le cas prévisible</u>, $X_o = X(., 0)$ <u>est une variable</u> $\mathcal{B}_o$<u>-mesurable</u>.

<u>Cette condition peut aussi s'exprimer sous la forme équivalente suivante :</u> <u>pour tout</u> $n \in \mathbb{N}$, <u>il existe un processus réel</u> $X'^{(n)}$ $\mathcal{B}_n \otimes \mathcal{R}_+$<u>- mesurable tel que</u>

(6') $$X = \sum_{\mathbb{N}} X'^{(n)} 1_{[\![T_n, T_{n+1} [\![}$$

$$\text{resp. } X = X_o 1_{[\![0]\!]} + \sum_{\mathbb{N}} X'^{(n)} 1_{]\!] T_n, T_{n+1}]\!]}$$

<u>où en outre dans le cas prévisible</u> X_o <u>est</u> $\mathcal{B}_o$<u>-mesurable.</u>

<u>Démonstration</u>

a) Pour établir que ces conditions sont suffisantes, montrons d'abord que T_n est un temps optionnel de la famille $(\mathcal{F}_t, t \in R_+)$ et que $\mathcal{B}_n \subset \mathcal{F}_{T_n}$, quel que soit $n \in \mathbb{N}$. Or, si $n \in \mathbb{N}$ et $B \in \mathcal{B}_n$, pour tous $t \in R_+$ et $p \in \mathbb{N}$, les relations

$$B\{T_n \leqslant t\} \cap \{t < T_{p+1}\} \begin{cases} = \emptyset & \text{si } p < n, \\ \in \mathcal{B}_p \cap \{t < T_{p+1}\} & \text{si } n \leqslant p \end{cases}$$

montrent que $B\{T_n \leqslant t\} \in \mathcal{F}_t$. Pour $B = \Omega$, cela établit que T_n est un temps optionnel et pour B quelconque dans $\mathcal{B}_n$, que $\mathcal{B}_n \subset \mathcal{F}_{T_n}$.

Les propositions 3.2 et 3.6 impliquent alors que pour tout $n \in \mathbb{N}$ et tout processus réel $\mathcal{B}_n \otimes \mathcal{R}_+$ mesurable Z, les processus $Z 1_{[\![T_n, \infty [\![}$ et $Z 1_{]\!] T_n, \infty [\![}$ sont respectivement optionnel et prévisible. Chacune des deux conditions de la proposition est donc suffisante pour que X soit optionnel ou prévisible ; en effet la première condition implique la seconde en prenant simplement $X'^{(n)} = X^{(n)}$ tandis que la seconde condition dans le cas optionnel par exemple, permet d'écrire X sous la forme

$$X = \sum_{\mathbb{N}} (X'^{(n)} 1_{[\![T_n, \infty[\![}) \cdot (1 - 1_{[\![T_{n+1}, \infty[\![})$$

où tous les termes du second membre sont optionnels.

b) Il est un peu plus difficile de démontrer que les conditions de la proposition dans le cas optionnel sont nécessaires. Par définition de la tribu $\mathcal{J}_t$, pour tout $F \in \mathcal{J}_t$ et tout $n \in \mathbb{N}$, il existe au moins un $B \in \mathcal{B}_n$ tel que $F\{t < T_{n+1}\} = B_n\{t < T_{n+1}\}$. Par les arguments habituels, ce résultat s'étend aux processus réels $\mathcal{J}_t \otimes \mathcal{R}_+$ mesurables : pour un tel processus sur R_+, soit Z, il existe au moins un processus $\mathcal{B}_n \otimes \mathcal{R}_+$ mesurable, soit U, tel que $Z = U$ sur $[\![0, T_{n+1}[\![$.

Considérons ensuite un processus réel optionnel positif X et fixons-nous un entier $n \in \mathbb{N}$. D'après la proposition 3.2, le processus $X 1_{[\![0, t]\!]}$ est $\mathcal{J}_t \otimes \mathcal{R}_+$ mesurable sur $\Omega \times R_+$ pour tout $t \in R_+$ et d'après ce que nous venons de dire, il existe des processus réels positifs $\mathcal{B}_n \otimes \mathcal{R}_+$ mesurables U_t $(t \in R_+)$ tels que

$$X(\omega, s) 1_{(s \leq t)} = U_t(\omega, s) \qquad \text{si } t < T_{n+1}(\omega)$$

sur $\Omega \times R_+$. Nous allons montrer qu'il existe alors un processus réel positif $\mathcal{B}_n \otimes \mathcal{R}_+$ mesurable U tel que pour tout $t \in Q_+$ au moins

$$U_t(\omega, s) = U(\omega, s) 1_{(s \leq t)} \qquad \text{si } t < T_{n+1}(\omega) ;$$

cela permettra d'écrire que $X(\omega, s) = U(\omega, s)$ si $s \leq t < T_{n+1}(\omega)$, donc puisque t est arbitraire dans Q_+, si $s < T_{n+1}(\omega)$. Nous aurons ainsi démontré que la première condition de la proposition est nécessaire dans le cas optionnel.

Les U_t que nous venons d'introduire ne varient pas de manière arbitraire avec t : pour tout couple $t < u$ dans R_+, nous avons

$$U_t(\omega, s) = X(\omega, s) 1_{(s \leq t)} = U_u(\omega, s) 1_{(s \leq t)} \qquad \text{si } u < T_{n+1}(\omega).$$

Malheureusement T_{n+1} n'est pas $\mathcal{B}_n$ - mesurable ; aussi pour définir U, nous devrons introduire les ensembles

$$A_u = \{(\omega, s) : U_t(\omega, s) = U_u(\omega, s)\, 1_{(s\leq t)} \text{ pour tout } t \in Q \cap [0, u[\,\}$$

$(u \in Q_+)$. Il est évident que $A_u \in \mathcal{B}_n \otimes \mathcal{R}_+$, que les A_u décroissent en u sur Q_+ et que $A_u \supset \{(\omega, s) : u < T_{n+1}(\omega)\}$. La formule $U = \sup_{Q_+} (U_u 1_{A_u})$ définit alors un processus réel positif $\mathcal{B}_n \otimes \mathcal{R}_+$ qui a les propriétés désirées, comme il est facile de le vérifier.

La première condition de la proposition pour que X soit optionnelle est donc nécessaire ; la seconde l'est donc aussi puisque la première l'implique. Pour éviter des répétitions, nous ne démontrerons la nécessité de la condition de la proposition pour que X soit prévisible que après avoir établi le corollaire suivant. □

Corollaire III.11

Soit $(D_n, n \in N)$ une suite de variables réelles positives adaptée à la suite $(\mathcal{B}_n, n \in N)$ telle que pour tout $n \in \mathbb{N}$:

$$D_n < T_{n+1} \Longleftrightarrow D_{n+1} < T_{n+1} \Longrightarrow D_{n+1} = D_n$$

$$\text{resp. } D_n \leq T_{n+1} \Longleftrightarrow D_{n+1} \leq T_{n+1} \Longrightarrow D_{n+1} = D_n$$

La formule

$$(7) \qquad S = \begin{cases} D_n & \text{si } D_n < T_{n+1}, \text{ resp. } D_n \leq T_{n+1} \qquad (n \in N) \\ +\infty & \text{si } D_n \geq T_{n+1}, \text{ resp. } D_n > T_{n+1} \quad \text{pour tout } n \in N \end{cases}$$

définit alors sans ambiguité un temps optionnel resp. prévisible et tout temps optionnel resp. prévisible est de cette forme.

La tribu $\mathcal{J}_S$ des évènements antérieurs à un temps optionnel S est donnée par chacune des deux formules suivantes :

$$(8) \qquad \begin{aligned} \mathcal{J}_S &= \{F : F \in \mathcal{J} \text{ et } F\{S < T_{n+1}\} \in \mathcal{B}_n \cap \{S < T_{n+1}\} \quad \forall n \in N\} \\ \mathcal{J}_S &= \sum_N \mathcal{B}_n \cap \{T_n \leq S < T_{n+1}\} \end{aligned}$$

<u>La tribu $\mathcal{F}_S^-$ des évènements strictement antérieurs à un temps prévisible est donnée par les formules analogues</u>

$$\mathcal{F}_S^- = \{F : F \in \mathcal{F} , F \cap \{S=0\} \in \mathcal{B}_o \ \underline{\text{et}}\ F\,\{0<S\leq T_{n+1}\} \in \mathcal{B}_n \cap \{0<S\leq T_{n+1}\}\}$$

(8')

$$= \mathcal{B}_o \cap \{S=0\} + \sum_N \mathcal{B}_n \cap \{T_n < S \leq T_{n+1}\} .$$

La construction des temps optionnels et prévisibles que ce corollaire donne peut se comprendre de la manière suivante. Le temps s'écoulant à partir de l'instant 0, attendons le premier des deux instants D_o et T_1. Si D_o se produit avant T_1 (strictement dans le cas optionnel , non strictement dans le cas prévisible), prenons le comme valeur de S. Sinon, ne nous occupons plus de l'instant D_o et attendons le premier des deux instants ultérieurs D_1 et T_2 ; notons que si $D_o \geq T_1$ resp. $D_o > T_1$, par hypothèse $D_1 \geq T_1$ resp. $D_1 > T_1$. Si l'instant D_1 se produit avant T_2 , prenons le pour valeur de S ; sinon attendons le premier des instants ultérieurs D_2 et T_3 et ainsi de suite.

Dans le corollaire précédent, les temps constants t qui sont prévisibles correspondent aux suites $D_n \equiv t$ $(n \in N)$. D'autre part si U est une variable réelle positive $\mathcal{B}_p$ - mesurable telle que $U \geq T_p$, resp. $U > T_p$, sur Ω , en prenant $D_n = +\infty$ si $n < p$ et $D_n = U$ si $n \geq p$, nous trouvons bien que U est un temps optionnel, resp. prévisible. Enfin si S est un temps prévisible le corollaire montre que $\{S = T_{n+1}\} = \{D_n = T_{n+1}\}$ pour tout $n \in N$; en particulier T_{n+1} n'est prévisible que si et seulement s'il est $\mathcal{B}_n$ - mesurable.

<u>Démonstration</u>

a) Si $(D_n, n \in N)$ est une suite de variables satisfaisant aux hypothèses du corollaire, notons d'abord que si $D_n < T_{n+1}$, resp $D_n \leq T_{n+1}$ alors $D_n = D_{n+1} = D_{n+2} = \ldots$; en effet si $D_n < T_{n+1}$ resp. $D_n \leq T_{n+1}$ alors $D_n = D_{n+1}$ et par conséquent $D_{n+1} < T_{n+2}$, ce qui établit une récurrence. Cette remarque montre que la formule définissant S ne présente aucune ambiguité.

Il résulte de cette formule de définition de S que

$$[\![S, \infty [\![\; = \; \sum_N [\![D_n, \infty [\![\cap [\![T_n, T_{n+1} [\![$$

dans le premier cas et que

$$[\![S, \infty [\![\; = \{D_o = 0\} \times \{0\} + \sum_N [\![D_n, \infty [\![\cap]\!] T_n, T_{n+1}]\!]$$

dans le second cas ; d'après la condition suffisante de la proposition III.10, cela établit que S est optionnel dans le premier cas et que S est prévisible dans le second.

b) Réciproquement si S est un temps optionnel, d'après cette même proposition il existe pour tout $n \in N$, un ensemble $A_n \in \mathcal{B}_n \otimes \mathcal{R}_+$ tel que

$$[\![S, \infty [\![\cap [\![0, T_{n+1} [\![\; = A_n \cap [\![0, T_{n+1} [\![\quad \text{dans } \Omega \times R_+ .$$

La formule $D_n(\omega) = \inf (r : r \in Q,\ r > 0 \text{ et } (r, \omega) \in A_n)$ définit alors une variable réelle positive $\mathcal{B}_n$ - mesurable (car $D_n = \inf_{Q_+} U_r$ où $U_r(\omega) = r$ si $(r, \omega) \in A_n$, $= \infty$ sinon) qui étant donné l'égalité précédente vérifie les relations

$$D_n < T_{n+1} \iff S < T_{n+1} \implies D_n = S$$

Les variables D_n $(n \in N)$ vérifient alors les premières hypothèses du corollaire et la formule (7) de ce corollaire donne bien la variable S précédente.

Ceci permet d'achever la démonstration de la proposition III.10 : il restait à montrer qu'un processus réel prévisible X vérifie bien les conditions de cette proposition. Or, par les arguments habituels, il suffit d'après la proposition III.6 (b) de le montrer lorsque X est de la forme $1_{F \times \{0\}}$ avec $F \in \mathcal{F}_o$, ce qui est immédiat, et lorsque X est de la forme $1_{]\!] S,\infty [\![}$ pour un temps optionnel S. Or, dans ce dernier cas, d'après ce que nous venons d'établir à l'alinéa précédent, pour tout $n \in N$

$$]\!] S, \infty [\![\cap [\![0, T_{n+1}]\!] \; =]\!] D_n, \infty [\![\cap [\![0, T_{n+1}]\!] \qquad (n \in N).$$

La première condition de la proposition III.10 pour que $]\!] S,\infty [\![$ soit prévisible est ainsi vérifiée et elle implique la seconde.

c) Soit S un temps prévisible. D'après la proposition III.10, pour tout $n \in N$, il existe un ensemble $A'_n \in \mathcal{B}_n \otimes \mathcal{R}_+$ tel que

$$[\![S, \infty [\![\; \cap \;]\!] 0, T_{n+1}]\!] = A'_n \cap]\!] 0, T_{n+1}]\!] .$$

Comme précédemment, introduisons la variable réelle positive $\mathcal{B}_n$-mesurable $D'_n(\omega) = \inf (r : r \in Q, r > 0 \text{ et } (r, \omega) \in A'_n)$ mais introduisons aussi l'ensemble $B_n = \{\omega : (\omega, D'_n(\omega)) \in A_n\}$ qui appartient à $\mathcal{B}_n$ puisque $1_{B_n}(\omega) = 1_{A'_n}[\omega, D'_n(\omega)]$. L'égalité ci-dessus montre que :

a) sur $\{D'_n < T_{n+1}\} + \{D'_n = T_{n+1}\} \cap B_n$: $S(\omega) = D'_n(\omega)$ et $\omega \in B_n$,

b) sur $\{D'_n > T_{n+1}\} + \{D'_n = T_{n+1}\} \cap B^c_n$: $S(\omega) > T_{n+1}(\omega)$.

Les nouvelles variables $\mathcal{B}_n$-mesurables définies par $D_n = D'_n + 1_{B^c_n}$ jouissent alors des propriétés plus simples suivantes

$$D_n \leq T_{n+1} \iff S \leq T_{n+1} \implies S = D_n .$$

Ces variables vérifient alors les secondes hypothèses du corollaire et la formule (7) de ce corollaire redonne bien la variable S précédente.

d) Il reste à identifier les tribus $\mathcal{J}_S$ et $\mathcal{J}^-_S$. Si S est un temps optionnel et si Z est une variable réelle $\mathcal{J}_S$ mesurable, le processus réel $Z\, 1_{[\![S,\infty [\![}$ est optionnel et il existe donc des processus réels $\mathcal{B}_n \otimes \mathcal{R}_+$-mesurables $X^{(n)}$ $(n \in \mathbb{N})$ tels que

$$Z\, 1_{[\![S, \infty [\![} = \sum_N X^{(n)} \, 1_{[\![T_n, T_{n+1} [\![}$$

Alors, comme $S = D_n$ lorsque $S < T_{n+1}$, nous pouvons écrire que

$$Z = \sum_N X^{(n)}_S \, 1_{\{T_n \leq S < T_{n+1}\}} = \sum_N X^{(n)}_{D_n} \, 1_{\{T_n \leq S < T_{n+1}\}}$$

et puisque $X^{(n)}_{D_n}$ est $\mathcal{B}_n$-mesurable pour tout $n \in N$, nous avons montré que

$$\mathcal{F}_S \subset \sum_{\mathbb{N}} \mathcal{B}_n \cap \{T_n \leq S < T_{n+1}\} .$$ N.III.24

Remarquons que pour les temps optionnels T_n pour lesquels nous avons déjà établi que $\mathcal{B}_n \subset \mathcal{F}_{T_n}$, cela établit que $\mathcal{F}_{T_n} = \mathcal{B}_n$ $(n \in N)$.

Réciproquement si z_n est pour chaque $n \in N$ une variable $\mathcal{B}_n$-mesurable, la formule $z = \sum_N z_n 1_{\{T_n \leq S < T_{n+1}\}}$ définit une variable $\mathcal{F}_S$-mesurable puisqu'elle est la valeur au temps optionnel S du processus réel $\sum_N z_n 1_{[\![T_n, T_{n+1}[\![}$ qui est optionnel.

L'identification de la tribu $\mathcal{F}_S^-$ associée à un temps prévisible S se fait de manière entièrement analogue. □

III.2. PROJECTIONS OPTIONNELLES ET PREVISIBLES

Aux données $[\Omega , \mathfrak{F} ; (\mathfrak{F}_t , t \in R_+)]$ du paragraphe précédent, ajoutons celle d'une probabilité P sur $(\Omega, \mathfrak{F})$, ce qui nous permettra notamment de considérer les espérances conditionnelles $E^{\mathfrak{F}.}$ par rapport aux diverses tribus $\mathfrak{F}_T$ et $\mathfrak{F}_T^-$ introduites dans le premier paragraphe. Il sera commode de supposer comme nous le ferons, que les ensembles négligeables de $\mathfrak{F}$ pour P appartiennent à la sous-tribu $\mathfrak{F}_o$ et donc à toutes les tribus $\mathfrak{F}_T$ et $\mathfrak{F}_T^-$.

Pour toute variable réelle intégrable Z définie sur $(\Omega, \mathfrak{F} , P)$, la théorie des martingales nous apprend que grâce à la continuité à droite de la famille $(\mathfrak{F}_t , t \in R_+)$ il existe un processus réel cadlag M qui soit une version de la martingale $(E^{\mathfrak{F}_t} (Z), t \in R_+)$ et qui vérifie plus précisément les deux familles de relations suivantes

a) pour tout temps optionnel T : $M_T = E^{\mathfrak{F}_T} (Z)$ P p.s.,

b) pour tout temps prévisible S : $M_S^- = E^{\mathfrak{F}_S^-} (Z)$ P p.s.,

si M^- désigne le processus régularisé à gauche de M, soit $M_s^- = M_{s-o}$ si s > 0 et $M_o^- = M_o$.

Ce résultat de théorie des martingales s'étend aux processus de la manière suivante.

Théorème III.12

A tout processus réel positif $X : \Omega \times R_+ \to \overline{R}_+$ mesurable par rapport à la tribu $\mathfrak{F} \otimes \mathcal{R}_+$ sont associés deux processus réels positifs resp. optionnel et prévisible que nous noterons X^o et X^p , tels que

(9)
a) pour tout temps optionnel T : $X_T^o = E^{\mathfrak{F}_T} (X_T)$ P p.s. sur $\{T < \infty\}$

b) pour tout temps prévisible S : $X_S^p = E^{\mathfrak{F}_S^-} (X_S)$ P p.s. sur $\{S < \infty\}$

Ces deux processus X^o et X^p sont caractérisés par ces propriétés à une indistinguabilité près (Deux processus Y et Z sont dits indistinguables si $P^* \{\omega : Y (\omega, .) \neq Z (\omega, .)\}) = 0$).

Démonstration

Si X est de la forme $Z\,1_{[t,\infty[}$ pour un Z $\mathcal{F}$-intégrable et positif et pour un $t \in R_+$, les résultats de théorie des martingales rappelés ci-dessus entraînent que les processus $X^o = M\,1_{[\![t,\infty[\![}$ et $X^p = M^-\,1_{[\![t,\infty[\![}$ ont les propriétés a et b respectivement. L'existence dans le cas général de processus X^o et X^p se démontre alors par les arguments habituels.

L'unicité de X^o et X^p sont des conséquences des théorèmes de sections : deux processus optionnels resp. prévisibles sont indistinguables d'après ces théorèmes si et seulement s'ils sont égaux en chaque temps optionnel resp. prévisible. □

Les opérations de projections optionnelle et prévisible ont évidemment les propriétés habituelles des espérances conditionnelles. En nous limitant toujours à ne considérer que des processus réels positifs, nous pouvons écrire en notant $(\ )^{\bullet}$ l'une ou l'autre des deux projections :

1) $(a_1 X_1 + a_2 X_2)^{\bullet} = a_1 X_1^{\bullet} + a_2 X_2^{\bullet}$ $\qquad (a_1, a_2 \geq 0)$,

2) $X_1^{\bullet} \leq X_2^{\bullet}$ si $X_1 \leq X_2$,

3) pour toute suite croissante $(X_n, n \in N)$: $(\lim_n \uparrow X_n)^{\bullet} = \lim_n \uparrow X_n^{\bullet}$ et donc pour toute suite $(X_n, n \in N)$:

$$(\sum_N X_n)^{\bullet} = \sum_N X_n^{\bullet}, \quad (\liminf_{n\to\infty} X_n)^{\bullet} \leq \liminf_{n\to\infty} X_n^{\bullet}.$$

En outre quel que soit le processus réel positif optionnel, resp. prévisible, Z :

(10) $\qquad (X Z)^o = X^o Z$ resp. $(X Z)^p = X^p Z$;

ceci entraîne en particulier que si T est un temps optionnel et S un temps prévisible :

$$(X\,1_{[\![0,T]\!]})^o = X^o\,1_{[\![0,T]\!]}, \quad (X\,1_{[\![0,T]\!]})^p = X^p\,1_{[\![0,T]\!]},$$

$$(X\,1_{[\![0,T[\![})^o = X^o\,1_{[\![0,T[\![}, \quad (X\,1_{[\![0,S[\![})^p = X^p\,1_{[\![0,S[\![}.$$

Remarquons aussi que pour tout processus réel positif $\mathcal{F} \otimes \mathcal{R}_+$ mesurable X

(11) $(X^o)^p = X^p$

Cela résulte en effet immédiatement des inclusions $\mathfrak{F}_S^- \subset \mathfrak{F}_S$ et de l'unicité de la projection prévisible puisque pour tout temps prévisible (donc optionnel) S nous avons

$$(X^o)^p{}_S = E^{\mathfrak{F}_S^-}[X^o{}_S] = E^{\mathfrak{F}_S^-} E^{\mathfrak{F}_S} X_S = E^{\mathfrak{F}_S^-} X_S = X^p{}_S .$$

Exemple

Comment s'expriment les projections optionnelle et prévisible dans le cas de l'exemple de la fin du dernier paragraphe ? Comme nous allons le voir, dans le cas de cet exemple, il n'est pas nécessaire d'invoquer le théorème général précédent pour construire ces projections.

Rappelons d'abord le lemme suivant, qui n'est que le cas particulier du théorème précédent où la famille $(\mathfrak{F}_t, t \in R_+)$ est constante.

Lemme III.13

Si $\mathcal{B}$ est une sous-tribu de $\mathfrak{F}$ dans l'espace de probabilité $(\Omega, \mathfrak{F}, P)$, pour tout processus réel positif X $\mathfrak{F} \otimes \mathcal{R}_+$- mesurable il existe un processus réel positif Y $\mathcal{B} \otimes \mathcal{R}_+$ -mesurable tel que $Y_D = E^{\mathcal{B}}(X_D)$ p.s. sur $\{D < \infty\}$ pour tout temps D : $\Omega \to \overline{R}_+$ $\mathcal{B}$-mesurable. Nous noterons $\overline{E}^{\mathcal{B}} X$ ce processus Y qui est unique à une indistinguabilité près.

Démonstration

Si X est de la forme $1_{F \times [t,\infty[}$ pour un $F \in \mathfrak{F}$ et un $t \in R_+$, le processus $Y = E^{\mathcal{B}}(1_F) \cdot 1_{[t,\infty[}$ convient car $E^{\mathcal{B}}(X_D) = E^{\mathcal{B}}(1_F\, 1_{t\leq D}) = E^{\mathcal{B}}(1_F)\, 1_{(t\leq D)}$ $= Y_D$ si D est un temps $\mathcal{B}$-mesurable. Par les arguments habituels, ce résultat d'existence s'étend à tous les processus réels positifs X. L'unicité de Y s'obtient par le théorème de section de Von Neumann : pour tout $B \in \mathcal{B} \otimes \mathcal{R}_+$ il existe au moins un temps $\mathcal{B}$-mesurable D tel que $P[\omega : (\omega, D(\omega)) \in B] = P^*[\text{proj}(B)]$. □

Considérons donc la famille $(\mathcal{F}_t\ ,\ t \in R_+)$ associée aux deux suites $(\mathcal{B}_n,\ n \in N)$ et $(T_n,\ n \in N)$ de l'exemple cité ; nous supposerons en outre que les ensembles négligeables de $\mathcal{F}$ appartiennent à $\mathcal{B}_o = \mathcal{F}_o$. Pour tout $n \in N$, il existe un processus réel à valeurs dans $[0,\ 1]$, décroissant, cadlag et $\mathcal{B}_n \otimes \mathcal{R}_+$ mesurable, soit π_n , tel que

$$\pi_n\ (.,\ t) = P^{\mathcal{B}_n}\ (t < T_{n+1})\ \text{p.s.} \qquad (n \in N,\ t \in R_+)$$

(il suffit par exemple de poser $\pi_n\ (\omega,\ t) = \sup\ (Z_r\ (\omega)\ ;\ r \in Q,\ r > t)$ où les variables $Z_r\ (r \in Q_+)$ à valeurs dans $[0,\ 1]$ sont dans les classes d'équivalence $P^{\mathcal{B}_n}\ (r < T_{n+1})$) ; notons que les régularisés à gauche de ces processus valent

$$\pi_n^-\ (t) = \pi_n\ (t\text{-}o) = P^{\mathcal{B}_n}\ (t \leqslant T_{n+1}) \qquad (n \in N,\ t \in R_+)$$

(pour $t = 0$: $\pi_n^-\ (0) = \pi_n\ (0) = 1$ p.s. puisque $T_{n+1} > 0$).

Voici alors les formules donnant les projections optionnelle et prévisible d'un processus réel positif $\mathcal{F}\otimes\mathcal{R}_+$- mesurable quelconque X :

$$X^o = \sum_N \frac{1}{\pi_n}\ \overline{E}^{\mathcal{B}_n}\ (X\ 1_{[\![T_n,\ T_{n+1}[\![})\ .\ 1_{[\![T_n,\ T_{n+1}[\![}$$

(12)

$$X^p = E^{\mathcal{B}_o}\ (X_o)\ 1_{[\![0]\!]}\ + \sum_N \frac{1}{\pi_n^-}\ \overline{E}^{\mathcal{B}_n}\ (X\ 1_{]\!]T_n,T_{n+1}]\!]}),\ 1_{]\!]T_n,T_{n+1}]\!]}.$$

Pour démontrer la première de ces deux formules par exemple, calculons la valeur du processus X^o défini par cette formule en un temps optionnel S. D'après le corollaire 3.9, il existe des temps $\mathcal{B}_n$-mesurables $D_n\ (n \in N)$ tels que $S = D_n$ sur $\{S < T_{n+1}\} = \{D_n < T_{n+1}\}$. Le lemme III.13 nous permet alors d'écrire que :

$$X_S^o = \sum_N \frac{1}{\pi_n\ (D_n)}\ \overline{E}^{\mathcal{B}_n}\ (X\ 1_{[\![T_n,T_{n+1}[\![})\ (D_n) \quad 1_{\{T_n \leqslant S < T_{n+1}\}}$$

$$= \sum_N \frac{E^{\mathcal{B}_n}\ [X_{D_n}\ 1_{\{T_n \leqslant D_n < T_{n+1}\}}]}{P^{\mathcal{B}_n}\ (T_n \leqslant D_n < T_{n+1})} \quad 1_{\{T_n \leqslant S < T_{n+1}\}}$$

$$= \sum_N \frac{E^{\mathcal{B}_n}\left[X_S\, 1_{\{T_n \leqslant S < T_{n+1}\}}\right]}{P^{\mathcal{B}_n}(T_n \leqslant S < T_{n+1})}\, 1_{\{T_n \leqslant S < T_{n+1}\}}.$$

Mais comme $\mathcal{F}_S = \sum_N \mathcal{B}_n \cap \{T_n \leqslant S < T_{n+1}\}$, le dernier membre vaut encore l'espérance conditionnelle $E^{\mathcal{F}_S}(X_S)$ [le démontrer !] et cela établit puisque S est un temps optionnel arbitraire, que X^o est bien la projection optionnelle de X. La démonstration de la formule donnant X^p est entièrement analogue.□

Avant d'étudier les projections de mesures aléatoires, introduisons la notion d'accessibilité d'un temps optionnel qui nous sera utile dans la suite

Définition III.14

Un temps optionnel T est dit accessible s'il existe une suite de temps prévisibles S_n , deux à deux disjoints, c'est-à-dire tels que $\{S_m = S_n < \infty\} = \emptyset$ si $m \neq n$, pour lesquels

(13) $\{T < \infty\} = \sum_n \{S_n = T < \infty\}$.

Un temps optionnel T est dit totalement inaccessible si pour tout temps prévisible S : $P(T = S) = 0$.

Il est immédiat de vérifier sur la définition de la projection prévisible d'un processus réel que pour tout temps optionnel T les trois conditions suivantes sont équivalentes :

a) T est totalement inaccessible,

b) $(1_{[\![T]\!]})^p = 0$,

c) $(1_{[\![T,\infty[\![})^p = 1_{]\!]T,\infty[\![}$.

Dans le cas où la famille $(\mathcal{F}_t, t \in R_+)$ est quasi continue à gauche, un temps optionnel est accessible si et seulement s'il est prévisible. En effet si T est accessible au sens de la définition précédente

$$1_{[\![T,\infty[\![} = \sum_n 1_{\{T = S_n\}}\, 1_{[\![S_n,\infty[\![}$$

et le deuxième membre est prévisible d'après la proposition III.6 puisque $\{T = S_n\} \in \mathfrak{F}_{S_n} = \mathfrak{F}^-_{S_n}$ pour tout n.

Lemme III.15

A tout temps optionnel T est associée une partie essentiellement unique A_T de $\{T < \infty\}$ dans $\mathfrak{F}_T$ telle que les temps optionnels

(14) $$T'(\omega) = \begin{cases} T(\omega) & \text{si } \omega \in A_T \\ +\infty & \text{sinon} \end{cases}, \quad T''(\omega) = \begin{cases} T(\omega) & \text{si } \omega \in \{T < \infty\} - A_T \\ +\infty & \text{sinon} \end{cases}$$

soient respectivement accessible et totalement inaccessible.

Démonstration

Considérons la borne supérieure essentielle $B = \text{ess sup } \{T = S < \infty\}$ où S parcourt les temps prévisibles ; comme toute borne supérieure essentielle, B est la classe d'équivalence d'une réunion dénombrable $A = \bigcup_n \{T = S_n < \infty\}$ (S_n prévisibles) qui a les propriétés de l'ensemble A_T du lemme. En effet il est clair que pour tout temps prévisible S : $\{T = S < \infty\} \subset A$ p.s. et donc que $P(T'' = S < \infty) = 0$; autrement dit T" est totalement inaccessible. D'autre part posons

$$S'_n = S_n \text{ sur } \{S_m \neq S_n < \infty \ \forall m < n\}, \ = +\infty \text{ ailleurs}$$

Les S'_n sont des temps prévisibles (noter que $\{S_m \neq S_n < \infty\} \in \mathfrak{F}^-_{S_n}$ car $1_{[\![S_n]\!]} \in \mathcal{P}$ si bien que $1_{\{S_m = S_n < \infty\}} = 1_{[\![S_m]\!]}(S_n)$ est $\mathfrak{F}^-_{S_n}$ mesurable), les S'_n sont deux à deux disjoints par construction et enfin $\sum_n \{T = S'_n < \infty\} = A$; il s'ensuit bien que T' est accessible.□

Ce lemme permet de donner un critère simple pour qu'une mesure aléatoire soit prévisible à une équivalence près.

Proposition III.16

Une mesure optionnelle positive est presque sûrement égale à une mesure prévisible positive si et seulement si

a) $N(\{T\}) = 0$ p.s. <u>pour tout temps optionnel totalement inaccessible</u> T,

b) $N(\{T\})$ est $\mathcal{F}_T^-$ - <u>mesurable pour tout temps prévisible</u> T.

<u>Démonstration</u>

Si N est une mesure optionnelle positive de primitive A, le processus ΔA défini par $\Delta A_t = A_t - A_{t-o} = N(\{t\})$ $(t \in R_+ , A_{o-} = 0)$ est optionnel comme différence du processus optionnel A et du processus prévisible A^-. Pour tout $\varepsilon > 0$, la mesure aléatoire positive $N^\varepsilon = 1_{\{\Delta A \geqslant \varepsilon\}} . N$ formée par les sauts $\geqslant \varepsilon$ de N est donc optionnelle ; comme elle s'écrit $\sum_{N^*} Z_n \, \varepsilon_{T_n}$ pour une suite de T_n croissant vers $+\infty$ et des Z_n positifs tels que $\{Z_n > 0\} = \{T_n < \infty\}$, les T_n sont des temps optionnels et les Z_n sont des variables $\mathcal{F}_{T_n}$ mesurables. Mais $N(\{T_n\}) = Z_n > 0$ sur $\{T_n < \infty\}$; l'hypothèse (a) impose donc à la partie totalement inaccessible de T_n d'être p.s. égale à $+\infty$; quitte à modifier les T_n sur des ensembles négligeables nous pourrons donc les supposer accessibles.

Soient S_n^k des temps prévisibles 2 à 2 disjoints lorsque k varie, tels que $\{T_n < \infty\} = \sum_k \{T_n = S_n^k < \infty\}$ et soit $(S'_p , p \in N)$ une suite de temps prévisibles deux à deux disjoints telle que

$$\sum_p [\![S'_p]\!] = \bigcup_n \sum_k [\![S_n^k]\!] .$$

Il est alors clair que

$$N^\varepsilon \underset{\text{p.s.}}{=} \sum_p Y_p \, \varepsilon_{S'_p} \qquad \text{avec} \qquad Y_p = \Delta A_{S'_p} \, 1_{\{\Delta A S'_p \geqslant \varepsilon\}} .$$

Puisque les S'_p sont des temps prévisibles, l'hypothèse (b) entraîne que $\Delta A_{S'_p}$ et donc que Y_p sont $\mathcal{F}_{S'_p}^-$ mesurables ; mais la mesure aléatoire $\sum_p Y_p \, \varepsilon_{S'_p}$ est alors prévisible.

En faisant tendre $\varepsilon \downarrow 0$, nous trouvons que la mesure aléatoire positive $N^* = \lim_{\varepsilon \downarrow 0} \uparrow N^\varepsilon$ est presque sûrement égale à une mesure prévisible et positive. Mais d'autre part la mesure aléatoire positive

$$N - N^* = N - 1_{(\Delta A > 0)} . N$$

est optionnelle et diffuse, donc prévisible. Il s'en suit bien que la mesure aléatoire $N = N^* + (N - N^*)$ est p.s. égale à une mesure prévisible sous les hypothèses de la proposition.

Nous démontrerons que les conditions de la proposition sont nécessaires pour que N soit à une équivalence près une mesure prévisible, en même temps que la proposition III.17 ci-dessous.□

Dans le cas où la famille $(\mathcal{F}_t, t \in R_+)$ est quasi-continue à gauche, la condition (b) peut être omise dans la proposition précédente. Cette condition est en effet vérifiée dès que N optionnelle (puisque $N(\{T\}) = \Delta A_T$ est $\mathcal{F}_T$-mesurable pour tout T optionnel). D'autre part la démonstration précédente se simplifie un peu puisque les T_n de cette démonstration, modifiés pour devenir accessibles, seront alors déjà prévisibles et qu'il suffira donc de définir les S'_n comme ces T_n modifiés.

La proposition suivante simple, bien qu'un peu longue à démontrer, constitue le résultat technique fondamental pour la définition des projections optionnelle et prévisible de mesures.

Proposition III.17

Si $N : \Omega \to M_+(R_+)$ est une mesure aléatoire positive sur R_+, la mesure positive sur $(\Omega \times R_+, \mathcal{F} \otimes \mathcal{R}_+)$ qui lui est associée par la formule $\nu(d\omega\ dt) = P(d\omega)\, N(\omega, dt)$ jouit des propriétés suivantes :

a) $\nu(A) = 0$ pour tout $A \in \mathcal{F} \otimes \mathcal{R}_+$ tel que $P^*[\text{proj}_\Omega(A)] = 0$,

b) $\nu([\![0, T_n[\![) < \infty$ pour une suite croissante de temps $\mathcal{F}$-mesurables $T_n : \Omega \to \overline{R}_+$ tels que $\lim_n \uparrow T_n = +\infty$.

Réciproquement toute mesure positive ν sur $\Omega \times R_+$ qui vérifie les conditions a et b est de la forme précédente pour une mesure aléatoire positive essentiellement unique N.

De plus pour que la mesure aléatoire N soit optionnelle resp. soit presque sûrement égale à une mesure prévisible, il faut et il suffit que pour tout processus réel positif $\mathcal{I} \otimes \mathcal{R}_+$ -mesurable X

(15) $\nu(X) = \nu(X^o)$, resp. $\nu(X) = \nu(X^p)$.

Lorsque cette condition est satisfaite, la mesure ν est dite optionnelle resp. prévisible et les temps T_n ($n \in N$) de la condition (b) peuvent être pris optionnels, resp. prévisibles.

Démonstration

La mesure positive $\nu(d\omega\, dt) = P(d\omega)\, N(\omega, dt)$ associée à une mesure aléatoire positive N jouit de la propriété (a) car pour tout $A \in \mathcal{I} \otimes \mathcal{R}_+$, $\nu(A)$ est l'espérance de la variable positive $Z = \int N(., dt)\, 1_A(., t)$ qui est telle que $\{Z \neq 0\} \in \mathcal{I}$ soit contenu dans $\mathrm{proj}_\Omega(A)$. Ensuite pour tout $a \in R_+$, posons $T_a = \min(t : A_t \geq a)$ en désignant par A le processus réel croissant cadlag $A_t = N([0, t])$; comme dans $\Omega \times R_+$

$$[\![T_a, \infty[\![= \{A \geq a\} \in \mathcal{I} \otimes \mathcal{R}_+ \text{, resp. } \mathcal{O} \text{, resp. } \mathcal{P}$$

suivant que N est une mesure aléatoire positive générale, resp. optionnelle, resp. prévisible, nous voyons que les T_a sont des temps $\mathcal{I}$-mesurables, optionnels ou prévisibles . Mais d'autre part $N([0, T_a[) \leq a$, si bien que la condition (b) et ses variantes sont établies. De plus la formule élémentaire

$$\int_{R_+} X(\omega, t)\, N(\omega, dt) = \int_{R_+} X_{T_a}(\omega)\, 1_{(T_a(\omega) < \infty)}\, da \qquad (\omega \in \Omega)$$

est valable pour tout processus réel positif $\mathcal{I} \otimes \mathcal{R}_+$-mesurable X et entraîne que pour un tel processus

$$\nu(X) = \int_{R_+} E\left[X_{T_a}\, 1_{(T_a < \infty)}\right] da$$

Si N est optionnel, resp. prévisible, les temps T_a sont optionnels resp. prévisibles de sorte que le second membre ne change pas si X est remplacé par sa projection optionnelle X^o , resp. par sa projection prévisible X^p ; donc $\nu(X) = \nu(X^o)$ resp. $\nu(X) = \nu(X^p)$ suivant le cas. Les parties directes de la proposition sont ainsi établies.

Réciproquement si ν est une mesure positive sur $\mathcal{F} \otimes \mathcal{R}_+$ vérifiant les conditions a et b de la proposition, il existe un processus réel positif et fini, croissant et cadlag, soit A tel que $\nu(. \times [0, t]) = A_t . P$ sur $(\Omega, \mathcal{F})$ pour tout $t \in R_+$. En effet les hypothèses montrent que pour tout $t \in R_+$ fixé, la mesure positive $\nu(. \times [0, t])$ est nulle sur les ensembles P-négligeables de $\mathcal{F}$ et finie sur les ensembles $\{t < T_n\}$ qui croissent vers Ω lorsque $n \uparrow \infty$. D'après le théorème de Radon-Nikodym, cette mesure est donc de la forme $A'_t . P$ pour une variable $\mathcal{F}$- mesurable positive et finie A'_t ; de plus $A'_u \geq A'_t$ p.s. si $u \geq t$. Le processus réel positif A défini par $A_t(\omega) = \inf(A'_u(\omega)$; $u \in Q, u > t)$ est alors croissant et cadlag. De plus pour tout $F \in \mathcal{F}$, l'intégrale $\int_F A_t \, dP$ est la limite décroissante des intégrales $\int_F A'_u \, dP$ lorsque $u \in Q$, $u > t$ et $u \to t$, ce qui entraîne qu'elle vaut $\nu(F \times [0, t])$; ainsi $\nu(. \times [0, t]) = A_t . P$ pour tout $t \in R_+$.

Comme pour tout $\omega \in \Omega$, la fonction réelle positive $A.(\omega)$ est croissante et cadlag sur R_+ il existe une mesure positive $N(\omega, .)$ unique sur $\mathcal{R}_+$ telle que $A_t(\omega) = N(\omega, [0, t])$ $(t \in R_+)$. Il est alors facile de vérifier que N est une mesure aléatoire telle que $\nu(d\omega\, dt) = P(d\omega)\, N(\omega, dt)$ sur $\mathcal{F} \otimes \mathcal{R}_+$. L'unicité de la mesure N ainsi construite à une équivalence près se démontre aussi aisément : si N' est une autre mesure aléatoire positive de mesure associée ν, son processus croissant A' vérifie $A'_t = A_t$ p.s. pour tout $t \in R_+$; comme deux fonctions cadlag coïncident sur R_+ dès qu'elles sont égales sur Q_+, il s'en suit que

$$\{\omega : N'(\omega, .) \neq N(\omega, .)\} = \{\omega : A'(\omega, .) \neq A(\omega, .)\} = \bigcup_{t \in Q_+} \{A'_t \neq A_t\}$$

est négligeable.

Supposons maintenant que la mesure positive ν vérifie l'égalité $\nu(X) = \nu(X^o)$ pour tout processus réel positif $\mathcal{F} \otimes \mathcal{R}_+$ mesurable X. Pour tout temps optionnel T, la projection optionnelle du processus $Y\, 1_{[\![0, T]\!]}$ étant la même que celle du processus $E^{\mathcal{F}_T}(Y)\, 1_{[\![0, T]\!]}$ quelle que soit la variable positive $\mathcal{F}$- mesurable Y [utiliser les relations (4)]. Nous trouvons que

$$E\,(Y\,A_T) = \nu\,(Y\,1_{|0,\,T|}) = \nu\,(E^{\mathcal{F}_T}\,Y\,.\,1_{[\![0,\,T]\!]}) = E\left[E^{\mathcal{F}_T}\,Y\,.\,A_T\right]$$

grâce à l'hypothèse faite sur ν. L'égalité des membres extrêmes quelle que soit Y n'est possible que si A_T est $\mathcal{F}_T$ - mesurable, compte tenu de ce que $\mathcal{F}_T$ est complète dans $\mathcal{F}$. Pour les temps T constants cela montre que A est adapté et donc que N est optionnelle.

Supposons enfin que la mesure positive ν vérifie l'égalité $\nu\,(X) = \nu\,(X^p)$ pour tout processus réel positif $\mathcal{F} \otimes \mathcal{R}_+$ mesurable X. En vertu de l'égalité des projections prévisibles de X et X^o (formule 11), la mesure ν vérifie en particulier l'hypothèse de l'alinéa précédent et la mesure aléatoire N est donc optionnelle. Pour montrer que N est équivalente à une mesure prévisible, appliquons la proposition III.16. Ses hypothèses sont vérifiées car d'une part pour tout temps optionnel totalement inaccessible T : $(1_{[\![T]\!]})^p = 0$ puisque $1_{[\![T]\!]}(S) = 1_{\{T = S < \infty\}} = 0$ p.s. pour tout temps prévisible S, de sorte que $E\,(N\,(\{T\})) = \nu\,(1_{[\![T]\!]}) = 0$. D'autre part pour tout temps prévisible T, la projection prévisible du processus $Y\,1_{[\![0,\,T]\!]}$ est la même que celle du processus $E^{\mathcal{F}_T^-}(Y)\,1_{[\![0,\,T]\!]}$, quelle que soit la variable positive $\mathcal{F}$- mesurable Y [utiliser les relations (4)] ; donc en raisonnant comme ci-dessus dans le cas optionnel, nous trouverons que A_T est $\mathcal{F}_T^-$ - mesurable. Mais le processus A^- étant prévisible, la variable A_T^- est toujours $\mathcal{F}_T^-$ mesurable de sorte que nous avons montré que $\Delta\,A_T$ est $\mathcal{F}_T^-$ - mesurable.

La proposition III.17 est ainsi complètement démontrée. Remarquons que nous avons aussi établi la condition nécessaire de la proposition III.16 : en effet, si N est équivalente à une mesure prévisible, la première partie de la démonstration montre que $\nu\,(X) = \nu\,(X^p)$ pour tout processus réel positif mesurable X et la dernière partie de la démonstration montre que les conditions (a-b) de la proposition III.16 sont alors satisfaites. □

Le deuxième théorème fondamental de ce paragraphe se déduit immédiatement de la proposition précédente.

Théorème III.18

Soit N une mesure aléatoire positive sur R_+ pour laquelle il existe une suite croissante de temps optionnels T_n $(n \in \mathbb{N})$ telle que $\lim_n \uparrow T_n = \infty$ et que

$$E\,(N\,([\![0, T_n[\![)) < \infty \qquad \text{resp. } E\,(N\,([\![0, T_n]\!])) < \infty \quad .$$

Il existe alors une mesure optionnelle, resp. prévisible, positive, unique à une équivalence près que nous noterons N^o, resp. N^p et qui vérifie les égalités

$$(16) \quad E\left[\int_{R_+} X \, d\, N^o\right] = E\left[\int_{R_+} X^o \, d\, N\right] \text{ resp. } \left[E\int_{R_+} X\, d\, N^p\right] = \left[E\int_{R_+} X^p\, dN\right]$$

pour tout processus réel positif $\mathcal{J} \otimes \mathcal{R}_+$ - mesurable X.

La mesure aléatoire N^o s'appelle la projection optionnelle de N et la mesure aléatoire N^p s'appelle la projection prévisible de N. Lorsque N est elle-même optionnelle, sa projection prévisible est aussi appelée sa compensatrice.

Démonstration

Soit ν la mesure positive sur $\Omega \times R_+$ associée à N par $\nu\,(d\omega\, dt) = P\,(d\omega)\, N\,(\omega, dt)$. Les propriétés des projections optionnelle et prévisible, leur σ-additivité notamment, montrent que les formules $\nu^o\,(A) = \nu(1_A^o)$, $\nu^p\,(A) = \nu\,(1_A^p)$ $(A \in \mathcal{J} \otimes \mathcal{R}_+)$ définissent deux mesures positives sur $\Omega \times R_+$ telles que plus généralement

$$\nu^o\,(X) = \nu\,(X^o)\ , \quad \nu^p\,(X) = \nu\,(X^p)$$

pour tout processus réel positif et mesurable X. Il s'ensuit évidemment que $\nu^o\,(X) = \nu^o\,(X^o)$ et que $\nu^p\,(X) = \nu^p\,(X^p)$ pour tout X. D'autre part ces mesures ν^o et ν^p vérifient les conditions a - b de la proposition III.17. En effet d'une part si $A \in \mathcal{J} \otimes \mathcal{R}_+$ est tel que $P^*\,(\text{proj}_\Omega\, A) = 0$, les projections optionnelle et prévisible de 1_A sont nulles et $\nu^o\,(A) = 0 = \nu^p\,(A)$. D'autre part pour tout temps optionnel $\nu^o\,([\![0, T[\![) = \nu\,([\![0, T[\![)$ et

ν^p (⟦0, T⟧) = ν (⟦0, T⟧) car ⟦0, T⟦ et ⟦0, T⟧ sont respectivement optionnel et prévisible ; les hypothèses du théorème entraînent donc la validité de la condition (b) de la proposition III.17. L'application de cette proposition permet alors de conclure. □

Les projections optionnelle et prévisible des mesures aléatoires ne doivent évidemment pas être confondues avec celles des processus ; aussi, compte tenu des formules (16), sont elles souvent appelées projections optionnelle et prévisible <u>duales</u>. Néanmoins pour tout processus réel Z positif, $\mathcal{F} \otimes \mathcal{R}_+$ mesurable et localement intégrable par rapport à la mesure de Lebesgue λ sur R_+, nous avons les formules

(17) $$(Z \,.\, \lambda)^o = Z^o \,.\, \lambda \qquad , \qquad (Z \,.\, \lambda)^p = Z^p \,.\, \lambda$$

où les projections sont des projections de mesures dans les membres de gauche et des projections de processus dans les membres de droite. Là vérification de cette propriété est facile et d'ailleurs le lemme suivant (alinéa b) montre que cette propriété reste vraie lorsque la mesure de Lebesgue est remplacée par une mesure positive, optionnelle ou prévisible suivant le cas.

<u>LEMME III.19</u>

<u>Soient</u> N <u>une mesure aléatoire positive et</u> Y <u>un processus réel mesurable positif tels que</u>

$$E\left[\int_{[0,\, T_n[} (1 + Y_t)\, d\, N\, (t)\right] < \infty$$

<u>pour une suite de temps optionnels</u> T_n $(n \in \mathbb{N})$ <u>croissant vers</u> $+\infty$. <u>Alors</u>

(18) a) $(Y \,.\, N)^o = Y \,.\, N^o$ <u>si le processus</u> Y <u>est optionnel,</u>

b) $(Y \,.\, N)^o = Y^o \,.\, N$ <u>si la mesure</u> N <u>est optionnelle.</u>

<u>Des résultats analogues sont valables dans le cas prévisible à condition dans l'hypothèse de remplacer</u> $[0, T_n[$ par $[0, T_n]$.

Démonstration

Si le processus Y est optionnel, tout processus réel mesurable positif X est tel que $(X\,Y)^o = X^o\,Y$ et alors la formule

$$E\left[\int_{R_+} X^o \, d\,(Y\,.\,N)\right] = E\left[\int_{R_+} (X\,Y)^o \, d\,N\right] = E\left[\int_{R_+} X\,Y\,d\,N^o\right]$$

établit bien que $(Y\,.\,N)^o = Y\,.\,N^o$.

Si N est optionnelle, si A est sa primitive et si $T_a = \inf\,(t : A_t \geqslant a)$ les temps T_a sont optionnels puisque $[\![T_a, \infty[\![\; = \{A \geqslant a\}$ de sorte que $E\,(X^o_{T_a}\; Y_{T_a}\; 1_{(T_a<\infty)}) = E\,(X_{T_a}\; Y^o_{T_a}\; 1_{(T_a<\infty)})$ pour tout $a \in R_+$ puisque $X^o_{T_a} = E^{\mathcal{F}_{T_a}}\,(X_{T_a})$ sur $\{T_a < \infty\}$. Cela entraîne que

$$E\left[\int_{R_+} X^o \, d\,(Y\,N)\right] = \int_{R_+} \lambda\,(da)\; E\,(X^o_{T_a}\; Y_{T_a}\; 1_{(T_a<\infty)})$$

$$= \int_{R_+} \lambda\,(da)\; E\,(X_{T_a}\; Y^o_{T_a}\; 1_{(T_a<\infty)}) = E\left(\int_{R_+} X\,Y^o\,dN\right)$$

pour tout X et montre donc que $(Y\,.\,N)^o = Y^o\,.\,N$.

Les démonstrations dans le cas prévisible se font de la même manière. □

Voici un autre lien entre les projections optionnelles ou prévisibles de mesures aléatoires qui facilite certains calculs.

Proposition III.20

Etant donné un temps $T : \Omega \to \overline{R}_+$ $\mathcal{F}$ - mesurable et une variable positive $\mathcal{F}$ - mesurable Z, la projection optionnelle, resp. prévisible, du processus réel positif $X = Z\,1_{[\![T]\!]}$ s'exprime à partir de celle de la mesure aléatoire positive $N = Z\,\varepsilon_T$ par la formule

$$X^o\,(\omega, t) = N^o\,(\omega, \{t\}) \text{ , resp. } X^p\,(\omega, t) = N^p\,(\omega, \{t\}).$$

En particulier un temps optionnel T est totalement inaccessible si et seulement si la compensatrice N^p de la mesure aléatoire $N = \varepsilon_T$ est diffuse.

Démonstration

Si S est un temps optionnel, resp. prévisible et si Y est une variable positive $\mathcal{F}_S$ - mesurable, resp. $\mathcal{F}_S^-$ mesurable, la formule

$$E\left[\int Y\, 1_{[\![S]\!]}\, dN\right] = E\left[\int Y\, 1_{[\![S]\!]}\, dN^q\right] \text{ s'écrit ici}$$

$E\left[Y\, Z\, 1_{S=T}\right] = E\left[Y\, N^q\, (\{S\})\right]$ (q = 0 resp. p), et implique donc que :

$$E^{\mathcal{F}_S}\left[Z\, 1_{[\![T]\!]}(S)\right] = E^{\mathcal{F}_S}\left[N^o\, (\{S\})\right] \text{ pour tout temps optionnel S,}$$

$$E^{\mathcal{F}_S^-}\left[Z\, 1_{[\![T]\!]}(S)\right] = E^{\mathcal{F}_S^-}\left[N^p\, (\{S\})\right] \text{ pour tout temps prévisible S.}$$

Mais $N^o\,(\omega, \{t\}) = \Delta A_t^o\,(\omega)$ est un processus optionnel de sorte que $N^o\,(\{S\})$ est $\mathcal{F}_S$ - mesurable pour tout temps optionnel ; d'après la définition de la projection optionnelle d'un processus réel il s'ensuit bien que $(Z\, 1_{[\![T]\!]})^o = N^o\,(\omega, \{t\})$. De même $N^p\,(\omega, \{t\}) = \Delta A_t^p\,(\omega)$ est un processus prévisible de sorte que $N^p\,(\{S\})$ est $\mathcal{F}_S^-$ - mesurable pour tout S prévisible (ou même seulement optionnel) et il s'ensuit aussi que $(Z\, 1_{[\![T]\!]})^p = N^p\,(\omega, \{t\})$.

Pour qu'un temps optionnel T soit totalement inaccessible, nous avons remarqué qu'il était nécessaire et suffisant que $(1_{[\![T]\!]})^p = 0$; or cette projection prévisible est égale pour tout temps optionnel au processus des sauts de la mesure N^p si $N = \varepsilon_T$. $\square$

Exemple

Il n'est pas difficile dans le cas de l'exemple de la fin du premier paragraphe de trouver les projections optionnelle et prévisible de mesures aléatoires. Comme dans l'étude des projections de processus nous commencerons par énoncer un lemme qui n'est qu'un cas particulier de la proposition III.17.

Lemme III.21

Si N est une mesure aléatoire sur R_+ définie sur l'espace de probabilité $(\Omega, \mathcal{F}, P)$, si $\mathcal{B}$ est une sous-tribu de $\mathcal{F}$ et s'il existe une suite croissante de temps $\mathcal{B}$- mesurables U_n $(n \in N)$ croissant vers $+\infty$, tels que $E\,(N\,([0, U_n[)) < \infty$, il existe une mesure aléatoire positive $\mathcal{B}$ - mesurable sur R_+, unique à une équivalence près, soit $\tilde{N}$ telle que

$$\tilde{N}(A) = E^{\mathcal{B}}[N(A)]$$

pour tout borélien A de R_+ et plus généralement telle que

$$E\left[\int X\, d\tilde{N}\right] = E\left[\int \overline{E}^{\mathcal{B}} X \; dN\right]$$

pour tout processus réel mesurable positif X. Cette mesure $\tilde{N}$ sera notée $E^{\mathcal{B}}N$.

A partir de ce lemme, des formules de définition (16) et des formules (12) donnant les projections des processus dans l'exemple considéré, nous trouvons que

$$N^o = \sum_{\mathbb{N}} \frac{1_{[\![T_n, T_{n+1}[\![}}{\Pi_n(.)} \, \overline{E}^{\mathcal{B}_n}\left[1_{[\![T_n, T_{n+1}[\![} \, . N\right]$$

(19)

$$N^p = E^{\mathcal{B}_o}\left[N(\{0\})\right] \varepsilon_o + \sum_{\mathbb{N}} \frac{1_{]\!]T_n, T_{n+1}]\!]}}{\Pi_n^-(.)} \, \overline{E}^{\mathcal{B}_n}\left(1_{]\!]T_n, T_{n+1}]\!]} \, . N\right)$$

En effet, si N^o par exemple est défini par la formule précédente, l'égalité $E(\int X\, d\, N^o) = E(\int X^o\, d\, N)$ qui montre que N^o est bien la projection optionnelle de N s'obtient en additionnant en n les égalités

$$E\int_{R_+} \frac{1}{\Pi_n} \, 1_{[\![T_n, T_{n+1}[\![} X \; d\,\overline{E}^{\mathcal{B}_n}(1_{[\![T_n, T_{n+1}[\![} \, . N)$$

$$= E \int_{R_+} \frac{1}{\Pi_n} \, \overline{E}^{\mathcal{B}_n}\left(1_{[\![T_n, T_{n+1}[\![} X\right) d\left(1_{[\![T_n, T_{n+1}[\![} \, . N\right);$$

ces dernières égalités s'obtiennent par la dernière formule du lemme précédent compte tenu de ce que Π_n est $\mathcal{B}_n$ - mesurable.

En particulier, calculons le compensateur N^p de la mesure optionnelle $N = \sum_{N^*} Z_n \, \varepsilon_{T_n}$ associée à une suite de variables positives et respectivement $\mathcal{B}_n$ - intégrables Z_n ($n \in N$). Si ν_n désigne la mesure aléatoire positive $\mathcal{B}_n$ - mesurable $\overline{E}^{\mathcal{B}_n}(Z_{n+1} \, \varepsilon_{T_{n+1}})$ dont la primitive vaut

$$\nu_n(. \,; [0, t]) = E^{\mathcal{B}_n}\left[Z_{n+1} \, 1_{(T_{n+1} \leq t)}\right] \qquad (t \in R_+)$$

la formule (19) ci-dessus montre que sur $\mathbb{R}_+$, pour tout $\omega \in \Omega$

$$(20) \qquad N^p(\omega, dt) = \sum_{\mathbb{N}} 1_{\{T_n(\omega)<t\leq T_{n+1}(\omega)\}} \frac{\nu_n(\omega\,;\,dt)}{\Pi_n^-(\omega, t)}$$

où, rappelons-le, $\Pi_n^-(\omega, t) = P^{\mathcal{B}_n}(T_{n+1} \geq t)$.

En particulier la projection prévisible de la mesure optionnelle $\varepsilon_{T_{n+1}}$ vaut

$$\frac{\mu_n(\omega, dt)}{\mu_n(\omega, [t, \infty])} 1_{\{t \leq T_{n+1}(\omega)\}} \qquad \text{si } \mu_n(\omega, .) = \bar{E}^{\mathcal{B}_n}\left[\varepsilon_{T_{n+1}}\right] ;$$

d'après la proposition III.20, pour que T_{n+1} soit totalement inaccessible il est donc nécessaire et suffisant que μ_n soit diffuse. □

L'extension des résultats de ce paragraphe aux mesures aléatoires "marquées" ne présente aucune difficulté ; faisons-la brièvement. Si E est un espace localement compact à base dénombrable muni de sa tribu borélienne $\mathcal{E}$, un processus réel optionnel, resp. prévisible sur $R_+ \times E$ sera défini comme une application $X : \Omega \times R_+ \to R$ mesurable pour la tribu $\mathcal{O} \otimes \mathcal{E}$ resp. $\mathcal{P} \otimes \mathcal{E}$; la projection optionnelle resp. prévisible d'un processus réel positif $\mathcal{F} \otimes \mathcal{R}_+ \otimes \mathcal{E}$ mesurable X défini sur $\Omega \times R_+ \times E$ sera alors définie comme le processus réel positif optionnel X^o resp. prévisible X^p sur $R_+ \times E$ qui pour tout $x \in E$ fixé soit la projection optionnelle resp. prévisible de $X(., x)$; son existence est assurée par les arguments habituels et son unicité s'entend à une indistinguabilité près en ω, t pour tout x.

Soit $N : \Omega \to M_+(R_+ \times E)$ une mesure aléatoire positive $\mathcal{F}$ - mesurable sur $R_+ \times E$; pour tout borélien relativement compact de E, soit $F \in \mathcal{E}_c$, la formule $N_F(\omega, dt) = N(\omega\,;\,dt \times F)$ définit alors une mesure aléatoire positive sur R_+ . Par les arguments habituels, il est facile de vérifier que les formes suivantes de la définition du caractère optionnel, resp. prévisible de N sont équivalentes

a) $N(\{o\}) = 0$ et pour tout $F \in \mathcal{E}_c$, N_F est une mesure aléatoire positive optionnelle resp. prévisible sur F ;

b) pour tout processus réel positif optionnel, resp. prévisible sur $R_+ \times E$, soit X, le processus réel $Y(\omega, t) = \int_{[0,t] \times E} X(\omega, s, x)\, N(\omega ; ds\, dx)$ est optionnel resp. prévisible sur R_+ ;

c) pour tout processus réel positif X sur $R_+ \times E$

$$E\left[\int_{R_+ \times E} X(t,x)\, N(dt\, dx)\right] = E\left[\int_{R_+ \times E} X^q(t,x)\, N(dt\, dx)\right]$$

où q = 0 ou p.

Enfin la projection optionnelle resp. prévisible d'une mesure aléatoire positive sur $R_+ \times E$ peut être définie lorsque pour une suite au moins d'ouverts relativement compacts G_p croissant vers E il existe des temps optionnels T_n^p $(n, p \in \mathbb{N})$ tels que $T_n^p \uparrow \infty$ lorsque $n \uparrow \infty$ et que $E\, N([0, T_n^p[\times G_p) < \infty$ resp. $E\, N([0, T_n^p] \times G_p) < \infty$; c'est la mesure aléatoire positive optionnelle resp. prévisible sur $R_+ \times E$, unique à une équivalence près, que nous noterons N^o resp. N^p et qui soit telle que

$$E\left[\int_{R_+ \times E} X(t,x)\, N^q(dt\, dx)\right] = E\left[\int_{R_+ \times E} X^q(t,x)\, N(dt\, dx)\right]$$

(q = o ou p)

pour tout processus réel positif mesurable X.

Si $\overline{N}(\omega, dt) = N(\omega, dt \times E)$ est une mesure aléatoire sur R_+ (c'est-à-dire si $\overline{N}(\omega, .)$ est de Radon pour tout $\omega \in \Omega$), il existe une probabilité de transition $a(\omega, t ; dx)$ de $(\Omega \times R_+, \mathcal{J} \otimes \mathcal{R}_+)$ vers $(E, \mathcal{E})$ telle que

$$N(\omega ; dt\, dx) = \overline{N}(\omega ; dt)\, a(\omega, t ; dx) \quad \text{sur } R_+ \times E$$

pour P presque tout ω ($a(\omega, t ; F)$ est défini comme la densité de Radon-Nikodym de $P(d\omega)\, N(\omega ; dt \times F)$ par rapport à $P(d\omega)\, \overline{N}(\omega ; dt)$ pour tout $F \in \mathcal{E}$; et on choisit les fonctions $a(. ; F)$ dans leurs classes d'équivalence pour qu'elles soient positives et σ - additives en F ; voir [ ]). En outre si N est optionnelle, resp. prévisible sur $R_+ \times E$ la mesure aléatoire $\overline{N}$ l'est aussi sur R_+ et la probabilité de transition a peut être choisie $\mathcal{O}$-mesurable resp. $\mathcal{P}$- mesurable en (ω, t) sur $\Omega \times R_+$ (on montre dans la construction précédente qu'on peut prendre pour a (. ; F) des densités de Radon-Nikodym sur $\mathcal{O}$, resp. sur $\mathcal{P}$).

Construction de la loi d'un processus ponctuel de projection prévisible donnée

Nous terminerons ce paragraphe par la construction de la loi d'un processus ponctuel marqué dont la projection prévisible relativement à ses tribus propres est donnée. Mais avant d'aborder ce problème, faisons une digression sur la notion d'intensité de probabilité".

Soit T une variable réelle positive représentant une durée de vie ou une durée d'attente, dont nous supposerons d'abord qu'elle est entière auquel cas nous poserons $p(t) = P(T = t)$ $(t \in N)$ ou qu'elle possède une densité, soit f, sur R_+. L'intensité de la variable T est alors définie comme la probabilité conditionnelle lorsque $T \geqslant t$ pour que $T = t$ dans le cas entier ou "pour que $t \leqslant T < t + dt$" dans le cas continu, soit

$$\theta(t) = \frac{p(t)}{\sum\limits_{s \geqslant t} p(s)} \quad \text{resp. } \theta(t)\, dt = \frac{f(t)}{\int_t^\infty f(s)\, ds}\, dt.$$

Cette intensité représente un taux de mort ou un taux d'attente qui s'estime expérimentalement par la proportion d'individus ou d'objets terminant leur vie ou leur attente à l'instant t ou dans l'intervalle (t, t + dt) parmi ceux ayant vécu ou attendu au moins jusqu'à t. La loi de probabilité de la variable T peut être retrouvée à partir de l'intensité θ par les deux formules que le lecteur n'aura aucun mal à établir

$$p(t) = \prod_{0 \leqslant s < t} [1 - \theta(s)] \,.\, \theta(t) \quad \text{resp.} \quad f(t) = \theta(t) \exp\left[- \int_0^t \theta(s)\, ds\right]$$

$(t \in N$ resp. $R_+)$. Après ces préliminaires, voici un résultat un peu plus général, où pour la commodité nous nous placerons sur $\overline{R}_+$ plutôt que sur R_+.

Lemme III.21

Soit μ une probabilité sur $\overline{R}_+ = [0, \infty]$ et soit $I = [0, d|$ l'intervalle ouvert ou fermé à droite sur lequel $\mu([., \infty]) > 0$. La formule

$$(21) \qquad \theta(dt) = \frac{\mu(dt)}{\mu([t, \infty])} \qquad \text{sur } I$$

définit alors une mesure de Radon positive sur I telle que $\theta(\{t\}) < 1$ sauf en $t = d$, si $d \in I$, où $\theta(\{d\}) = 1$ et telle que si $d \notin I$: $\theta(I) = +\infty$. Inversement à une telle mesure θ définie sur un intervalle $I = [0, d|$, la formule

$$(22) \qquad \mu(dt) = \begin{cases} \exp\left[-\int_0^t \theta^*(ds)\right] \cdot \prod_{s<t} \left[1 - \theta(\{s\})\right] \cdot \theta(dt) & \text{sur } I \\ 0 & \text{sur } I^c \end{cases}$$

associe une mesure de probabilité μ sur $\overline{R}_+$ et les deux formules précédentes sont inverses l'une de l'autre. (θ^* désigne la partie diffuse de θ).

La démonstration est élémentaire et laissée au lecteur. Dans la suite de ce paragraphe, les lettres θ et μ affectées des mêmes indices désigneront toujours des mesures liées par les formules du lemme précédent ; la mesure θ sera appelée l'intensité de la loi μ ou de toute variable de loi μ.

Dans le problème que nous allons étudier, nous appellerons mesure ponctuelle sur $R_+ \times E$ une mesure de la forme $m = \sum_{\mathbb{N}^*} \varepsilon_{(t_n, x_n)}$ où $(t_n, n \in \mathbb{N}^*)$ est une suite croissante dans $]0, \infty]$ ne tendant pas nécessairement vers $+\infty$, où $(x_n, n \in N^*)$ est une suite de points dans $\overline{E} = E + \{\delta\}$ pour un point auxiliaire δ et où : $x_n \in E$ et $t_n < t_{n+1}$ si $t_n < \infty$, $x_n = \delta$ si $t_n = +\infty$. L'espace de ces mesures ponctuelles sera pris pour espace Ω, les applications $m \to t_n$ de Ω dans $]0, \infty]$ et $m \to x_n$ de Ω dans $\overline{E}$ seront notées T_n et ξ_n respectivement ($n \in \mathbb{N}^*$) ; nous noterons $\mathcal{B}_n$ ($n \in N$) les tribus de parties de Ω engendrées respectivement par les applications T_m, ξ_m ($1 \leqslant m \leqslant n$) $\left[\mathcal{B}_0 = \{\emptyset, \Omega\}\right]$.
Pour tout $n \in N$, donnons nous une probabilité de transition $\mathcal{B}_n$ - mesurable sur Ω et $R_+ \times \overline{E}$, soit $\mu_n(\omega ; dt\, dx)$, telle que $\mu_n(\omega ; .)$ soit pour tout ω portée par $]T_n(\omega), +\infty[\times E + \{(+\infty, \delta)\}$. Le théorème de Tulcea

(Neveu [2] p.153) permet alors de construire une probabilité d'ailleurs unique P sur Ω telle que

$$P^{\mathcal{B}_n}((T_{n+1}, \xi_{n+1}) \in .) = \mu_n(\omega ; .) \text{ sur } R_+ \times \overline{E}$$

pour P presque tout ω et tout $n \in N$; le théorème de Tulcea construit en fait la probabilité P sur $\left[ ]0, \infty] \times \overline{E} \right]^{N^*}$ mais il est facile de voir que cette probabilité P est portée par la partie de l'espace produit précédent qui est correspondance bijective avec Ω.

La mesure aléatoire $N = \sum_{\mathbb{N}^*} \varepsilon_{(T_m, \xi_m)}$ définie sur Ω comme l'application identique est optionnelle pour la famille de tribus $(\mathcal{F}_t, t \in R_+)$ associée aux deux suites $(\mathcal{B}_n, n \in N)$ et $(T_n, n \in N)$. Sa projection prévisible vaut

$$N^p = \sum_{\mathbb{N}^*} 1_{]\!]T_n, T_{n+1}]\!]} \cdot \theta_n(\omega; dt\, dx) \text{ où } \theta_n(\omega; dt dx) = \frac{\mu_n(\omega ; dt\, dx)}{\mu_n(\omega, [t, \infty] \times \overline{E})} ;$$

d'après la formule (20); en effet, pour tout borélien F de E, la projection prévisible de $N(\omega ; dt \times F) = \sum_{\mathbb{N}^*} 1_F(\xi_n) \varepsilon_{T_n}$ vaut

$$\frac{E^{\mathcal{B}_n}\left[1_F(\xi_{n+1}) \varepsilon_{T_{n+1}}\right](\omega, dt)}{P^{\mathcal{B}_n}(T_n \geq t)} = \frac{\mu_n(\omega ; dt \times F)}{\mu_n(\omega ; [t, \infty] \times \overline{E})} \text{ sur }]\!]T_n, T_{n+1}]\!] .$$

Montrons maintenant que la mesure prévisible N^p caractérise la probabilité P , c'est-à-dire qu'elle permet de retrouver les probabilités de transition $\mu_n(\omega ; dt\, dx)$. Compte tenu de la définition ci-dessus des θ_n et compte tenu du lemme III.21, la donnée des θ_n et celle des μ_n sont équivalentes : en effet puisque $\theta_n(\omega ; dt \times \overline{E}) = \dfrac{\mu_n(\omega ; dt \times \overline{E})}{\mu_n(\omega ; [t, \infty] \times E)}$, la formule 22 du lemme s'applique et permet de trouver $\mu_n(\omega ; dt \times \overline{E})$ à partir de θ_n et alors aussi $\mu_n(\omega ; dt\, dx)$ qui est égale à $\mu_n(\omega ; [t, \infty] \times \overline{E})\, \theta_n(\omega ; dt\, dx)$. Le seul problème est donc de montrer que :

$$1_{]\!]T_n, T_{n+1}]\!]} \theta_n(\omega ; dt\, dx) \quad \text{détermine } \theta_n(\omega ; dt\, dx)$$

et comme par hypothèse les mesures $\mu_n(\omega ; .)$ et $\theta_n(\omega ; .)$ sont portées par $]T_n(\omega), +\infty] \times \overline{E}$, la difficulté est seulement du côté de T_{n+1}.

III.3. MARTINGALES DISCONTINUES

Commençons par quelques préliminaires destinés à justifier l'étude à cet endroit des martingales dites discontinues.

Relativement à la famille $(\mathcal{F}_t, t \in R_+)$ que nous supposons croissante, continue à droite et complète dans l'espace de probabilité $(\Omega, \mathcal{F}, P)$, un processus réel cadlag adapté M est appelé une martingale si pour tout $t \in R_+$ la variable M_t est intégrable et si $M_s = E^{\mathcal{F}_s}(M_t)$ lorsque $s < t$ dans R_+ (ce n'est pas une réelle restriction d'exiger qu'une martingale soit cadlag puisque d'après un théorème, toute martingale qui ne serait pas supposée cadlag peut être modifiée en tout $t \in R_+$ sur un ensemble négligeable dépendant de t, pour le devenir, grâce à la continuité à droite de la famille $(\mathcal{F}_t, t \in R_+)$. Pour toute martingale M uniformément intégrable, la limite $M_\infty = \lim_{t\to\infty} M_t$ existe non seulement presque partout (ce pourquoi il suffit que $\sup_{R_+} E|M_t| < \infty$) mais aussi dans L^1 de sorte que $M_s = E^{\mathcal{F}_s}(M_\infty)$ pour tout $s \in R_+$ et que plus généralement

$$M_T = E^{\mathcal{F}_T}(M_\infty) \qquad \text{(où } M_T = M_\infty \text{ sur } \{T = \infty\}\text{)}$$

pour tout temps optionnel T.

Inversement pour toute variable $\mathcal{F}$- intégrable Z, la martingale $(M_s = E^{\mathcal{F}_s}(Z), s \in R_+)$ est uniformément intégrable et la notion de "martingale uniformément intégrable" apparait donc comme particulièrement importante.
Comme pour toute martingale M, la famille $(M_{t \wedge s}, s \in R_+) = (E^{\mathcal{F}_s}(M_t), s \in R_+)$ est uniformément intégrable pour tout $t \in R_+$ fixé, une martingale peut être définie comme un processus réel M tel que pour tout $t \in R_+$, le processus réel arrêté $(M_{t\wedge s}, s \in R_+)$ soit une martingale uniformément intégrable !
La notion de martingale fait donc jouer à la famille des temps optionnels constants t un rôle particulier et c'est sans doute pourquoi il est apparu nécessaire de la généraliser comme le fait la définition suivante.

Définition III.23

Un processus réel cadlag M est appelé une martingale locale s'il existe une suite de temps optionnels T_n $(n \in N)$ croissant vers $+\infty$, soit $\lim_n \uparrow T_n = +\infty$ sur Ω, telle que pour tout $n \in N$, le processus $(M_{T_n \wedge t},\ t \in R_+)$ soit une martingale uniformément intégrable.

Une martingale locale est nécessairement adaptée et donc optionnelle puisque $M_t = \lim_n M_{T_n \wedge t}$ pour tout $t \in R_+$. Si M est une martingale locale, la limite $M_\infty = \lim_{t\to\infty} M_t$ existe sur l'ensemble $\bigcup_N \{T_n = \infty\}$ (qui est peut-être vide !) et alors l'égalité $M_S = E^{\mathcal{F}_S}(M_T)$ a lieu sur $\{S \leqslant T\}$ pour tout couple S, T de temps optionnels majorés sur Ω par l'un des T_n au moins ; en particulier $(M_{T\wedge t},\ t \in R_+)$ est une martingale uniformément intégrable pour tout temps optionnel T majoré sur tout Ω par l'un des T_n.

Les martingales interviennent de deux manières dans l'étude des mesures aléatoires et en particulier des processus ponctuels ; d'une part dans les formules de changement de loi, d'autre part dans l'étude des compensateurs (= projections prévisibles) des mesures aléatoires optionnelles. De manière plus précise, nous avons les deux résultats suivants.

Lemme III.24

Si P' est une probabilité sur $(\Omega, \mathcal{F})$ absolument continue par rapport à P sur la tribu $\mathcal{F}_\infty$, resp. sur les tribus $\mathcal{F}_{T_n}$ associées à une suite $(T_n,\ n \in N)$ de temps optionnels croissant vers $+\infty$, il existe une martingale uniformément intégrable, resp. une martingale locale, M telle que $P' = M_T \cdot P$ sur $\mathcal{F}_T$ pour tout temps optionnel T, resp. pour tout temps optionnel T majoré sur tout Ω par l'un des T_n.

Démonstration

Soit Z la dérivée de Radon-Nikodym de P' par rapport à P sur $\mathcal{F}_\infty$ lorsque $P' \ll P$ sur $\mathcal{F}_\infty$ et soit M une version cadlag de $(E^{\mathcal{F}_t} Z,\ t \in R_+)$: M est une martingale intégrable, $M_T = E^{\mathcal{F}_T}(Z)$ pour tout temps optionnel T et alors

$P' = Z \, . \, P = M_T \, . \, P$ sur $\mathcal{F}_T$ puisque $\mathcal{F}_T \subset \mathcal{F}_\infty$.

Lorsque P' n'est absolument continue par rapport à P que sur $\mathcal{F}_{T_n}$ $(n \in \mathbb{N})$, la martingale uniformément intégrable $M^{(n)}$ associée à la dérivée de Radon-Nikodym $\dot{Z}^{(n)}$ de P' par rapport à P sur $\mathcal{F}_{T_n}$ vérifie encore $M_T^{(n)} = E^{\mathcal{F}_T}(Z_T^{(n)})$ pour tout temps optionnel T de sorte que $P' = Z^{(n)} \, . \, P = M_T^{(n)} \, . \, P$ pour tout temps optionnel T majoré par T_n (et donc tel que $\mathcal{F}_T \subset \mathcal{F}_{T_n}$). Par application aux temps $t \wedge T_n$, il s'ensuit que $M_t^{(n+1)} = M_t^{(n)}$ P p.p sur $\{t \leq T_n\}$; grâce à la propriété cadlag des martingales $M^{(n)}$ et $M^{(n+1)}$ il existe donc un ensemble P-négligeable $N^{(n)}$ tel que $M^{(n+1)}(\omega, t) = M^{(n)}(\omega, t)$ si $t \leq T_n(\omega)$ et $\omega \notin N^{(n)}$. Cela permet de définir un processus réel cadlag M sur $\Omega \times \mathbb{R}_+$ tel que $M^{(n)}(\omega, t) = M(\omega, t)$ si $t \leq T_n(\omega)$ et $\omega \notin N$ où $N = \bigcup_n N^{(n)}$ est P-négligeable, donc tel que les processus arrêtés $M_{T_n \wedge \, .}$ soient indistinguables des $M^{(n)}$ $(n \in \mathbb{N})$. Ceci achève la démonstration. □

Proposition III.25

Soit N une mesure optionnelle positive sur $\mathbb{R}_+$ telle que $E[N(\mathbb{R}_+)] < \infty$, resp. telle que $E\Big[N([0, T_n])\Big] < \infty$ pour une suite $(T_n, n \in \mathbb{N})$ de temps optionnels croissant vers $+\infty$. Le compensateur N^p de N est alors l'unique mesure prévisible positive sur $\mathbb{R}_+$ telle que le processus cadlag défini sur $\Omega \times \mathbb{R}_+$ par

$$M(\omega, t) = N(\omega, [0, t]) - N^p(\omega, [0, t])$$

soit une martingale uniformément intégrable, resp. une martingale locale, nulle en 0.

De plus pour tout temps optionnel totalement inaccessible T le saut de M en T vaut $\Delta M_T = N(\{T\})$ p.s. tandis que pour tout temps prévisible S ce saut vaut $\Delta M_S = N(\{S\}) - E^{\mathcal{F}_{S^-}}[N(\{S\})]$. En particulier si la mesure optionnelle N ne change aucun temps prévisible, sa compensatrice est une mesure aléatoire diffuse et la martingale M admet les mêmes sauts que N, soit $\Delta M_t = N(\{t\})$ $\forall t \in \mathbb{R}_+$.

La martingale M de la proposition est la primitive de la mesure aléatoire $N - N^p$; nous dirons donc aussi que $N - N^p$ est une <u>différentielle de martingale</u>.

Démonstration

Soient A et A^p respectivement les primitives de N et de sa projection prévisible N^p dont l'existence est assurée par le théorème III.18. Le processus réel $M = A - A^p$ est alors un processus cadlag adapté différence de deux processus croissants dont la variation vérifie l'inégalité suivante pour tout temps optionnel T

$$E\left[\int_{[\![0,T]\!]} |d M_t|\right] \leqslant E\left[N([\![0,T]\!]) + N^p([\![0,T]\!])\right] = 2\,E\left[N([\![0,T]\!])\right] ;$$

suivant les hypothèses faites nous aurons donc

$$E\left[\int_{R_+} |d M_t|\right] < \infty \qquad \text{ou} \qquad E\left[\int_{[\![0,T_n]\!]} |d M_t|\right] < \infty \qquad (n \in N).$$

Pour tout temps prévisible S, la variable positive $N(\{S\})$ est $\mathfrak{F}_S$ - mesurable et son espérance conditionnelle par rapport à $\mathfrak{F}_S^-$ vaut $N^p(\{S\})$. En effet d'une part $N(\{S\}) = A_S - A_S^-$ et $N^p(\{S\}) = A_S^p - A_S^{p,-}$ sont respectivement $\mathfrak{F}_S$ et $\mathfrak{F}_S^-$ - mesurables (A est optionnel tandis que A^p, A^- et $A^{p,-}$ sont prévisibles), et d'autre part comme le processus $Z\,1_{[\![S]\!]} = Z\,1_{[\![S,\infty[\![} - Z\,1_{]\!]S,\infty[\![}$ est prévisible dès que Z est $\mathfrak{F}_S^-$ - mesurable, nous avons $E\left[Z\,N^p(\{S\})\right] = E\left[Z\,N(\{S\})\right]$ pour tout Z $\mathfrak{F}_S^-$ - mesurable positif. Nous avons donc établi que

$$\Delta M_S = N(\{S\}) - E^{\mathfrak{F}_S^-}\left[N(\{S\})\right] \qquad \text{(S prévisible)}.$$

Pour S = 0, puisque $\mathfrak{F}_0^- = \mathfrak{F}_0$ par définition cela montre que $M_0 = N(\{0\}) - N^p(\{0\}) = 0$. Le processus M ou les processus arrêtés $M_{T_n \wedge .}$ suivant le cas sont donc uniformément intégrables puisque

$$\sup_{R_+} |M_t| \leqslant \int_{R_+} |d M_t| \text{ , resp. } \sup_{t \leqslant T_n} |M_t| \leqslant \int_{[0,T_n]} |d M_t| .$$

De plus quels que soient s < t dans R_+ et quelle que soit la variable Z $\mathfrak{F}_s$ - mesurable positive, le processus réel $Z\,1_{]s,t]}$ est prévisible de sorte que $E(\xi N(]s,t])) = E(\xi N^p(]s,t]))$ par définition de N^p ; cela montre, au moins lorsque $E[N(R_+)] < \infty$, que $E^{\mathcal{F}_s}(M_t - M_s) = 0$ et donc d'après ce qui précède que M est une martingale uniformément intégrable. Sous l'hypothèse $E(N([0, T_n])) < \infty$, en appliquant ce résultat à la mesure aléatoire $1_{[\![0, T_n]\!]} . N$ dont la compensatrice vaut $1_{[\![0, T_n]\!]} . N^p$, nous trouvons que $M_{T_n \wedge .}$ est une martingale uniformément intégrable, donc que M est une martingale locale.

Si T est un temps optionnel totalement inaccessible, $N^p(\{T\}) = 0$ p.s. d'après la proposition III.16 et par conséquent $\Delta M_T = N(\{T\})$ p.s. Si la mesure aléatoire N ne charge aucun temps prévisible, sa compensatrice N^p a la même propriété puisque nous venons d'établir que $N^p(\{S\}) = E^{\mathcal{F}_{S^-}}[N(\{S\})]$ si S est prévisible ; la démonstration de la proposition III.16 montre dans ce cas que N^p est diffuse p.s. (ici les variables Y_p de cette démonstration sont nulles).

Enfin si N' est une mesure aléatoire positive et prévisible telle que $N'([0, t]) = N([0, t]) - M_t$ $(t \in R_+)$ pour une martingale uniformément intégrable M, resp. pour une martingale locale M, pour tout processus réel de la forme $X = \xi\, 1_{]s,t]}$ où $s < t$ dans R_+ et ξ est une variable positive bornée $\mathcal{F}_s$ - mesurable, nulle éventuellement sur l'un des ensembles $\{T_n \leqslant s\}$ $(n \in N)$

$$E\left[\int_{R_+} X\, dN'\right] = E\left[\int_{R_+} X\, dN\right] - E\left[\xi(M_t - M_s)\right] = E\left[\int_{R_+} X\, dN^p\right].$$

L'égalité des deux membres extrêmes s'étend à tous les processus réels positifs prévisibles par les arguments habituels (puisque $\{T_n \leqslant s\} \downarrow \emptyset$ lorsque $n \uparrow \infty$, quel que soit $s \in R_+$) ; d'après le résultat d'unicité de la proposition III.17, ceci montre que $N' = N^p$ presque sûrement. □

L'extension de la proposition précédente à des différences de mesures optionnelles positives ne présente aucune difficulté. Le principal intérêt de cette extension réside dans le corollaire suivant.

Corollaire III.26

Soit M une martingale locale nulle en 0 et telle que $E\left[\int_{[0,T_n]} |d M_t|\right] < \infty$ pour une suite $(T_n, n \in N)$ de temps optionnels croissant vers $+\infty$. La formule $N = \sum_{R_+} \Delta M_t \varepsilon_t$ définit alors une mesure optionnelle telle que $M_t = N([0,t]) - N^p([0,t])$ $(t \in R_+)$ si N^p désigne la compensatrice de N.

Démonstration

Pour tout processus réel cadlag adapté X, $T_X^\varepsilon = \inf(t : \Delta X_t \geqslant \varepsilon)$ est un temps optionnel en vertu de la formule facile à établir

$$\left\{T_X^\varepsilon \leqslant t\right\} = \inf_{n,\, p \in \mathbb{N}^*} \sup\left[\left\{X_v - X_u \geqslant \varepsilon - \frac{1}{p}\right\} ;\ u, v \in Q \cup \{t\},\ 0 \leqslant u < v \leqslant t,\ v - u \leqslant \frac{1}{n}\right].$$

La suite strictement croissante $(T^{\varepsilon,n}, n \in \mathbb{N}^*)$ des instants auxquels $\Delta X_t \geqslant \varepsilon$ est par conséquent une suite de temps optionnels puisque $T^{\varepsilon,n+1} = T^\varepsilon_{X^n}$ si $X^n_t = (X_t - X_{T^{\varepsilon,n}})\, 1_{(T^{\varepsilon,n} < t)}$ $(t \in R_+)$ et ces temps optionnels croissent vers $+\infty$; comme ΔX est optionnel, il s'ensuit que

$$N^\varepsilon = \sum_{t \in R_+} \Delta X_t\, 1_{\{\Delta X_t \geqslant \varepsilon\}}\, \varepsilon_t = \sum_{\mathbb{N}^*} (\Delta X)_{T^{\varepsilon,n}}\, \varepsilon_{T^{\varepsilon,n}}$$

est une mesure optionnelle positive. En faisant tendre $\varepsilon \downarrow 0$, nous trouvons sous les hypothèses du corollaire que la formule $N^{\pm} = \sum_{t \in R_+} (\Delta M_t)^{\pm}\, \varepsilon_t$ définit deux mesures optionnelles positives telles que $E(N^+([0,T_n]) + N^-([0,T_n])) < \infty$ pour tout $n \in \mathbb{N}$.

Considérons d'autre part les deux mesures optionnelles positives $\dot{M}^{\pm}$ définies par $\dot{M}^{\pm}([0,t]) = \int_0^t [d M_s(\omega)]^{\pm}$ $(t \in R_+)$; il s'agit bien de mesures optionnelles car cette définition entraîne que $\dot{M}^{\pm}([0,t]) = \lim_n \sum_k \left[M_{t \wedge (k+1) 2^{-n}} - M_{t \wedge k 2^{-n}}\right]^{\pm}$. Il est clair que $E(\dot{M}^+([0,T_n]) + M^-([0,T_n])) < \infty$ pour tout $n \in N$.

D'autre part comme $\dot{M}^+([0,t]) - \dot{M}^-([0,t]) = M_t$ $(t \in R_+)$ est une martingale locale nulle en 0, la proposition précédente entraîne que la projection prévisible de la mesure optionnelle $\dot{M} = \dot{M}^+ - \dot{M}^-$, soit $(\dot{M}^+)^p - (\dot{M}^-)^p$ est nulle. La mesure optionnelle $N - \dot{M}$ où $N = N^+ - N^-$ a donc même projection prévisible que N, soit N^p. Mais par construction les deux mesures optionnelles $\dot{M}^+ - N^+$ et $\dot{M}^- - N^-$ sont positives et diffuses, donc prévisibles ; la mesure $N - \dot{M}$ est donc prévisible et comme nous venons de démontrer que sa projection prévisible vaut N^p, nous avons établi que $N - \dot{M} = N^p$. □

BIBLIOGRAPHIE

ACQUAVIVA A.
Théorème de relèvement et répartitions ponctuelles à points distincts en nombre localement fini. C.R. Acad. Sci. Paris 281 (1975) 297-300

ADAMOPOULOS L.
Some counting and interval properties of the mutually-exciting processes J. Appl. Proba. 12 (1975) 78-86

AMBARTZUMIAN R.V.
[1] Correlation properties of the intervals in the superpositions of independent stationary recurrent point processes. Studia Sci. Hung 4 (1969) 161-170

[2] Random plane mosaics. Dokl Akad Nauk SSSR 200 (1971) 255-258 ; Soviet Math Dokl 12 (1971) 1349-1353

[3] Palm distributions and superpositions of independent point processes in R^n. Stochastic point processes. PAW Lewis Ed. Wiley (1972) 626-645

[4] The solution of the Buffon-Sylvester problem in R^3. ZfW 27 (1973) 53-74

AMBROSE W.
Representation of ergodic flows. Ann. Math 42 (1941) 723-739

AMBROSE W. et KAKUTANI S.
Structure and continuity of measurable flows. Duke Math J. 9 (1942) 25-42

AMBROSE W., HALMOS P.R. et KAKUTANI S.
Decomposition of measures. Duke Math J. 9 (1942) 43-47

BARTLETT M.S.
<u>An introduction to stochastic processes</u>. Cambridge Univ. Press (1966) 2ème édition.

BENÈS V.E.
<u>General stochastic processes in the theory of queues</u>. Addison - Wesley Mass. 1963

BENVENISTE A.
Processus stationnaires et mesures de Palm du flot spécial sous une fonction. Thèse Paris.

BENVENISTE A. et JACOD J.
Intensité stochastique d'un processus ponctuel stationnaire, reconstruction du processus ponctuel à partir de son intensité stochastique C.R. Acad. Sci.Paris 280 (1975) 821-5

BEUTLER F.J. et LENEMAN OAZ
The theory of stationary point processes. Acta Math. 116 (1966) 159-197

BILLINGSLEY P.
<u>Convergence of probability measures</u>. Wiley (1968)

BLANCHARD F.
Processus de points marqués et processus ramifiés. Ann. IHP 9 (1973) 259-276

BOEL R., VARAYIA P. et WONG E.
Martingales of jump processes, I and II. Siam J. Control 13 (1975) 999-1061

BOROVKOV A.A.
[1] On the first passage time for a class of processes with independent increments. Teor. Veraj. Prim. 10 (1965) 360-364

[2] Some limit theorems in queuing theory. Teor. Ver°j. Prim. 9 (1964) 608-625 ; 10 (1965) 409-437

[3] Asymptotic analysis of some queuing systems. Teor. Ver°j. Prim. 11 (1966) 675-682

[4] Stochastic processes in queuing theory. Springer (1976)

BREIMAN L.
The Poisson tendency in traffic distribution. Ann. Math. Stat. 34 (1963) 308-311

BREMAUD P.
[1] The martingale theory of point processes over the real half-line admitting an intensity. Proc. IRIA Conf., Lect. Notes Op. Res. and Math. Systems, Springer 107 (1974) 519-542

[2] An extension of Watanabe's theorem of characterization of Poisson processes over the positive real half line. J. Appl. Prob. 12 (1975) 396-399

[3] Estimation de l'état d'une file d'attente et du temps de panne d'une machine par la méthode des semi-martingales. Adv. Appl. Proba. 7 (1975)

[4] On the information carried by a stochastic point process. Revue Cethedec 43 (1975)

[5] Bang-Bang controls of point processes. Adv. Appl. Proba 8 (1976) 385-394

BREMAUD P. et JACOD J.
Processus ponctuels et martingales

BRILLINGER D.
The identification of point process systems. Ann. Proba 3 (1976) 909-929

BROWN M.
[1] Sampling with random jitter. J. Soc. Ind. Appl. Math. 11 (1963) 460-473

[2] An invariance property of Poisson processes. J. Appl. Proba 6 (1969) 453-458

[3] A property of Poisson processes and its applications to macroscopic equilibrum of particle systems. Ann. Math. Stat. 41 (1970) 1935-41

[4] Discrimination of point processes. Ann. Math. Stat. 42 (1971)

CARTER D.S. et PRENTER P.M.
Exponential spaces and counting processes. ZfW 21 (1972) 1-19

CHOQUET G.
Theory of capacities. Ann. Fourier 5 (1953) 131-295

CHOU C.S. et MEYER P.A.
Représentation des martingales comme intégrales stochastiques dans les processus ponctuels. C.R. Acad. Sci. Paris A 278 (1974) 1561-3

CHIMG K.L.
Crudely stationary counting processes. Amer. Math. Monthly 79 (1972) 867-877

CINLAR E.
Superposition of point processes. Stochastic point processes, P. Lewis Ed., Wiley (1972) 549-606

COHEN J.W.
[1] <u>The single server queue</u>. North-Holland, Amsterdam 1969

[2] Asymptotic relations in queuing theory. Stoch. Prob. Appli 1 (1973) 107-124

COX D.R.
<u>Renewal theory</u>. Methuen's monograph, London 1962

COX D.R. et LEWIS P.A.W.
[1] <u>The statistical Analysis of series of Events</u>. Methuen, London (1966) ; Trad. fr. Dunod 1969

[2] Multivariate point processes. Proc. 6th Berkeley Symp 3 (1972) 401-425.

CRAMER H. et LEADBETTER M.R.
<u>Stationary and related stochastic processes</u>. Wiley (1967)

CRAMER H., LEADBETTER M.R. et SERFLING
On distribution function moment relationships in a stationary point process. ZfW 18 (1971) 1-8

DALEY D.J.
[1] The correlation structure of the output process of some single server queuing systems. Ann. Math. Stat 39 (1968) 1007-19

[2] Asymptotic properties of stationary point processes with generalized clusters. ZfW 21 (1972) 65-76

[3] Markovian processes whose jump epochs constitute a renewal process Quart. J. Math. Oxford 24 (1973) 97-105

[4] Poisson and alternating renewal processes with superposition a renewal process. Math. Nachr 57 (1973) 359-369

[5] Various concepts of orderliness for point-processes. Stochastic Geometry, Wiley (1974), 148-161

[6] Queuing output processes. Adv. Appl. Proba 8 (1976) 395-415

DALEY D.J. et MILNE R.K.
Orderliness, intensities and Palm-Khinchin equations for multivariate processes. J. Appl. Proba. 12 (1975) 383-389

DALEY D.J. et OAKES D.
Random walk point processes. ZfW 30 (1974) 1-16

DALEY D.J. et VERE-JONES D.
A summary of the theory of point processes. Stochastic point processes, P. Lewis Ed. Wiley (1972) 299-383

DAVIDSON R.
Some arithmetic and geometry in probability theory. Thesis Cambridge 1968. Stochastic geometry, a memorial volume, Wiley 1974.

DAVIS M.H.A. et ELLIOTT R.J.
Representation of martingales of jump processes. Siam J. Control (1976)

DAVIS M.H.A. et
Optimal control of a jump process (1975)

DAVIS M.H.A., KAILATH T. et SEGALL A.
Non-linear filtering with counting observations. IEEE IT 21 (1975) 143-150

DEBES H., KERSTAN J., LIEMANT A. et MATTHES K.
Verallgemeinerung eines Satzes von Dobrušin. Math Nachrichten 47 (1970) 183-244, 50 (1971) 99-139 et 51 (1971) 149-188.

DELLACHERIE C.
[1] Capacités et processus stochastiques. Springer 1972

[2] Intégrales stochastiques par rapport aux processus de Wiener et de Poisson. Sem. Proba. Strasbourg, Lect. Notes Math 381 (1974) 25-26 et 465 (1975) 494

DELLACHERIE C. et MEYER P.A.
Probabilités et Potentiel, Hermann (1976)

DISNEY R.L. et CHERRY W.P.
Some topics in queuing network theory. Proc. Conf. Western Michigan Univ 1973. Lecture Notes in Economics 98, 23-44

DOLEANS - DADE C.
[1] Quelques applications de la formule de changement de variables pour les semi-martingales. ZfW 16 (1970) 181-194

[2] Existence and Unicity of stochastic integral equations. ZfW 36 (1976) 93-101

DOLEANS - DADE C. et MEYER P.A.
Intégrales stochastiques par rapport aux martingales locales.
Sém. de Proba, Strasbourg 4, Lecture Notes Springer 124 (1970) 77-107

DRISCOLL M.F. et WEISS N.A.
Random translations of stationary point processes. J. Math. Anal. Appl. 48 (1974) 423-433

ELKAROUI N. et LEPELTIER J.P.
Représentation des processus ponctuels multivariés à l'aide de processus de Poisson. C.R. Acad. Sci. Paris A (1975)

ELLIOTT R.J.
Stochastic integrals for martingales of a jump process. Report Hull Univ.

Martingales of a jump process with partially accessible jump times. Report Hull Univ.

Levy systems and absolutely continuous changes of measures for a jump process. Report Hull Univ.

Levy functionals and jump process martingales. Report Hull Univ.

Innovation projections of a jump process and local martingales. Report Hull Univ.

FELLER W.
<u>An introduction to probability and its applications.</u> Vol. 1 and 2. Wiley

FICHTNER K.H.
Charakterisierung Poissonscher zufülliger Punktfolgen und infinitesimale Verdünnungsschemata. Math. Nachr 68 (1975) 93-104

FIEGER W.
[1] Zwer Verallgemeinerungen der Palmschen Formeln. Trans. 3nd Prague Conf (1964), 107-122

[2] Eine für beliebige Call-Prozesse geltende Verallgemernerung der Palmschen Formeln. Math. Scan 16 (1965) 121-147

[3] Die Anzahl der γ-niveau Kreuzungspunkte von stochastischen Prozessen ZfW 18 (1971) 227-260

FISCHER L.
A survey of the math theory of multi-dimensional point processes. Stochastic point processes, PAW Lewis ed. Wiley (1972) 468-513

FORTET R.
[1] Sur les répartitions ponctuelles aléatoires. Ann IHP 4 (1968) 99-112

[2] Définition et lois de probabilité des répartitions ponctuelles aléatoires. Zastosowania Matematyki 10 (1969) 57-73

FORTET R. et KAMBOUZIA M.
[1] Lois de probabilité des répartitions ponctuelles markoviennes et des répartitions ponctuelles cumulatives. CR. Acad. Sci. 268 (1969) 644-5

[2] Recouvrement par un ensemble aléatoire. CR. Acad. Sci. 281 (1975) 397-8

[3] Ensembles aléatoires induits par une répartition ponctuelle aléatoire CR. Acad. Sci. 280 (1975) 1447-50

FRANKEN P.
Approximation durch Poissonsche Prozesse. Math Nachr 26 (1963) 101-114

FRANKEN P., LIEMANT A. et MATTHES K.
Stationäre zufällige Puntkfolgen I-III. J.Ber Deutsch Math Verein 65 (1963) 66-79, 66 (1964) 106-118 et 67 (1965) 183-202.

FRIEDMAN N.A.
Introduction to Ergodic theory. Van Nostrand (1970)

FRISCH H.L. et HAMMERSLEY J.M.
Percolation processes and related topics. J. Soc. Ind. Appl. Math II (1963) 894-918

FUJISAKI M., KALLIANPUR G. et KURITA H.
Stochastic differential equations for the non-linear filtering problem Osaka J. Math 9 (1972) 19-40

FURSTENBERG H. et TZKONI I.
Spherical functions and integral geometry. J. Israel Math 10 (1971) 327-338

GEMAN D.
[1] Horizontal-window conditioning and the zeros of stationary processes

[2] On the variance of the number of zeros of a stationary Gaussian process. Ann. Math. Stat.

GEMAN D. et HOROWITZ J.
[1] Remarks on Palm measures. Ann IHP 9 (1973) 215-232

[2] Random shifts which preserve measure. Proc AMS 49 (1975) 143-150

[3] Polar sets and Palm measures in the theory of flows. Trans AMS 208 (1975) 141-159

[4] Time conditionning of random processes

[5] Transformation of flows by discrete random measures. Indiana J. Math 24 (1975) 291-306

GIGER et HADWIGER
Uber Treffzahlwahrscheinlickkeiten in Eikörperfeld. ZfW 10 (1968) 329-334

GIRSANOV I.
On transforming a certain class of stochastic processes by absolutely continnous substitution of measures. Teor. Veroj i Prim. 5 (1960) 285-301

GNEDENKO B.V. et KOVALENKO I.N.
Introduction to queuing theory. Nauka, Moscou (1966). Trad. anglaise Jesuralem (1968)

HINCIN A Ya (ou KHINTCHINE)
[3] On Poisson streams of random events. Teor. Veroj. 1 (1956) 320-7 ; Th. Proba. 1 (1956) 291-7

HOFFMAN-JORGENSEN J.
Markov sets. Math. Scand. 24 (1969) 145-166

ISHAM V.
On a point process with independent locations. J. Appl. Proba. 12 (1975) 435-446

ITO K.
Poisson point processes attached to Markov processes. Proc. 6th Berkeley Symp (1972) III, 225-239

JACOD J.
[1] Two dependent Poisson processes whose sum is still a Poisson process. J. Appl. Proba 12 (1974) 1-12

[2] Multivariate point processes : predictable projection, Radon-Nikodym derivatives, representation of martingales. Z.f.W. 31 (1975) 235-253

[3] Un théorème de représentation pour les martingales discontinues. Z.f.W. 34 (1976) 225-244

JACOD J. et MEMIN J.
[1] Caractéristiques locales et conditions de continuité pour les semi-martingales. Z.f.W. 35 (1976) 1-37

[2] Un théorème de représentation des martingales pour les ensembles régénératifs. Sem. Proba. Strasbourg X. Lecture Notes Springer 511 (1976) 24-39

JAGERS P.
[1] On the weak convergence of superpositions of point processes. Z.f.W. 22 (1972) 1-7

[2] On Palm probabilities. Z.f.W. 26 (1973) 17-32

[3] Aspects of random measures and point processes. Adv. Proba. 3, ed. Ney. M. Dekker (1974) 179-239

[4] Convergence of general branching processes and functionals thereof J. Appl. Proba. 11 (1974) 471-8

JAGERS P. et LINDVALL T.
Thinning and rare events in point processes. Z.f.W. 28 (1974) 89-98 et 29 (1974) 272

JIŘINA M.
Branching process with measure-valued states. Trans. 3rd Prague Conf. (1964) 333-357

KABANOV I.
Représentation comme intégrales stochastiques des fonctionnelles d'un Wiener ou d'un Poisson. Teor. Veraj. Prim 18 (1973) 376-380

KABANOV I., LIPSER R. et SHIRYAEV A.
Martingale methods in the theory of point processes. Proc. Steklov Inst. (1975) 269-354

KAC M. et STEPIAN D.
Large excursions of Gaussian processes. Ann. Math. Stat. 30 (1959) 1215-1228

KAILATH T. et SEGALL A.
Radon-Nikodym derivatives with respect to measures induced by discontinuous independent increment processes. Ann. Proba 3 (1975) 449-464

KALLENBERG O.
[1] Characterization and convergence of random measures and point processes. ZfW 27 (1973) 9-21

[2] Canonical representations and convergence criteria for processes with interchangeable increments. ZfW 27 (1973) 23-36

[3] Extremality of Poisson and sample processes. Stoch. Prob. Appli. 2 (1974) 73-83

[4] On symmetrically distributed random measures. Trans. AMS 202 (1975) 105-121

[5] Limits ofcompound and thinned point processes. J. Appl. Prob. 12 (1975) 269-278

KENDALL D.G.
[1] Stochastic processes occuring in the theory of queues and their analysis by the method of the imbedded Markov chains. Ann. Math. Stat. 24 (1953) 338-354

[2] Foundations of a theory of random sets. Stochastic geometry. Harding et Kendall, ed. Wiley (1974) 322-376

KERSTAN J. et MATTHES K.
[1] Verallgemernerungeines Satzes von Sliwnjak. Revue Romaine Math Pures et appliquées 9 (1964) 811-829

[2] Ergodische unbegrenzt teilbare stationäre zufällige Punktfolgen. Trans. 4th Prague Conf (1967) 399-415

KERSTAN J. MATTHES K. ET MECKE J.
Unbegrenzt teilbare Punktprozesse. Akademie Verlag. Berlin (1974)

KIEFER J. et WOLFOWITZ J.
[1] On the theory of queues with many servers. Trans. AMS 78 (1955) 1-18

[2] On the characteristics of the general queuing process with applications to random walks. Ann. Math. Stat 27 (1956) 147-161

KINGMAN J.F.C.
[1] On doubly stochastic Poisson processes. Proc Cambridge Phil Soc 60 (1964) 923-930

[2] The stochastic theory of regenerative events. ZfW 2 (1964) 180-224

[3] Completely random measures. Pacific J. Math 21 (1967) 59-78

[4] Markov population processes. J. Appl. Proba. 6 (1969) 1-18

[5] Regenerative Phenomena. Wiley 1972

KOSTEN L.
Stochastic theory of service systems. Pergamon Press, Oxford (1973)

KRENGEL U.
Darstellungen von Strömungen I, II. Math. Annalen 176 (1968) 181-190

KRICKEBERG K.
[1] The Cox processes. Symp. Math. Roma 9 (1972) 151-167

[2] Theory of hyperplane processes. Conf. Stoch. point processes I.B.M. Wiley (1972) 514-521

[3] Moments of point processes. Lecture Notes in Mathematics, Springer 246 (1973) 70-101

[4] Invariance properties of the correlation measure of line processes. Stochastic geometry, ed. Kendall ; Wiley (1974) 76-88

KRYLOV N.V. et YUSKENI A.A.
Markov random sets. Terr. Verojatn Prim. 9 (1964) 738-743

KUMMER G. et MATTHES K.
Verallgemeinesung eines Satzes von Sliwnyak. Rev. Roumaine Math Pures Appl. 15 (1970) 845-870 et 1631-1642

KUNITA H.
[1] Asymptotic behavior of the non-linear filtering errors of Markov processes. J. Multiv. Anal. 1 (1971) 365-393

[2] Cours de 3ème cycle, Paris VI (1974)

KUNITA H. et WATANABE S.
Square integrable martingales. Nagoya J. Math 30 (1967) 209-245

KURTZ T.G.
[1] Limit theorem for sequences of jump Markov processes approximating ordinary differential processes. J. Appl. Proba. 8 (1971) 344-356

[2] Point processes and completely monotone set functions. ZfW 31 (1974) 57-67

LAWRANCE A.J.
[1] Selective interaction of a point process and a renewal process. J. Appl. Proba 7 (1970) 359-372, 7 (1970) 483-9, 8 (1971) 170-183, 8 (1971) 731-744

[2] Stationary series of univariate events. Stochastic point processes. P. Lewis, Ed. Wiley (1972) 199-256

LAZARO J. de Sam et MEYER P.A.
[1] Méthodes des martingales et théorie des flots. ZfW 18 (1971) 116-140

[2] Questions de théorie des flots. Sém. Proba Strasbourg (1972-3)

LEADBETTER M.R.
[1] On three basic results in the theory of stationary point processes. Proc. AMS 19 (1968) 115-7

[2] On basic results of point process theory. Proc. 6th Berkeley Symp 3 (1972) 449-462

[3] Point processes generated by level crossings. Conf. Stoch. point processes I.B.M. Wiley (1972) 436-467

LECAM L.
An approximation theorem for the Poisson binomial distribution Pacific J. Math 10 (1960) 1181-1197

LEE P.M.
[1] Infinitely divisible stochastic processes. ZfW 7 (1967) 147-160

[2] Some examples of infinitely divisible point processes. Studia Sc. Math Hung 3 (1968) 219-224

LEGALL
Les systèmes avec ou sans attente et les processus stochastiques. Dunod 1962

LEONOV V.P.
Applications of the characteristic functional and semi-invariants to the ergodic theory of stationary processes. Dokl Akad Nauk SSSR (1960) 523-6

LEWIS P.A.W.
[1] Asymptotic properties and equilibrium conditions for branching Poisson processes. J. Appl. Proba 6 (1969) 355-371

[2] Asymptotic properties of branching renewal processes

LIND D.A.
Locally compact measure preserving flows. Trans Amer Math Soc (1973)

LITTLE D.V.
A third note on recent research in general critical probability. Adv. Appl. Proba 6 (1974) 103-130

MACCHI O.
[1] Etude d'un processus ponctuel par ses multicoïncidences. C.R. Acad. Sci. Paris 268 (1969) 1616, 271 (1970) 660

[2] The coïncidence approach to stochastic point processes. Adv. Appl. Proba 7 (1975), 83-122

MAISONNEUVE B.
[1] Temps local d'un fermé droit aléatoire : cas d'un ensemble régénératif. C.R. Acad. Sci. Paris A 270 (1970) 1526-8

[2] Un résultat de renouvellement pour des processus de Markov généraux C.R. Acad. Sci. Paris A 272 (1971) 964-6

[3] Ensembles régénératifs, temps locaux et subordinateurs. Sém. Proba. Strasbourg, Lecture Notes Mathematics Springer 191 (1971)

MAISONNEUVE B. et MORANDO Ph.
Temps locaux pour les ensembles régénératifs. CR. Acad. Sci. Paris A 269 (1969) 523-5

MARTINS-NETTO et WONG E.
A martingale approach to queuing theory (1975)

MARUYAMA
Transformation of flows. J. Math. Soc. Japan 18 (1966) 303-330.

MATHERON G.

[1] Ensembles aléatoires, ensembles semi-markoviens et polyèdres poissoniens. Adv. Appl. Proba. 4 (1972) 508-541

[2] Un théorème d'unicité pour les hyperplans poissoniens. J. Appl. Proba 11 (1974) 184-9

[3] Hyperplans poissoniens et compacts de Steiner. Adv. Appl. Proba 6 (1974) 563-579

[4] Random sets and integral geometry. Wiley 1975

MATTHES K.

[1] Stationäre zufüllige Punktfolgen. Jahresbericht der DMV 66 (1963) 66-79

[2] Zur theorie der Bedienungs prozesse. Trans. 3rd Prague Conf (1964) 513-518

[3] Eine charackterisierung der kontinuierlichen unbegrenzt teilbaren Verteinlungsgesetze zufälliger Punktfolgen.Revue Roumaine Math.Pures et appliquées 14 (1969) 1121-1127

[4] Infinitely divisible point processes. Stochastic point processes, P. Lewis Ed. Wiley (1972) 384-404

Mc FADDEN J.A.
On the lengths of intervals in a stationary point processes. J.R. Stat. Soc. B 24 (1962) 364-382

Mc FADDEN J. et WEISSBLUM W.
Higher order properties of a stationary point process. J.R. Stat. Soc. B 25 (1963) 413-431

MECKE J.

[1] Stationäre zufällige Masse auf lokalkompakten Abelschen Gruppen. ZfW 9 (1967) 36-58

[2] Eine kharakteristischeEigenschaft der doppelt stochastischen Poissonschen Prozesse. ZfW 11 (1968) 74-81

[3] Invariance Eigenschaften allgemeiner Palmscher Masse. Math. Nachrichten

[4] Dac Erlangsche Modell mit abhängigen Bedienungszeiten. Math. op. Forsch u Stat 3 (1972) 453-464

[5] Stationäre Verteilungen für das Erlangsche Modell. Math. Nachr

[6] A result on the output of stationary Erlang processes

[7] A characterization of mixed Poisson processes. J. Appl. Proba.

MEYER P.A.

[1] Processus de Poisson ponctuels suivant Ito. Sém. Proba. Strasbourg V. Lecture Notes Springer 191 (1971)

[2] Ensembles aléatoires markoviens homogènes. Sém. Proba. Strasbourg 8, Lecture Notes Springer 381 (1974) 172-261

[3] Un cours sur les intégrales stochastiques. Sém. Proba. Strasbourg X, Lecture Notes Springer 511 (1975) 245-400

[4] Generation of σ-fields by step processes. Sem Strasbourg X. Lecture Notes Springer 511 (1976) 118-124

MILES R.E.

[1] Poisson flats in euclidean space. Adv. Appl. Proba. 1 (1969) 211-237 3 (1971) 1-43

[2] A synopsis of Poisson flats in euclidean spaces. Izv. Akad Nauk Armianskoi SSR 3 (1970) 263-285

MILNE R.K.

Simple proofs of some theorems on point processes. Ann. Math Stat 42 (1971) 368-372

MILNE R.K. et WESTCOTT M.

Further results for Gauss-Poisson processes. Adv. Appl. Proba. 4 (1972) 151-176

MOGYÓRODI J.

On the rarefaction of renewal processes. Studia Sci. Math. Hung 7 (1972) 285-305 et 8 (1973) 21-38, 193-209

MÖNCH G.

Verallgeneinerung eines Satzes von A. Renyi. Studia Sci Math Hung 6 (1971) 81-90

MORAN P.A.P.

[1] A non-markovian quasi-Poisson process. Studia Sci. Math Hung 2 (1967) 425-429

[2] A second note on recent research in geometrical probability. Adv. Appl. Proba 1 (1969) 73-89

MORI T.

Ergodicity and identifiability for random translations of stationary point processes. J. Appl. Proba. 12 (1975) 734-743

MOYAL J.E.

[1] The general theory of stochastic population processes. Acta Math 108 (1962) 1-31

[2] Multiplicative population processes. J. Appl. Proba. 1 (1964) 267-283

[3] Particle populations and number operators in quantum theory Adv. Appl. Proba. (1972) 39-80

NEVEU J.

[1] Une généralisation des processus à accroissements positifs indépendants. Abh. Math. Sem. Hamburg 25 (1961) 36-61

[2] Bases mathématiques des Probabilités. Masson 1964 et 1970

[3] Sur la structure des processus ponctuels stationnaires. CR. Acad. Sci. Paris A 267 (1968) 561-4

[4] Martingales à temps discret. Masson 1972

[5] Sur les mesures de Palm de deux processus ponctuels stationnaires. ZfW 34 (1976) 189-203

NEWMAN D.S.

A new family of point processes which are characterized by their second moment properties. J. Appl. Proba. 7 (1970) 338-358

NEY P.

Convergence of a random distribution function associated with a branching process. J. Math. Anal. Appl. 12 (1965) 316-327

NGUYEN Xuan Xanh et ZESSIN H.

Punktprozesse mit Wechselwirkung. ZfW 37 (1976) 91-126

OAKES D.

[1] Synchronous and asynchronous distributions for Poisson cluster processes. J. Roy Stat. Soc. B 37 (1975) 238-247

[2] The markovian self exciting process. J. Appl. Proba. Appl. 12 (1975) 69-77

[3] Random overlopping intervals. A generalization of Erlang's loss formula. Ann. Proba. 4 (1976) 940-6

OREY S.

Radon-Nikodym derivatives of probability measures : martingale methods. Dpt of Math. Educ., Tokyo Univ. Educ. (1974) 1-38

PALM C.

Intensitätschwankungen in Fernsprechverkehr.Ericsson Technics 44 (1943) 1-189

PAPANGELOU

[1] The Ambrose-Kakutani theorem and the Poisson process. Lecture Notes Springer 160 (1970)

[2] Integrability of expected increments of point processes and a related change of scale. Trans. AMS 165 (1972) 483-506

[3] Summary of some results on point and line processes. Stochastic point processes, etc ... I.B.M. Conf. Wiley (1972) 522-532

[4] Stochastic geometry. Trans. AMS 165 (1972) 483-506

[5] On the Palm probabilities of processes of points and processes of lines. Stochastic geometry, ed. Kendall. Wiley (1974) 114-147

[6] The conditional intensity of general point processes and an application to line processes. ZfW 28 (1974) 207-226

PARTHASARATHY K.R.

Probability measures on metric spaces. Academic Press, New York 1967

POLLACZEK F.

Problèmes stochastiques posés par le phénomène de formation d'une queue d'attente à un guichet et par des phénomènes apparentés Mém. Sc. Math Paris (1957)

PORT S.C.

[1] Equilibrium systems of recurrent Markov processes. J. Math. Anal Appl. 12 (1965) 555-569

[2] Equilibrium processes Trans Amer Math Soc 124 (1966) 168-184

[3] Limit theorems involving capacities. J. Math. Mech 15 (1966) 805-832

PORT S.C. et STONE C.J.

Infinite particle systems. Trans. Amer. Math. Soc. 178 (1973) 307-340

PYKE R.

Markov renenval processes. Ann. Math. Stat. 32 (1961) 1231-1259

RÅDE L.

Limit theorems for thinning of renewal point processes. J. Appl. Proba 9 (1972) 847-851

RENYI A.

Remarks on the Poisson process. Studia Sci Math Hung 2 (1967) 119-123

RENSHAW A.

Interconnected population processes. J. Appl. Proba. 10 (1973) 1-14

RIORDAN J.

Stochastic service systems. Wiley (1962)

RIPLEY R.D.

[1] Locally finite random sets ; foundations for point process theory. Ann. Proba. 4 (1976) 983-994

[2] The foundations of stochastic geometry. Ann. Proba. 4 (1976) 995-998

[3] On stationary and superposition of point processes. Ann. Proba. 4 (1976) 999-1005

ROOT D.

A counterxample in renewal theory. Ann. Math. Stat. 42 (1971) 1763-6

RUBEN H. et REED W.J.

A more general form of a theorem of Crofton. J. Appl. Proba. 10 (1973) 479-482

RUBIN I.

Regular point processes and their detection. IEEE IT 18 (1972) 5

Regular point processes and their information processing. IEEE IT 20 (1974) 617-624

RUDEMO M.

[1] Point processes generated by transition of Markov chains. Adv. Appl. Proba 5 (1973) 262-286

[2] Multivariate point processes generated by transitions of Markov chains. J. Appl. Proba (1973)

[3] Some non stationary point processes with stationary forward recurrence time distribution. J. Appl. Proba 12 (1975) 167-9

RYLL NARDZEWSKI C

Remarks on processes of calls. Proc 4th Berkeley Symp II (1961) 455-465

SAMUELS S.M.

A characterization of the Poisson process. J. Appl. Proba 11 (1974) 72-85

SERFLING R.J.

Research in point processes, with applications to reliability and biometry. Proc. Conf. Florida. SIAM (1974) 109-127

SERFOZO R.F.

[1] Conditional Poisson processes. J. Appl. Proba 9 (1972) 288-302

[2] Semi-stationary processes. ZfW 23 (1972) 125-132

[3] Compositions, inverses and thinnings of random measures

SERFOZO R. et STIDHAM S.

Semi-stationary clearing processes

SKOROHOD A.

Studies in the theory of random processes. Trad. du russe. Addison-Wesley (1965)

SLIVNYAK

Some properties of stationary streams of homogeneous random events. Teor. Veraj. Prim. 7 (1962) 347-352

SNYDER D.

[1] Filtering and detection for doubly stochastic Poisson processes. IEEE Trans IT 18 (1972) 97-102

[2] Smoothing for doubly stochastic Poisson processes. IEEE Trans IT 18 (1972) 558-562

[3] Information processing for observed jump processes. Info et Control 22 (1973)

SRINIVASAN S.K.

Stochastic point processes and their applications. Griffin's Stat. Monographs and Courses 34 (1974)

STONE C.J.

[1] On a theorem of Dobrusin. Ann Math Stat 39 (1968) 1391-1401

[2] An supper bound for the renewal function. Ann Math Stat 43 (1972) 2050-2052

STÖRMER H.

Zur Uberlagerung von Erneuerungs prozessen. ZfW 13 (1969) 9-24

STRAUSS D.J.

A model for clustering. Biometrika 62 (1975) 467-475

SYSKI R.

Introduction to Congestion theory in telephone systems. Oliver et Boyd, London (1962)

SZASZ D.

[1] On the general branching process with continuous time-parameter. Studia Sci. Math Hung 2 (1967) 227-247

[2] Once more on the Poisson process. Studia Sci. Math. Hung 5 (1970) 441-4

[3] On the convergence of sums of point processes with integer marks Conf. on Stochastic point processes, etc. I.B.M. Wiley (1972) 607-615

TAKACS L.

[1] Introduction to the theory of queues. Oxford Univ. Press (1962)

[2] Sojourn time problems. Ann. Proba. 2 (1974) 420-431

TENHOOPEN M. and REUVER HA

[1] Selective interaction of two independent recurrent processes. J. Appl. Proba 2 (1965) 286-292

[2] Interaction between two independent recurrent time series. Info et Control 10 (1967) 149-158

THEDEEN T.

[1] A note on the Poisson tendency in traffic distribution. Annals Math Stat 35 (1964) 1823-4

[2] Convergence and invariance questions for point systems in R_1 under random motion. Arkiv för Math 7 (1967) 211-239

TOMKO J.

On the rarefaction of multivariate point processes. European Meeting of Statisticians, Budapest (1972) 843-860

TOTOKI H.

Time changes of flows. Mem Fac. Sci.Kyushu Univ. A 20 (1966) 27-55

TORTRAT A.

Sur les mesures aléatoires dans les groupes non abéliens. Ann IHP 5 (1969) 31-47

VAN SCHUPPEN J. et WONG E.
Translation of local martingales under a change of law. Ann. Proba. 2 (1974) 879-888

VARAYIA P.
The martingale theory of jump processes. IEEE AC 20 (1975) 34-42

VERE-JONES D.
[1] Stochastic models for earthquake occurence. JR Stat Soc B 32 (1970) 1-62

[2] A renewal equation for point processes with Markov dependent intervals. Math. Nachr 68 (1975) 133-9

VON WALDENFELS W.
[1] Charakteristische funktionale zufälliger Masse ZfW 10 (1968) 279-283

[2] Taylor expansion of a Poisson measure. Sem Strasbourg. Lecture Notes Springer 381 (1974) 344-354

WATANABE S.
On discontinuous additive functionals and Levy measures of a Markov process. Japanese J. Math 34 (1964) 53-70

WEISS N.A.
Infinite particle systems of recurrent random walks. J. Math Anal. Appl. 41 (1973) 565-574

Limit theorems for infinite particle systems. ZfW

WESTCOTT M.
[1] On existence and mixing results for cluster point processes. J.R. Stat. Soc. B 33 (1971) 290-300

[2] The probability generating functional. J. Austr. Math. Soc. 13 (1972) 448-466

[3] Results in the asymptotic and equilibrum theory of Poisson cluster processes. J. Appl. Proba. 10 (1973) 807-823

[4] Some remarks on a property of the Poisson process. Sankhya A 35 (1973) 29-34

[5] Simple proof of a result on thinned point process. Ann. Proba. 4 (1976) 89-90

WHITT W.
[1] The continuity of queues. Adv. Appl. Proba. 6 (1974) 175-183

[2] Representation and convergence of point processes on the line. Ann. Proba. 2 (1974)

WISMIEWSKI T.K.M.
Bivariate stationary point processes. Adv. Appl. Proba. 4 (1972) 296-317

WONG E.
Recent progresses in stochastic processes ; a survey. IEEE Trans. IT 19 (1973) 262-275

WOLFF R.W. et WRIGHTSON C.W.
An extension of Erlang's loss formula. J. Appl. Proba. 13 (1976) 628-632

YASHIN
Filtering of jump processes. Artomatika i Telemektanica 5 (1970)

YLVISAKER N.D.
The expected number of zeros of a stationary Gaussian process. Ann. Math. Stat. 36 (1965) 1043-6

YOR M.
[1] Sur les intégrales stochastiques optionnelles et une suite remarquable de formules exponentielles. Sém. Proba. Strasbourg, Lecture Notes Springer (1975)

[2] Représentation des martingales de carré intégrable relative aux processus de Wiener et de Poisson à n paramètres. C.R. Acad. Sci. Paris 281 (1975) 111-113

YOEURP CH.
Décomposition de martingales locales et formules exponentielles. Sém. Proba. Strasbourg, Lecture Notes Springer.

INDEX